T0215172

Lecture Notes in Artificial Intelligence　9067

Subseries of Lecture Notes in Computer Science

More information about this series at http://www.springer.com/series/1244

Tsuyoshi Murata · Koji Mineshima
Daisuke Bekki (Eds.)

New Frontiers
in Artificial Intelligence

JSAI-isAI 2014 Workshops, LENLS, JURISIN, and GABA
Kanagawa, Japan, October 27–28, 2014
Revised Selected Papers

 Springer

Editors
Tsuyoshi Murata
Tokyo Institute of Technology
Tokyo
Japan

Daisuke Bekki
Ochanomizu University
Tokyo
Japan

Koji Mineshima
Ochanomizu University
Tokyo
Japan

ISSN 0302-9743   ISSN 1611-3349   (electronic)
Lecture Notes in Artificial Intelligence
ISBN 978-3-662-48118-9   ISBN 978-3-662-48119-6   (eBook)
DOI 10.1007/978-3-662-48119-6

Library of Congress Control Number: 2015945600

LNCS Sublibrary: SL7 – Artificial Intelligence

Printed on acid-free paper

Springer-Verlag GmbH Berlin Heidelberg is part of Springer Science+Business Media
(www.springer.com)

Preface

JSAI-isAI 2014 was the 6th International Symposium on Artificial Intelligence supported by the Japanese Society for Artificial Intelligence (JSAI). JSAI-isAI 2014 was successfully held during November 23–24 at Keio University in Kanagawa, Japan. In all, 98 people from 14 countries participated. The symposium took place after the JSAI SIG joint meeting. As the total number of participants for these two co-located events was about 500, it was the second-largest JSAI event in 2014 after the JSAI annual meeting.

JSAI-isAI 2014 included three workshops, where five invited talks and 52 papers were presented. This volume, *New Frontiers in Artificial Intelligence: JSAI-isAI 2014 Workshops,* contains the proceedings of JSAI-isAI 2014. From the three workshops (LENLS11, JURISIN2014, and GABA2014), 23 papers were carefully selected and revised according to the comments of the workshop Program Committee. About 40 % of the total submissions were selected for inclusion in the conference proceedings.

LENLS (Logic and Engineering of Natural Language Semantics) is an annual international workshop on formal semantics and pragmatics. LENLS11 was the 11th event in the series, and it focused on the formal and theoretical aspects of natural language. The workshop was chaired by Koji Mineshima (Ochanomizu University/JST CREST), Daisuke Bekki (Ochanomizu University/National Institute of Informatics/JST CREST), and Elin McCready (Aoyama Gakuin University).

JURISIN (Juris-informatics) 2014 was the eighth event in the series, organized by Satoshi Tojo (Japan Advanced Institute of Science and Technology). The purpose of this workshop was to discuss fundamental and practical issues for juris-informatics, bringing together experts from a variety of relevant backgrounds, including law, social science, information and intelligent technology, logic and philosophy (including the area of AI and law).

GABA (Graph-Based Algorithms for Big Data and Its Applications) 2014 was the first workshop on graph structures including string, tree, bipartite- and di-graph for knowledge discovery in big data. The purpose of this workshop was to discuss ideas for realizing big data integration, including algorithms with theoretical/experimental results. The workshop was chaired by Hiroshi Sakamoto (Kyushu Institute of Technology), Yoshinobu Kawahara (Osaka University), and Tetsuji Kuboyama (Gakushuin University).

It is our great pleasure to be able to share some highlights of these fascinating workshops in this volume. We hope this book will introduce readers to the state-of-the-art research outcomes of JSAI-isAI 2014, and motivate them to participate in future JSAI-isAI events.

April 2015

Tsuyoshi Murata
Koji Mineshima
Daisuke Bekki

Organization

JSAI-isAI 2014

Chair

Tsuyoshi Murata   Tokyo Institute of Technology, Japan

Advisory Committee

Daisuke Bekki   Ochanomizu University, JST CREST, National
                            Institute of Informatics, Japan
Ho Tu Bao   Japan Advanced Institute of Science and Technology
                            (JAIST), Japan
Ee-Peng Lim   Singapore Management University, Singapore
Thanaruk   SIIT, Thammasat University, Thailand
  Theeramunkong
Tsuyoshi Murata   Tokyo Institute of Technology, Japan
Yukiko Nakano   Seikei University, Japan
Ken Satoh   National Institute of Informatics, Japan
Vincent S. Tseng   National Cheng Kung University, Taiwan
Takashi Washio   Osaka University, Japan

LENLS 2014

Workshop Chair

Koji Mineshima   Ochanomizu University, JST CREST, Japan

Workshop Co-chairs

Daisuke Bekki   Ochanomizu University, JST CREST, National
                            Institute of Informatics, Japan
Elin McCready   Aoyama Gakuin University, Japan

Program Committee

Elin McCready   Aoyama Gakuin University, Japan
Daisuke Bekki   Ochanomizu University, JST CREST, National
                            Institute of Informatics, Japan
Koji Mineshima   Ochanomizu University, JST CREST, Japan
Alastair Butler   Tohoku University, Japan
Richard Dietz   University of Tokyo, Japan
Yoshiki Mori   University of Tokyo, Japan
Yasuo Nakayama   Osaka University, Japan

Katsuhiko Sano	Japan Advanced Institute of Science and Technology (JAIST), Japan
Katsuhiko Yabushita	Naruto University of Education, Japan
Tomoyuki Yamada	Hokkaido University, Japan
Shunsuke Yatabe	West Japan Railway Company, Japan
Kei Yoshimoto	Tohoku University, Japan

JURISIN 2014

Workshop Chair

| Satoshi Tojo | Japan Advanced Institute of Science and Technology (JAIST), Japan |

Organizing Committee

Katsumi Nitta	Tokyo Institute of Technology, Japan
Ken Satoh	National Institute of Informatics and Sokendai, Japan
Satoshi Tojo	Japan Advanced Institute of Science and Technology (JAIST), Japan
Katsuhiko Sano	Japan Advanced Institute of Science and Technology (JAIST), Japan

Advisory Committee

Trevor Bench-Capon	The University of Liverpool, UK
Tomas Gordon	Fraunfoher FOKUS, Germany
Henry Prakken	University of Utrecht and Groningen, The Netherlands
John Zeleznikow	Victoria University, Australia
Robert Kowalski	Imperial College London, UK
Kevin Ashley	University of Pittsburgh, USA

Program Committee

Satoshi Tojo	Japan Advanced Institute of Science and Technology (JAIST), Japan
Tokuyasu Kakuta	Nagoya University, Japan
Hanmin Jung	Korea Institute of Science and Technology Information (KISTI), Korea
Minghui Xiong	Sun Yat-sen University, China
Minghui Ma	Southwest University, China
NguyenLe Minh	Japan Advanced Institute of Science and Technology (JAIST), Japan
Katsuhiko Toyama	Nagoya University, Japan
Makoto Nakamura	Nagoya University, Japan
Marina De Vos	University of Bath, UK
Guido Governatori	NICTA, Australia
Paulo Novais	University of Minho, Portugal

Seiichiro Sakurai	Meiji Gakuin University, Japan
Ken Satoh	National Institute of Informatics and Sokendai, Japan
Thomas Agotnes	University of Bergen, Norway
Robert Kowalski	Imperial College London, UK
Katsuhiko Sano	Japan Advanced Institute of Science and Technology (JAIST), Japan
Katumi Nitta	Tokyo Institute of Technology, Japan
Katie Atkinson	University of Liverpool, UK
Philip T.H. Chung	Australasian Legal Information Institute (AustLII), Australia
Masahiro Kozuka	Okayama University, Japan
Fumihiko Takahashi	Meiji Gakuin University, Japan
Baosheng Zhang	China University of Political Science and Law, China
Akira Shimazu	Japan Advanced Institute of Science and Technology (JAIST), Japan
Mi-Young Kim	University of Alberta, Canada
Randy Goebel	University of Alberta, Canada

GABA 2014

Workshop Chair

| Hiroshi Sakamoto | Kyushu Institute of Technology, Japan |

Workshop Co-chairs

| Yoshinobu Kawahara | Osaka University, Japan |
| Tetsuji Kuboyama | Gakushuin University, Japan |

Program Committee

Hiroki Arimura	Hokkaido University, Japan
Kouichi Hirata	Kyushu Institute of Technology, Japan
Nobuhiro Kaji	The University of Tokyo, Japan
Yoshinobu Kawahara	Osaka University, Japan
Miyuki Koshimura	Kyushu University, Japan
Tetsuji Kuboyama	Gakushuin University, Japan
Yoshiaki Okubo	Hokkaido University, Japan
Hiroshi Sakamoto	Kyushu Institute of Technology, Japan
Takeshi Shinohara	Kyushu Institute of Technology, Japan
Yasuo Tabei	JST PRESTO, Japan
Akihiro Yamamoto	Kyoto University, Japan

Sponsored by

The Japanese Society for Artificial Intelligence (JSAI)

Contents

LENLS 11

Logic and Engineering of Natural Language Semantics (LENLS) 11   3
Koji Mineshima

Codability and Robustness in Formal Natural Language Semantics   6
Kristina Liefke

CI via DTS   23
Daisuke Bekki and Elin McCready

Formal Analysis of Epistemic Modalities and Conditionals Based on Logic
of Belief Structures   37
Yasuo Nakayama

A Type-Logical Account of Quantification in Event Semantics   53
Philippe de Groote and Yoad Winter

Towards a Wide-Coverage Tableau Method for Natural Logic   66
Lasha Abzianidze

Resolving Modal Anaphora in Dependent Type Semantics   83
Ribeka Tanaka, Koji Mineshima, and Daisuke Bekki

Canonical Constituents and Non-canonical Coordination: Simple Categorial
Grammar Account   99
Oleg Kiselyov

A Good Intensifier   114
Elena Castroviejo and Berit Gehrke

Strict Comparison and Weak Necessity: The Case of Epistemic *Yào*
in Mandarin Chinese   130
Zhiguo Xie

Computing the Semantics of Plurals and Massive Entities Using
Many-Sorted Types   144
Bruno Mery, Richard Moot, and Christian Retoré

On CG Management of Japanese Weak Necessity Modal *Hazu*   160
Shinya Okano and Yoshiki Mori

Using Signatures in Type Theory to Represent Situations.............. 172
 Stergios Chatzikyriakidis and Zhaohui Luo

Scope as Syntactic Abstraction............................... 184
 Chris Barker

Focus and Givenness Across the Grammar....................... 200
 Christopher Tancredi

JURISIN 2014

Eighth International Workshop on Juris-Informatics (JURISIN 2014)....... 225
 Satoshi Tojo

Classification of Precedents by Modeling Tool for Action and Epistemic
State: DEMO.. 227
 Tetsuji Goto and Satoshi Tojo

Legal Question Answering Using Ranking SVM and Syntactic/Semantic
Similarity.. 244
 Mi-Young Kim, Ying Xu, and Randy Goebel

Translating Simple Legal Text to Formal Representations 259
 Shruti Gaur, Nguyen H. Vo, Kazuaki Kashihara, and Chitta Baral

Analyzing Reliability Change in Legal Case..................... 274
 Pimolluck Jirakunkanok, Katsuhiko Sano, and Satoshi Tojo

GABA 2014

Workshop on Graph-Based Algorithms for Big Data and Its Applications
(GABA2014).. 293
 Yoshinobu Kawahara, Tetsuji Kuboyama, and Hiroshi Sakamoto

Anchored Alignment Problem for Rooted Labeled Trees 296
 Yuma Ishizaka, Takuya Yoshino, and Kouichi Hirata

Central Point Selection in Dimension Reduction Projection Simple-Map
with Binary Quantization................................. 310
 *Quming Jin, Masaya Nakashima, Takeshi Shinohara, Kouichi Hirata,
 and Tetsuji Kuboyama*

Mapping Kernels Between Rooted Labeled Trees Beyond Ordered Trees.... 317
 Kouichi Hirata, Tetsuji Kuboyama, and Takuya Yoshino

Finding Ambiguous Patterns on Grammar Compressed String.   331
    Koji Maeda, Yoshimasa Takabatake, Yasuo Tabei,
    and Hiroshi Sakamoto

Detecting Anomalous Subgraphs on Attributed Graphs via Parametric
Flow   340
    Mahito Sugiyama and Keisuke Otaki

Author Index   357

LENLS 11

Logic and Engineering of Natural Language Semantics (LENLS) 11

Koji Mineshima[1,2](✉)

[1] Ochanomizu University, Bunkyo, Tokyo, Japan.
[2] CREST, Japan Science and Technology Agency, Tokyo, Japan
mineshima.koji@ocha.ac.jp

1 The Workshop

On November 22–24, 2014 the Eleventh International Workshop of Logic and Engineering of Natural Language Semantics (LENLS 11) took place at Raiousha Building, Keio University, Kanagawa, Japan. This was held as a workshop of the Sixth JSAI International Symposia on AI (JSAI-isAI 2014), sponsored by The Japan Society for Artificial Intelligence (JSAI). This year, LENLS is being organized by an alliance of "Establishment of Knowledge-Intensive Structural Natural Language Processing and Construction of Knowledge Infrastructure" project, funded by JST CREST Programs "Advanced Core Technologies for Big Data Integration".

LENLS is an annual international workshop focusing on topics in formal semantics, formal pragmatics, and related fields. This year the workshop featured invited talks by Chris Barker, on the logic of scope and continuation, Matthew Stone, on the vagueness and flexibility of grounded meaning, and Chris Tancredi, on focus and givenness. A paper based on the first talk and one based on the third talk appear in the present volume. In addition, there were 23 presentations of talks selected by the program committee (see Acknowledgements) from the abstracts submitted for presentation.

As always with the LENLS workshops, the content of the presented papers was rich and varied. Topics represented included, on the empirical side, mass and plural expressions, modality, temporal expressions, honorifics, polarity items, questions and modifiers, and on the theoretical side, categorial grammar, type-logical semantics, dependent type theory, ludics, fuzzy logic, measurement theory, natural logic, and inquisitive semantics. The remainder of this introduction will briefly indicate the content of the papers selected to appear in the present volume.

2 Papers

The submitted papers in the LENLS part of the present volume fall into two classes. The first class of papers considers various linguistic phenomena from the perspective of type-logical semantics and/or categorial grammar. In "Codability

© Springer-Verlag Berlin Heidelberg 2015
T. Murata et al. (Eds.): JSAI-isAI 2014 Workshops, LNAI 9067, pp. 3–5, 2015.
DOI: 10.1007/978-3-662-48119-6_1

and robustness in formal natural language semantics", Kristina Liefke analyzes Partee's temperature puzzles using a typed language that is more parsimonious than the traditional type-logical semantics. The paper by Daisuke Bekki and Elin McCready, "CI via DTS", presents a proof system for capturing the interaction between conventional implicature and at-issue content in the framework of Dependent Type Semantics (DTS). The paper by Philippe de Groote and Yoad Winter, "A type-logical account of quantification in event semantics", proposes a solution to the problem of event quantification in Abstract Categorial Grammar (ACG). In "Towards a wide-coverage tableau method for natural logic", Lasha Abzianidze presents a natural-logic-based tableau system for textual entailment. The paper by Ribeka Tanaka, Koji Mineshima and Daisuke Bekki, "Resolving modal anaphora in dependent type semantics", addresses the problem of modal anaphora and subordination in the framework of Dependent Type Semantics. The paper by Oleg Kiselyov, "Canonical constituents and non-canonical coordination", presents a novel treatment of non-canonical coordination including the so-called "gapping" in the non-associative Lambek calculus. The paper by Bruno Mery, Richard Moot and Christian Retoré, "Computing the semantics of plurals and massive entities using many-sorted types", presents an analysis of mass nouns in a framework for rich lexical semantics based on a higher-order type-theory (System F). The paper by Stergios Chatzikyriakidis and Zhaohui Luo, "Using signatures in type theory to represent situations", applies the notion of signature in modern type theory to the formalization of the notion of situations in formal semantics.

The second class of papers focuses on modalities and intensional phenomena in natural languages from various perspectives. The paper by Yasuo Nakayama, "Formal analysis of epistemic modalities and conditionals based on logic of belief structures", proposes a logic of belief structures and applies it to the formalization of belief revision and semantics of conditionals. The paper by Elena Castroviejo and Berit Gehrke, "A GOOD intensifier", proposes a semantical analysis of Catalan modifiers *ben* and *bon*. The paper by Zhiguo Xie, "Strict comparison and weak necessity: the case of epistemic *Yào* in mandarin chinese", presents an analysis of the epistemic use of *Yào* in Mandarin Chinese. The paper by Shinya Okano and Yoshiki Mori, "On CG management of Japanese weak necessity modal *Hazu*", proposes an account of necessity modal *hazu* in Japanese.

The present volume also includes two invited papers, one by Chris Barker, "Scope as syntactic abstraction", which presents a substructural logic for scope-taking in natural language, and one by Christopher Tancredi, "Focus and Givenness Across the Grammar", which argues that focus and givenness need to be marked independently, against the standard assumption in formal semantics.

Acknowledgements. Let me acknowledge some of those who helped with the workshop. The program committee and organisers, in addition to myself, were Daisuke Bekki, Alastair Butler, Richard Dietz, Elin McCready, Yoshiki Mori, Yasuo Nakayama,

Katsuhiko Sano, Katsuhiko Yabushita, Tomoyuki Yamada, Shunsuke Yatabe and Kei Yoshimoto. Daisuke Bekki also was liaison with JSAI and together with Elin McCready organised and mentored many aspects of the workshop. Finally, the organisers would like to thank the JST CREST Program "Establishment of Knowledge-Intensive Structural Natural Language Processing and Construction of Knowledge Infrastructure" and JSAI for giving us the opportunity to hold the workshop.

Codability and Robustness in Formal Natural Language Semantics

Kristina Liefke[(⊠)]

Munich Center for Mathematical Philosophy, Ludwig-Maximilians-University,
Geschwister-Scholl-Platz 1, 80539 Munich, Germany
K.Liefke@lmu.de

Abstract. According to the received view of type-logical semantics (suggested by Montague and adopted by many of his successors), the correct prediction of entailment relations between lexically complex sentences requires many different types of semantic objects. This paper argues *against* the need for such a rich semantic ontology. In particular, it shows that Partee's temperature puzzle – whose solution is commonly taken to require a basic type for indices or for individual concepts – can be solved in the more parsimonious type system from [11], which only assumes basic individuals and propositions. We generalize this result to show the soundness of the PTQ-fragment in the class of models from [11]. Our findings support the robustness of type-theoretic models w.r.t. their objects' codings.

Keywords: PTQ-fragment · Individual concepts · Temperature puzzle · Entailment-preservation · Coding · Robustness

In choosing theories, we always invoke a principle of theoretical simplicity or parsimony: given two theories of equal explanatory power, the theory that postulates fewer irreducibly distinct kinds or types of entities is preferable.   [10, p. 51]

1 Introduction

It is a commonplace in ontology engineering that, to model *complex* target systems, we need to assume *many* different types of objects. The semantic ontology of natural language is no exception to this: To interpret a reasonably rich fragment of English, we assume the existence of individuals, propositions, properties of individuals, relations between individuals, situations, events, degrees, times, substances, kinds, and many other types of objects. These objects serve the interpretation of proper names, declarative sentences or complement phrases, common nouns or intransitive verbs, transitive verbs, neutral perception verbs, adverbial or degree modifiers, temporal adjectives, mass terms, bare noun phrases, etc.

I would Like to thank two anonymous referees for LENLS 11 for their comments and suggestions. Thanks also to Ede Zimmermann, whose comments on my talk at *Sinn und Bedeutung 18* have inspired this paper. The research for this paper has been supported by the *Deutsche Forschungsgemeinschaft* (grant LI 2562/1-1), by the LMU-Mentoring program, and by Stephan Hartmann's Alexander von Humboldt-professorship.

© Springer-Verlag Berlin Heidelberg 2015
T. Murata et al. (Eds.): JSAI-isAI 2014 Workshops, LNAI 9067, pp. 6–22, 2015.
DOI: 10.1007/978-3-662-48119-6_2

Traditional type-logical semantics (esp. [11,12]) tames this zoo of objects by assuming only a small set of primitive objects and constructing all other types of objects from these primitives via a number of object-forming rules. In this way, Montague reduces the referents of the basic fragment of English from [11] (hereafter, the *EFL-fragment*) to *two* basic types of objects: individuals (type ι) and propositions (analyzed as functions from indices to truth-values, type $\sigma \rightarrow t$;[1] abbreviated 'o'). From these objects, properties and binary relations of individuals are constructed as functions from individuals to propositions (type $\iota \rightarrow o$), resp. as curried functions from pairs of individuals to propositions (type $\iota \rightarrow (\iota \rightarrow o)$).

Since Montague's semantics *reduces* the number of basic objects in the linguistic ontology to a small set of primitives, we hereafter refer to this view of formal semantics as the *reduction view*. The latter is characterized below:

Reduction View. *Many (types of) objects in the linguistic ontology* **can be coded** *as constructions out of* a few *(types of) primitives. The coding relations between objects enable the compositional interpretation of natural language.*

In the last forty years, revisions and extensions of Montague's formal semantics have caused many semanticists to depart from the reduction view. This departure is witnessed by the introduction of a fair number of new basic types (including types for primitive propositions, situations, events, degrees, and times). The introduction of these new types is motivated by the coarse grain of set-theoretic functions (s.t. Montague's semantics sometimes generates wrong predictions about linguistic entailment), and by the need to find semantic values for 'new' kinds of expressions, which are not included in Montague's small linguistic fragments. Since many of these new values are either treated as primitives or are identified with constructions out of new primitives, we will hereafter refer to this view of formal semantics as the *ontology engineering view*. This view is captured below:

Ontology Engineering View. *Many (types of) objects in the linguistic ontology* **are** **not** **coded** *as constructions out of other objects. The compositionality of interpretation is only due to the coding relations between a small subset of objects.*

The ontology engineering view is supported by Montague's semantics from [11,12]. The latter interpret the EFL-fragment in models with primitive individuals and propositions, and interpret the fragment of English from [12] (called the *PTQ-fragment*) in models with primitive individuals, indices (i.e. possible worlds, or world-time pairs), and truth-values. Since the PTQ-fragment extends the lexicon of the EFL-fragment via intensional common nouns (e.g. temperature, price) and intensional intransitive verbs (e.g. rise, change), since PTQ-models extend the frames of EFL-models via individual concepts, and since PTQ-models interpret intensional nouns and intransitive verbs as functions *over individual concepts*, it is commonly assumed that any empirically adequate model for the PTQ-fragment requires a basic type for indices, or for individual concepts.

The ontology engineering view is further supported by its practical advantages. In particular, the availability of a larger number of basic types facilitates

---

[1] We follow the computer science-notation for function types. Thus, $\sigma \rightarrow t$ corresponds to Montague's type $\langle \sigma, t \rangle$ (or, given Montague's use of the index-type s, to $\langle s, t \rangle$).

work for the empirical linguist: In a rich type system, fewer syntactic expressions are interpreted in a non-basic type.[2] As a result, the compositional translations of many syntactic structures will be *simpler*, and will involve *less* lambda conversions than their 'reductive' counterparts.

However, the proliferation of basic types is not an altogether positive development: Specifically, the interpretation of new kinds of expressions in (constructions out of) *additional* basic types prevents an identification of the relation between the different type-theoretic models. As a result, we cannot check the relative consistency of the different models, transfer the interpretive success of some classes of models to other models, or identify the minimal semantic requirements on models for certain linguistic fragments. These problems are exemplified by the absence of a consequence-preserving translation between the terms of Montague's three-sorted logic IL from [12] (or of Gallin's logic TY_2 from [5]) and the terms of the two-sorted logic underlying [11], by the resulting impossibility of attributing the PTQ-model's solution of Partee's temperature puzzle to EFL-models, and by the related open question whether a primitive type for indices is really *required* for the interpretation of the PTQ-fragment.

This paper defends the reduction view with respect to the interpretation of the PTQ-fragment. In particular, it shows that the PTQ-fragment can be interpreted in EFL-models, which do not have a designated type for indices or individual concepts. This interpretation is sound: the EFL-interpretation of the PTQ-fragment preserves the entailment relation which is imposed on this fragment by its translation into Montague's logic IL. To illustrate this soundness, we show that our EFL-semantics blocks Partee's temperature puzzle from [12].

The plan of the paper is as follows: We first introduce Partee's temperature puzzle for extensional semantics of natural language and present Montague's solution to this puzzle (in Sect. 2). We then identify a strategy for the EFL-coding of indices and truth-values, which allows us to translate the linguistically relevant sublanguage of TY_2 into an EFL-language[3] (in Sect. 3). Finally, we apply this strategy to solve Partee's temperature puzzle (in Sect. 4). The paper closes with a summary of our results.

2   Partee's Temperature Puzzle and Montague's Solution

Partee's temperature puzzle [12, pp. 267–268] identifies a problem with extensional semantics for natural language, which regards their validation of the counterintuitive inference from $(\star)$:

---

[2] For example, since linguists often assign degree modifiers (e.g. very) the type for degrees δ (rather than the type for properties of individuals, $(\iota \to o) \to o$), gradable adjectives (e.g. tall) are interpreted in the type $\delta \to (\iota \to o)$, rather than in the type $((\iota \to o) \to o) \to (\iota \to o)$.

[3] In [11], Montague uses a *direct interpretation* of natural language into logical models, which does not proceed via the translation of natural language into the language of some logic. As a result, [11] does not identify a logical language with EFL-typed expressions. However, since such a language is easily definable (cf. Definition 7), we hereafter refer to any EFL-typed language as an 'EFL-language'.

The temperature is ninety.

$$\frac{\text{The temperature rises.}}{\text{Ninety rises.}} \qquad (\star)$$

The origin of this problem lies in the different readings of the phrase the temperature in the two premises of $(\star)$, and in the inability of extensional semantics (e.g. [1]) to accommodate one of these readings: In the second premise, the occurrence of the phrase the temperature is intuitively interpreted as a *function* (type $\sigma \to \iota$) from worlds and times (or 'indices', type σ) to the temperature at those worlds at the given times.[4] In the first premise, the occurrence of the temperature is interpreted as the *value* (type ι) of this function at the actual world at the current time. Since extensional semantics do not have a designated type for indices – such that they also lack a type for index-to-value functions –, they are unable to capture the reading of the phrase the temperature from the second premise.

The inference from the conjunction of the translations of the two premises of $(\star)$ to the translation of the conclusion of $(\star)$ in classical extensional logic is given below. There, constants and variables are subscripted by their semantic type:

$$\frac{\exists x_\iota \forall y_\iota.(\text{TEMP}_{\iota \to t}(y) \leftrightarrow x = y) \wedge x = \textit{ninety}_\iota}{\exists x_\iota \forall y_\iota.(\text{TEMP}_{\iota \to t}(y) \leftrightarrow x = y) \wedge \text{RISE}_{\iota \to t}(x)}{\text{RISE}_{\iota \to t}(\textit{ninety}_\iota)} \qquad (\text{ext-}\star)$$

The asserted identity of the temperature x with the value *ninety* in the first premise of $(\text{ext-}\star)$ justifies the (counterintuitive) substitution of the translation of the temperature from the second premise by the translation of the name ninety.

Montague's semantics from [12] blocks this counterintuitive inference by interpreting intensional common nouns (e.g. temperature) and intransitive verbs (e.g. rise) as (characteristic functions of) sets of *individual concepts* (type $(\sigma \to \iota) \to t$), and by restricting the interpretation of the copula is to a relation between the *extensions* of two individual concepts at the actual world @ at the current time (i.e. to a curried relation between *individuals*, type $\iota \to (\iota \to t)$). Since the first premise of $(\star)$ thus only asserts the identity of the individual 'the temperature at @' and the value ninety, it blocks the substitution of the individual concept-denoting phrase the temperature in the second premise of $(\star)$ by the name ninety.

The invalidity of the inference from the conjunction of the two premises of $(\star)$ to the conclusion of $(\star)$ in a streamlined version of Montague's Intensional Logic (cf. [5]) is captured in $(\text{PTQ-}\star)$. There, *ninety* denotes the constant function from indices to the type-ι denotation of the term *ninety* (s.t. $\forall i_\sigma.\, \textit{ninety}(i) = \textit{ninety}$):

$$\frac{\exists c_{\sigma \to \iota} \forall c1_{\sigma \to \iota}.(\textit{temp}_{(\sigma \to \iota) \to t}(c_1) \leftrightarrow c = c_1) \wedge c(@_\sigma) = \textit{ninety}_\iota}{\exists c_{\sigma \to \iota} \forall c1_{\sigma \to \iota}.(\textit{temp}_{(\sigma \to \iota) \to t}(c_1) \leftrightarrow c = c_1) \wedge \textit{rise}_{(\sigma \to \iota) \to t}(c)}{\textit{rise}_{(\sigma \to \iota) \to t}(\textit{ninety}_{\sigma \to \iota})} \qquad (\text{PTQ-}\star)$$

---

[4] As a result, this reading is sometimes called the *function reading* (cf. [8]). The reading of the phrase the temperature from the first premise is called the *value reading*.

Since Montague's EFL-models from [11] only assume basic types for individuals and propositions (s.t. they do not allow the construction of individual concepts), it is commonly assumed that these models are unable to block the inference from ($\star$). We show below that this assumption is mistaken.

3   Connecting PTQ and EFL

To demonstrate that Montague's models from [11] enable a solution to Partee's temperature puzzle, we first identify a strategy for the EFL-representation of indices and truth-values, which allows us to code every object in the class of models from [12] as an object in the class of models from [11] (in Sect. 3.1). We will see (in Sect. 3.3; cf. Sect. 3.2) that this strategy enables the translation of every linguistically relevant term[5] in a streamlined version of the language from [12] into a term in the language of a logic with basic types ι and o. We will then show that our translation avoids the emergence of Partee's temperature puzzle (in Sect. 4).

3.1   Coding PTQ-Objects as EFL-Objects

To enable the translation of Montague's PTQ-translations from [12] into terms of an EFL-typed language, we code indices and truth-values as type-o propositions. This coding is made possible by the interpretation of o as the type for functions from indices to truth-values, such that there are injective maps, $\lambda i_\sigma \lambda j_\sigma . j = i$ and $\lambda \theta_t \lambda i_\sigma . \theta$, from indices and truth-values to propositions. These maps enable the representation of indices via characteristic functions of the singleton sets containing these indices, and the representation of truth-values via constant functions from indices to these truth-values.

The existence of these maps suggests the possibility of replacing all non-propositional occurrences[6] of the types σ and t and all occurrences of the type $\sigma \to t$ by the type o. This replacement scheme converts the type for individual concepts into the type $o \to \iota$,[7] and converts the type for (characteristic functions of) sets of individuals into the type $\iota \to o$. The type for sets of individual concepts is then converted into the type $(o \to \iota) \to o$. However, this scheme fails to associate the types of some EFL-expressions from [12] with the ι- and o-based types from [11]. In particular, it associates the PTQ-type of common nouns, $\sigma \to (\iota \to t)$, with the type $o \to (\iota \to o)$, rather than with the type of common nouns from [11], $\iota \to o$. But this is undesirable.

---

[5] These are terms which are associated with PTQ-expressions.

[6] The latter are occurrences of index- and truth-value types which are not a constituent of the propositional type $\sigma \to t$. The need for the distinction between propositional and non-propositional occurrences of the types σ and t is discussed below.

[7] We will see that, since no other syntactic category of the PTQ-fragment receives an interpretation in a construction involving the type $o \to \iota$, semantic types involving this type still motivate the syntactic categories. This contrasts with the coding of degrees as equivalence classes of individuals (in [2]), which assigns adjectives (originally, type $\delta \to (\iota \to t)$) the type for verbal modifiers, $(\iota \to t) \to (\iota \to t)$.

To ensure the correct conversion of PTQ-types into EFL-types, we refine the above replacement scheme into the type-conversion rule from Definition 4. In the definition of this rule, the sets of TY_2 types and of EFL-types[8] (hereafter called 'TY_1 types'), and a TY_2 type's o-normal form are defined as follows:

Definition 1 (TY_2 Types). *The set* 2Type *of* TY_2 *types is the smallest set of strings such that* ι, σ, $t \in$ 2Type *and, for all* $\alpha, \beta \in$ 2Type, $(\alpha \to \beta) \in$ 2Type.

Definition 2 (TY_1 Types). *The set* 1Type *of* TY_1 *types is the smallest set of strings such that* ι, $o \in$ 1Type *and, for all* $\alpha, \beta \in$ 1Type, $(\alpha \to \beta) \in$ 1Type.

It is clear from the above and from the definition of o as $\sigma \to t$ that all TY_1 types are TY_2 types, but not the other way around. In particular, the TY_2 types σ, t, and constructions out of these types (esp. the types $\sigma \to \iota$, $(\sigma \to \iota) \to t$, and $\sigma \to (\iota \to t)$) are not TY_1 types.

Definition 3 (o-normal form). *An o-normal form, β, of a TY_2 type α is a TY_2 type that has been obtained from α via a unary variant, $^\Diamond$, of the permutation relation from [14, p. 119]. The former is defined as follows, where $0 \leq n \in \mathbb{N}$:[9]*

$$(i) \qquad\qquad \sigma^\Diamond = \sigma; \qquad\quad t^\Diamond = t; \qquad\quad \iota^\Diamond = \iota;$$
$$(ii) \quad (\sigma \to (\alpha_1 \to (\ldots \to (\alpha_n \to t))))^\Diamond = \alpha_1^\Diamond \to (\ldots \to (\alpha_n^\Diamond \to (\sigma \to t)));$$
$$(iii) \quad (\alpha \to \beta)^\Diamond = (\alpha^\Diamond \to \beta^\Diamond), \text{ if } (\alpha \to \beta) \neq (\sigma \to (\alpha_1 \to (\ldots \to (\alpha_n \to t))))$$

Definition 3 identifies the type for functions from individuals to propositions, $\iota \to (\sigma \to t)$ (i.e. the type for 'properties' of individuals), as the o-normal form of the type for parametrized sets of individuals, $\sigma \to (\iota \to t)$.

The conversion of TY_2 into TY_1 types is defined below:

Definition 4 (Type-conversion). *The relation ξ connects TY_2 types with TY_1 types via the following recursion:*

$$I. \ (i) \qquad \xi(\iota) = \iota;$$
$$(ii) \qquad \xi(\sigma) = \xi(t) = o, \quad \text{when } \sigma \text{ resp. } t \text{ does not occur in } (\sigma \to t);$$
$$II. \ (i) \ \xi(\sigma \to t) = o;$$
$$(ii) \ \xi(\alpha \to \beta) = \xi(\gamma \to \delta) = (\xi(\gamma) \to \xi(\delta)), \text{ where } (\gamma \to \delta) = (\alpha \to \beta)^\Diamond$$
$$\text{and } (\alpha \to \beta) \neq (\sigma \to t).$$

Clauses I and II.(i) capture the conversion of basic and propositional TY_2 types. Clause II.(ii) captures the conversion of all other complex TY_2 types. Specifically, the conjunction of this clause with the clauses for the conversion of basic TY_2 types enables the conversion of the type for individual concepts, $\boldsymbol{\sigma} \to \iota$, to the type for proposition-to-individual functions, $\boldsymbol{o} \to \iota$, and of the type for sets of individuals, $\iota \to t$, to the type for properties of individuals, $\iota \to \boldsymbol{o}$. The conjunction of clause II.(ii) with the converted TY_2 types for individual concepts and truth-values then enables the conversion of the type for sets of individual

---

[8] Since n-ary functions can be coded as unary functions of a higher type (cf. [16]), our definition of TY_1 types neglects n-ary function types, which are assumed in [11].

[9] Following Muskens, we write '$^\Diamond$' in postfix notation, such that '$\alpha^\Diamond$' denotes $^\Diamond(\alpha)$.

concepts, $(\sigma \to \iota) \to t$, to the type for properties of proposition-to-individual functions, $(o \to \iota) \to o$. Since the type $\iota \to (\sigma \to t)$ is the o-normal form of the type $\sigma \to (\iota \to t)$ (cf. clause II.(ii)), the type $\iota \to o$ is the converted type of both parametrized sets of individuals (type $\sigma \to (\iota \to t)$) and of properties of individuals (type $\iota \to (\sigma \to t)$).

Notably, the restriction of clauses I.(ii) and II.(ii) to non-propositional types, resp. to o-normal forms prevents the undesired conversion of the type $\sigma \to t$ into the type for properties of propositions, $o \to o$, and of the type for parametrized sets of type-α objects, $\sigma \to (\alpha \to t)$, to the type for functions from propositions to properties of type-α objects (type $o \to (\alpha \to o)$). The converted TY$_2$ types of all classes of expressions from the PTQ-fragment are listed in Table 1:[10]

Table 1. TY$_2$ and converted TY$_2$ (i.e. TY$_1$) types of PTQ-expressions.

Cat'y	$\alpha \in$ 2Type	$\xi(\alpha)$	Cat'y	$\alpha \in$ 2Type	$\xi(\alpha) \in$ 1Type
Name	ι	ι	NP	$(\sigma \to (\iota \to t)) \to t$	$(\iota \to o) \to o$
S	t	o	SCV	$(\sigma \to t) \to (\iota \to t)$ [6]	$o \to (\iota \to o)$
C, SAV	$(\sigma \to t) \to t$	$o \to o$	ADV	$(\sigma \to (\iota \to t)) \to (\iota \to t)$	$(\iota \to o) \to (\iota \to o)$
CN, IV	$\iota \to t$	$\iota \to o$ [2,3]	CN, IV	$(\sigma \to \iota) \to t$	$(o \to \iota) \to o$
TV [4]	$\iota \to (\iota \to t)$	$\iota \to (\iota \to o)$	ICV	$(\sigma \to (\iota \to t)) \to (\iota \to t)$	$(\iota \to o) \to (\iota \to o)$
TV [5]	$(\sigma \to ((\sigma \to (\iota \to t)) \to t)) \to (\iota \to t)$				$((\iota \to o) \to o) \to (\iota \to o)$
DET	$(\sigma \to (\iota \to t)) \to ((\sigma \to (\iota \to t)) \to t)$				$(\iota \to o) \to ((\iota \to o) \to o)$
P [8]	$\iota \to ((\sigma \to (\iota \to t)) \to (\iota \to t))$				$\iota \to ((\iota \to o) \to (\iota \to o))$
P	$(\sigma \to ((\sigma \to (\iota \to t)) \to t)) \to ((\sigma \to (\iota \to t)) \to (\iota \to t))$				$((\iota \to o) \to o) \to ((\iota \to o) \to (\iota \to o))$

This completes our discussion of the TY$_1$-coding of TY$_2$ types. To show the possibility of interpreting the PTQ-fragment in TY$_1$ models, we next describe the class of languages of the logics TY$_2$ and TY$_1$ (in Sect. 3.2), and translate all linguistically relevant TY$_2$ terms into terms of the logic TY$_1$ (in Sect. 3.3). We then observe that this translation is entailment-preserving.

3.2 The Languages of TY$_2$ and TY$_1$

The languages of the logics TY$_2$ and TY$_1$ are defined as countable sets $\cup_{\alpha \in 2\text{Type}} L_\alpha$, resp. $\cup_{\beta \in 1\text{Type}} L_\beta$, of uniquely typed non-logical constants. For every TY$_2$ type α and TY$_1$ type β, we further assume a countable set $\mathcal{V}^2_\alpha$, resp. $\mathcal{V}^1_\beta$ of uniquely typed variables, with '$\cup_{\alpha \in 2\text{Type}} \mathcal{V}^2_\alpha$' abbreviated as '$\mathcal{V}^2$' and '$\cup_{\beta \in 1\text{Type}} \mathcal{V}^1_\beta$' abbreviated as '$\mathcal{V}^1$'. From these basic expressions, we form complex terms inductively with the help of functional application, lambda abstraction, and the constants for *falsum*, $\perp$, and logical implication, $\to$.

---

[10] These type-assignments incorporate the type-ι interpretation of names and the meaning postulates from [12, pp. 263–264]. The latter are given in square brackets.

In the definition of TY_2 terms, the set CoType of conjoinable TY_2 types is defined as follows (cf. [15]):

Definition 5 (Conjoinable TY_2 Types). *The set* CoType *of conjoinable types of the logic* TY_2 *is the smallest set of strings such that, if* $\alpha_1, \ldots, \alpha_n \in$ 2Type, *then* $\alpha_1 \to (\ldots \to (\alpha_n \to t)) \in$ CoType, *where* $0 \le n \in \mathbb{N}$.

According to the above, a TY_2 term has a conjoinable type if its type is either the truth-value type t or a construction to the type t (via one or more applications of the rule from Definition 1). In these two cases, we say that the term is *conjoinable*.

Definition 6 (TY_2 Terms). *Let* $\alpha, \beta \in$ 2Type, *and let* $\epsilon \in$ CoType. *The set* T_α^2 *of* TY_2 *terms of type* α *is then defined as follows:*

(i) $L_\alpha^2, \mathcal{V}_\alpha^2 \subseteq T_\alpha^2, \quad \perp \in T_t^2$;
(ii) *If* $B \in T_{\alpha \to \beta}^2$ *and* $A \in T_\alpha^2$, *then* $(B(A)) \in T_\beta^2$;
(iii) *If* $A \in T_\beta^2$ *and* $x \in \mathcal{V}_\alpha^2$, *then* $(\lambda x.A) \in T_{\alpha \to \beta}^2$;
(iv) *If* $B, C \in T_\epsilon^2$, *then* $(B \to C) \in T_t^2$.

Clause (i) identifies all members of L_α^2 and $\mathcal{V}_\alpha^2$ as TY_2 terms. Clauses (ii) and (iii) identify the results of application and abstraction as TY_2 terms. Clause (iv) specifies the formation of complex TY_2 terms. From $\perp$ and $\to$, the familiar TY_2 connectives and quantifiers are standardly obtained (cf. [7]).

We next define the terms of the logic TY_1. Notably, since TY_1 does not have a type for truth-values, the TY_2 constants $\perp$ (type t) and $\to$ (type $\epsilon \to (\epsilon \to t)$) are not available in TY_1. The non-logical constants $\bigcirc$ (type o) and $\dot{\to}$ (type $\varepsilon \to (\varepsilon \to o)$) serve as their single-type stand-ins, where ε is in the proper subset, PropType $= \{\alpha_1 \to (\ldots \to (\alpha_n \to o)) \,|\, \alpha_1, \ldots, \alpha_n \in$ 1Type$\}$, of the set of conjoinable TY_2 types.[11] We hereafter call members of this set *propositional types*.

Definition 7 (TY_1 Terms). *Let* $\alpha, \beta \in$ 1Type, *and let* $\varepsilon \in$ PropType. *The set* T_α^1 *of* TY_1 *terms of type* α *is then defined as follows:*

(i) $L_\alpha^1, \mathcal{V}_\alpha^1 \subseteq T_\alpha^1, \quad \bigcirc \in T_o^1$;
(ii) *If* $B \in T_{\alpha \to \beta}^1$ *and* $A \in T_\alpha^1$, *then* $(B(A)) \in T_\beta^1$;
(iii) *If* $A \in T_\beta^1$ *and* $x \in \mathcal{V}_\alpha^1$, *then* $(\lambda x.A) \in T_{\alpha \to \beta}^1$;
(iv) *If* $B, C \in T_\varepsilon^1$, *then* $(B \dot{\to} C) \in T_o^1$.

The typing[12] of $\dot{\to}$, B, and C in clause (iv) suggests that the term $(B \dot{\to} C)$ be instead written as '$\dot{\to}(B)(C)$'. Our use of infix notation for $\dot{\to}$ (and similarly,

---

[11] [17] and [4] use a similar strategy for the introduction of propositional connectives.

[12] Since we only stipulate that $\varepsilon \in$ PropType, clause (iv) describes $\dot{\to}$ as a non-uniquely typed constant, which applies to pairs of arguments of all propositional TY_1 types. To avoid an extension of the TY_1 type system via polymorphic types, we assume a *schematic* (or *abbreviatory*) *polymorphism* of types. The latter is a syntactic device whereby a metatheoretical symbol is used to abbreviate a range of (monomorphic) types. Thus, in (iv), ε may be instantiated by any of the elements in PropType. The constant $\dot{\to}$ then represents a family, $\{\dot{\to}_{\varepsilon \to (\varepsilon \to o)} \,|\, \varepsilon \in$ PropType$\}$, of *distinct identical-looking constants*, one for each type.

for the TY_1 proxies of all other logical TY_2 constants; cf. Notation 1) is intended to remind the reader of their emulated logical role (cf. Definition 8).

From ⓛ and $\overset{\cdot}{\to}$, the TY_1 proxies of the other truth-functional connectives and quantifiers are easily obtained. In particular, the TY_1 proxies for the logical constants $\top, \forall, =, \neg, \wedge,$ and $\square$ (i.e. ⓣ, $\bigwedge, \overset{\cdot}{=}, \overset{\cdot}{\neg}, \overset{\cdot}{\wedge},$ and $\overset{\cdot}{\square}$) are obtained by variants of the definitions from [7]. Below, we let A, x and y, X (or B, C), and Y be variables (resp. constants) of the type o, α, $\alpha \to o$, resp. $(\alpha \to o) \to o$, where $\alpha \in 1\mathsf{Type}$:

Notation 1 *We write*

ⓣ	*for*	(ⓛ $\overset{\cdot}{\to}$ ⓛ);	$(\bigwedge x. A)$ *for*	$((\lambda x.ⓣ) \overset{\cdot}{\to} (\lambda x. A))$;
$B \overset{\cdot}{=} C$	*for*	$(\bigwedge Y. Y(B) \overset{\cdot}{\to} Y(C))$;	$\overset{\cdot}{\neg} B$ *for*	$(\lambda x. B(x) \overset{\cdot}{=} ⓛ)$;
$(B \overset{\cdot}{\wedge} C)$	*for*	$(\lambda x.(\lambda X. X(B \overset{\cdot}{=} C)) \overset{\cdot}{=} (\lambda X. X(ⓣ)))$		
$(B \overset{\cdot}{\vee} C)$	*for*	$\overset{\cdot}{\neg}(\overset{\cdot}{\neg} B \overset{\cdot}{\wedge} \overset{\cdot}{\neg} C)$;	$\overset{\cdot}{\square} A$ *for*	$(\bigwedge x. x \overset{\cdot}{=} ⓛ \overset{\cdot}{\vee} (x \overset{\cdot}{\to} A))$

The TY_1 stand-ins, $\overset{\cdot}{\neq}, \overset{\cdot}{\leftrightarrow}, \overset{\cdot}{\bigvee},$ and $\overset{\cdot}{\diamond}$, of the familiar symbols for inequality, (material) biimplication, the existential quantifier, and the modal diamond operator have their expected definitions.

The behavior of ⓛ, $\overset{\cdot}{\to}$, and of the defined constants from Notation 1 is governed by the constraints from Definition 8:

Definition 8 (Constraints on L^1-constants). *The interpretations of the TY_1 constants* ⓛ *and* $\overset{\cdot}{\to}$ *obey the following semantic constraints:*[13]

$$(C1) \quad ⓛ = (\lambda i_\sigma. \bot); \quad\quad (C2) \quad (B \overset{\cdot}{\to} C) = (\lambda i_\sigma \forall x. B(x)(i) \to C(x)(i))$$

The constraints (C1) and (C2) define the designated TY_1 constants ⓛ and $\overset{\cdot}{\to}$ as the results of lifting the TY_2 connectives $\bot$ and $\to$ to terms of the logic TY_1.[14] In particular, (C1) defines the constant ⓛ as the designator of the constant function from indices to *falsum*. From (C1), (C2), and Notation 1, the constraints for the remaining designated TY_1 constants are easily obtained. Since the TY_1 constants $\overset{\cdot}{\wedge}, \overset{\cdot}{\vee},$ and $\overset{\cdot}{\neg}$ are η-equivalent to their TY_2 constraints, we hereafter use instead the familiar connectives $\wedge$, $\vee$, and $\neg$.

3.3   Translating $\mathcal{L}^{TY_2}$ into $\mathcal{L}^{TY_1}$

To prepare the TY_1 translation of the PTQ-fragment, we next introduce the particular TY_2 and TY_1 languages, $\mathcal{L}^2$ and $\mathcal{L}^1$, whose constants are associated with the lexical elements of the PTQ-fragment. Following a streamlined presentation of Montague's PTQ-to-IL (or TY_2) translation from [12] (cf. [5]), we then identify a relation between $\mathcal{L}^2$- and $\mathcal{L}^1$-terms.

Tables 2 and 3 contain the non-logical constants of the designated languages $\mathcal{L}^2$, resp. $\mathcal{L}^1$. The small grey tables introduce our notational conventions for variables. In the tables, brackets contain the relevant meaning postulates from [12],

---

[13] These constraints are formulated in the TY_1 metatheory, TY_2 (cf. Sect. 3.3).

[14] This is reminiscent of the translation of dynamic to typed terms from [13, p. 9].

resp. the constants' interpretive domains from [11]. We will abbreviate x_1, x_2, and x_3 as 'x', 'y', resp. 'z', abbreviate i_1, i_2, and i_3 as 'i', 'j', resp. 'k', and abbreviate p_1, p_2, and p_3 and c_1, P_1, and Q_1 as 'p', 'q' resp. 'r' and 'c', 'P', resp. 'Q'.

Table 2. $\mathcal{L}^2$ constants and variables.

Constant				TY$_2$ Type		Var.	TY$_2$ Type
@ σ $\mid$ θ t $\mid$				$ninety$ $(\sigma \to \iota)$		$i_1, \ldots, i_n$	σ
$john, mary, bill, ninety$				ι		$x_1, \ldots, x_n$	ι
$man, woman, park, fish, pen, unicorn$				$\sigma \to (\iota \to t)$	[2]	$c_1, \ldots, c_n$	$\sigma \to \iota$
$run, walk, talk$				$\sigma \to (\iota \to t)$	[3]	$p_1, \ldots, p_n$	$\sigma \to t$
$temp, price, rise, change$				$\sigma \to ((\sigma \to \iota) \to t)$		$P_1, \ldots, P_n$	$\sigma \to (\iota \to t)$
$find, lose, eat, love, date$				$\iota \to (\sigma \to (\iota \to t))$ [4]		$\vec{T}$	$\sigma \to ((\sigma \to \iota) \to t)$
$believe, assert$				$(\sigma \to t) \to (\sigma \to (\iota \to t))$ [6]		$\vec{Q}$	$\sigma \to ((\sigma \to (\iota \to t)) \to t)$
$seek, conceive$				$(\sigma \to ((\sigma \to (\iota \to t)) \to t)) \to (\sigma \to (\iota \to t))$	[5]		
$rapidly, slowly, \ldots, allegedly, try, wish$				$(\sigma \to (\iota \to t)) \to (\sigma \to (\iota \to t))$	[7]		
in				$\iota \to ((\sigma \to (\iota \to t)) \to (\sigma \to (\iota \to t)))$	[8]		
$about$				$(\sigma \to ((\sigma \to (\iota \to t)) \to t)) \to ((\sigma \to (\iota \to t)) \to (\sigma \to (\iota \to t)))$			
All TY$_1$ constants from Table 3 are members of $\mathcal{L}^2$.							

Table 3. $\mathcal{L}^1$ constants and variables.

Constant			TY$_1$ Type		Var.	TY$_1$ Type	
$\top, \bot, @$ o $\mid$	$\wedge, \vee$		$(\alpha \to o) \to o$		$x_1, \ldots, x_n$ ι $\mid$	$\vec{p}$	o
$\Box, \Diamond$ $o \to o$ $\mid$	$\to, \doteq, \neq, \leftrightarrow$		$\alpha \to (\alpha \to o)$		$c_1, \ldots, c_n$	$o \to \iota$	
$john, mary, bill, ninety$			ι	[U_0]	$P_1, \ldots, P_n$	$\iota \to o$	
$man, woman, park, fish, pen, unicorn$			$\iota \to o$	[U_4]	$T_1, \ldots, T_n$	$(o \to \iota) \to o$	
$run, walk, talk$			$\iota \to o$	[U_2]	$Q_1, \ldots, Q_n$	$(\iota \to o) \to o$	
$temp, price, rise, change$			$(o \to \iota) \to o$				
$find, lose, eat, love, date$			$\iota \to (\iota \to o)$	[curried U_3]			
$believe, assert$			$o \to (\iota \to o)$	[U_5]			
$seek, conceive$			$((\iota \to o) \to o) \to (\iota \to o)$				
$rapidly, slowly, \ldots, allegedly, try, wish$			$(\iota \to o) \to (\iota \to o)$	[U_6]			
in			$\iota \to ((\iota \to o) \to (\iota \to o))$	[U_6]			
$ninety$ $(o \to \iota)$ $\mid$	$about$		$((\iota \to o) \to o) \to ((\iota \to o) \to (\iota \to o))$				

We denote the sets of TY$_2$ and TY$_1$ terms which are obtained from $\mathcal{L}^2$ and $\mathcal{L}^1$ via the operations from Definitions 6 and 7 by '$\mathcal{T}^2$', resp. '$\mathcal{T}^1$'.

Note that the language $\mathcal{L}^1$ adopts the individual constants, $john, mary, bill$, and $ninety$, of the language $\mathcal{L}^2$. To connect the designated languages of the logics TY$_2$ and TY$_1$, we further assume that each term from $\mathcal{L}^1$ is also a member of $\mathcal{L}^2$ (s.t. $\mathcal{L}^1$ is a sublanguage of $\mathcal{L}^2$), and that the designated TY$_2$ frame $\mathcal{F}^2$ and interpretation function $\mathcal{I}_{\mathcal{F}^2}$ embed the designated frame $\mathcal{F}^1$ and interpretation function $\mathcal{I}_{\mathcal{F}^1}$ of the logic TY_1, such that $\mathcal{F}^1 = \mathcal{F}^2 \restriction_{1\mathrm{Type}}$ and $\mathcal{I}_{\mathcal{F}^1} = \mathcal{I}_{\mathcal{F}^2 \restriction_{1\mathrm{Type}}}$.

We next give a streamlined presentation of Montague's PTQ-to-IL (or TY_2) translation from [12]:

We identify Logical Form (LF) with the component of syntactic representation which is interpreted in TY_2 models. Logical forms are translated into TY_2 terms via the process of type-driven translation (cf. [9]). The latter proceeds in two steps, by first defining the translations of lexical elements, and then defining the translations of non-lexical elements compositionally from the translations of their constituents.

Definition 9 (Basic TY_2 Translations). *The base rule of type-driven translation translates the lexical PTQ-elements[15] into the following TY_2 terms[16], where $X_1, \ldots, X_n$, R, and R_1 are TY_2 variables of the types $\alpha_1, \ldots, \alpha_n$, resp. $\alpha_1 \to (\ldots \to (\alpha_n \to t))$, and where t_n is the trace of a moved constituent in a logical form that is translated as a free variable:*

John $\rightsquigarrow$ *john;*	man $\rightsquigarrow$ *man;*	walks $\rightsquigarrow$ *walk;*	who $\rightsquigarrow$ $\lambda P.P$;
ninety $\rightsquigarrow$ *ninety;*	temp. $\rightsquigarrow$ *temp;*	rise $\rightsquigarrow$ *rise;*	finds $\rightsquigarrow$ *find;*
seeks $\rightsquigarrow$ *seek;*	tries to $\rightsquigarrow$ *try;*	about $\rightsquigarrow$ *about;*	in $\rightsquigarrow$ *in;*

that $\rightsquigarrow \lambda i \lambda p.p(i)$; believes $\rightsquigarrow$ *believe;* asserts $\rightsquigarrow \lambda p \lambda i \lambda x.\, assert\,(p)(i)(x) \wedge p$;

rapidly $\rightsquigarrow \lambda P \lambda i \lambda x.\, rapidly\,(P)(i)(x) \wedge P(i)(x)$;   allegedly $\rightsquigarrow$ *allegedly;*

and $\rightsquigarrow \lambda R_1 \lambda R \lambda \vec{X}.R(\vec{X}) \wedge R_1(\vec{X})$;   necessarily $\rightsquigarrow \lambda i \lambda p.\, \Box p(i)$;

a $\rightsquigarrow \lambda P_2 \lambda i \lambda P \exists x.P_2(i)(x) \wedge P(i)(x)$; $t_n/\mathsf{he}_n \rightsquigarrow x_n\ f.\ ea.\ n \in \mathbb{N}$;

the $\rightsquigarrow \lambda P_2 \lambda i \lambda P \exists x \forall y.(P_2(i)(y) \leftrightarrow x = y) \wedge P(i)(x)$;   is $\rightsquigarrow \lambda y \lambda i \lambda x. x = y$

On the basis of the above, we define the relation between $\mathcal{T}^2$ and $\mathcal{T}^1$ terms as follows:

Definition 10 (Embedding TY_2 in TY_1). *The relation $^\bullet$ connects the designated terms of the logic TY_2 with terms of the logic TY_1, such that*

I. (i) $john^\bullet = john$; $man^\bullet = \mathbf{man}$; $walk^\bullet = \mathbf{walk}$; $temp^\bullet = \mathbf{temp}$;

$ninety^\bullet = ninety$; $ninety^\bullet = \mathbf{ninety}$; $rise^\bullet = \mathbf{rise}$; $find^\bullet = \mathbf{find}$;

$seek^\bullet = \mathbf{seek}$; $allegedly^\bullet = \mathbf{allegedly}$; $about^\bullet = \mathbf{about}$; $in^\bullet = \mathbf{in}$;

$believe^\bullet = \mathbf{believe}$; $assert^\bullet = \mathbf{assert}$; $rapidly^\bullet = \mathbf{rapidly}$; $try^\bullet = \mathbf{try}$;

$@^\bullet = @$;

(ii) $x_k^\bullet = x_k$ for $1 \le k \in \mathbb{N}$; $c_k^\bullet = c_k$ for $1 \le k \in \mathbb{N}$;

$P_k^\bullet = \mathbf{P}_k$ for $1 \le k \in \mathbb{N}$; $p_k^\bullet = p_k$ for $1 \le k \in \mathbb{N}$;

$T_k^\bullet = \mathbf{T}_k$ for $1 \le k \in \mathbb{N}$; $i_k^\bullet = p_k$ for $1 \le k \in \mathbb{N}$;

---

[15] For reasons of space, we only translate some representative elements. Expressions of the same lexical (sub-)category receive an analogous translation.

[16] To perspicuate the compositional properties of our PTQ-translations, we assign lexical PTQ-elements variants of their TY_2 types from Table 1. Thus, the translation of extensional nouns as type-$(\sigma \to (\iota \to t))$ constants facilitates the application of translations of determiners to the translations of these expressions. To enable a compositional translation of other complex expressions (e.g. the application of verb-to name-translations), we use a permutation operation on the translations' lambdas.

II. (i) $(B_{\sigma \to \beta}(A_\sigma))^\bullet = B^\bullet$ *if $\beta = t$ or $\beta = (\gamma \to t)$, where $\gamma \in$ 2Type;*

      $(B_{\alpha \to \beta}(A_\alpha))^\bullet = (B^\bullet(A^\bullet))$ *otherwise;*

(ii) $(\lambda x_\sigma.A_\beta)^\bullet = (A[x := @])^\bullet$ *if $\beta = t$ or $\beta = (\gamma \to t)$, where $\gamma \in$ 2Type;*

       $(\lambda x_\alpha.A_\beta)^\bullet = (\lambda x^\bullet.A^\bullet)$ *otherwise, if $(\alpha \to \beta) = (\alpha \to \beta)^\Diamond$;*

       $(\lambda x_\alpha.A_\beta)^\bullet = (\lambda X_\gamma^\bullet.(\lambda x.A(X))^\bullet)$ *otherwise, granted $\beta := (\gamma \to \delta)$, w.*

                            X the 1st variable that doesn't occur free in A;

(iii) $\bot^\bullet = \textcircled{1};$ $(B \to C)^\bullet = (B^\bullet \dot{\to} C^\bullet)$

In the first item from II.(ii), '$A[x := @]$' denotes the result of replacing all bound occurrences of x in A by '@'.

The translation rules from Definition 10 respect the behavior of the type converter ξ from Definition 4. Thus, the relation $^\bullet$ translates individual and propositional TY_2 terms (e.g. *ninety*, x_k, p_k) into themselves, translates TY_2 terms for individual concepts (e.g. *ninety*, c_k) into TY_1 terms for proposition-to-individual functions, and translates TY_2 terms for parametrized sets of individuals (or of individual concepts) (e.g. *man*, P_k; resp. *temp*, T_k) into TY_1 terms for properties of individuals (resp. for properties of proposition-to-individual functions).

The translation rules from clause II ensure the correct translation of complex TY_2 terms. Specifically, the rules for the TY_1 translation of $\bot$ and $\to$ (cf. clause II.(iii)) associate the logical TY_2 constants with their propositional correspondents from TY_1. From the translations in clauses I and II.(iii), the rules for application and abstraction (clause II.(i), (ii)) enable the compositional TY_1 translation of all PTQ-translations from Definition 9. In these rules, the contraints on abstraction block the undesired translation of type-$(\sigma \to t)$ terms as TY_1 terms of the type $o \to o$. The constraints on application enable the translation of the result of applying a type-$(\sigma \to t)$ (or type-$(\sigma \to (\gamma \to t))$) term to a type-$\sigma$ term.

The translations of some example TY_2 terms are given below. In these translations, the TY_1 correlates of logical TY_2 constants other than $\bot$ and $\to$ are obtained from the TY_1 translations of $\bot$ and $\to$ via the definitions of the remaining logical TY_2 constants from [7] and the conventions from Notation 1. In particular, the TY_2 constant $\top$ is translated as follows:

$$\top^\bullet = (\bot \to \bot)^\bullet = (\bot^\bullet \dot{\to} \bot^\bullet) \quad \text{(by [7]; II.(iii))} \tag{1}$$
$$= (\textcircled{1} \dot{\to} \textcircled{1}) = \textcircled{T} \quad \text{(by II.(iii); Nota. 1)}$$

The translations of $\forall$, $\exists$, $=$, $\wedge$, and $\leftrightarrow$ are analogously obtained, such that

$$(\forall x.A)^\bullet = (\bigwedge x^\bullet.A^\bullet); \quad (B = C)^\bullet = B^\bullet \dot{=} C^\bullet; \quad (B \leftrightarrow C)^\bullet = (B^\bullet \dot{\leftrightarrow} C^\bullet);$$
$$(\exists x.A)^\bullet = (\bigvee x^\bullet.A^\bullet); \quad (B \wedge C)^\bullet = (B^\bullet \wedge C^\bullet).$$

From the above translations, the translations of the copula *is* and of the determiner *the* are obtained thus:

$$\text{is} \rightsquigarrow (\lambda y \lambda i \lambda x.x = y)^\bullet = (\lambda y^\bullet(\lambda i \lambda x.x = y)^\bullet) \quad \text{(by II.(ii))} \tag{2}$$
$$= (\lambda y^\bullet \lambda x^\bullet.(x = y)^\bullet) = (\lambda y^\bullet \lambda x^\bullet.x^\bullet \dot{=} y^\bullet) = (\lambda y \lambda x.x \dot{=} y) \text{ (by II.(i), etc.)}$$

$$\text{the} \rightsquigarrow (\lambda P_2 \lambda i \lambda P \, \exists x \forall y. (P_2(i)(y) \leftrightarrow x = y) \wedge P(i)(x))^\bullet \qquad (3)$$
$$= (\lambda P_2^\bullet (\lambda i \lambda P \, \exists x \forall y. (P_2(i)(y) \leftrightarrow x = y) \wedge P(i)(x))^\bullet \quad \text{(by II.(ii))}$$
$$= \lambda P_2^\bullet \lambda P^\bullet (\exists x \forall y. (P_2(@)(y) \leftrightarrow x = y) \wedge P(@)(x))^\bullet \quad \text{(by II.(i), (ii))}$$
$$= \lambda \boldsymbol{P_2} \lambda \boldsymbol{P} \bigvee x \bigwedge y. (\boldsymbol{P_2}(y) \not\leftrightarrow x \doteq y) \wedge \boldsymbol{P}(x) \qquad \text{(by I.(ii), [7]; all of II.)}$$

Since the relation $^\bullet$ respects the structure of each TY_2 term from Definition 9, the interpretation of the PTQ-fragment in the class of designated TY_1 models preserves the entailment relation which is imposed on this fragment by its translation into the logic TY_2. This observation is captured below:

Theorem 1 (Soundness of Translation). *Let Γ and Δ, and $\Gamma^\bullet := \{\gamma^\bullet \mid \gamma \in \Gamma\}$ and $\Delta^\bullet := \{\delta^\bullet \mid \delta \in \Delta\}$ be sets of designated TY_2 formulas and their TY_1 translations. Then,*
$$\Gamma^\bullet \vdash_{\mathrm{TY}_1} \Delta^\bullet \quad \textit{iff} \quad \Gamma \vdash_{\mathrm{TY}_2} \Delta.$$

In the above case, we say that the TY_2-to-TY_1 translation is sound.

Proof. The proof relies on the definition of $^\bullet$ and on the proof theories of TY_2 and TY_1.

4 Solving the Temperature Puzzle in EFL

To illustrate Theorem 1, we next show that the TY_1(-via-TY_2) translation of the PTQ-fragment blocks Partee's temperature puzzle. Since we use the strategy of "try[ing] simplest types first" (cf. Tables 1 and 2, Definition 9), the application of the TY_2 (or TY_1) translations of intensional expressions to the translations of other PTQ-expressions needs to be handled through type-shifting. In particular, to apply[17] the TY_2 translations of determiners (e.g. the; type $(\underline{\sigma \rightarrow (\iota \rightarrow t)}) \rightarrow (\sigma \rightarrow ((\sigma \rightarrow (\iota \rightarrow t)) \rightarrow t)))$ to the TY_2 translations of intensional common nouns (e.g. temperature; type $\sigma \rightarrow ((\sigma \rightarrow \iota) \rightarrow t))$, we introduce the *extensionalization* operator *ext*. This operator sends the designators of parametrized sets of *individual concepts* (type $\sigma \rightarrow ((\sigma \rightarrow \iota) \rightarrow t))$ to the designators of parametrized sets of *individuals* (type $\sigma \rightarrow (\iota \rightarrow t))$.

Definition 11 (Extensionalization). *The function $ext := \lambda T \lambda i \lambda x \exists c. T(i)(c) \wedge x = c(@)$ sends type-$(\sigma \rightarrow ((\sigma \rightarrow \iota) \rightarrow t))$ terms to type-$(\sigma \rightarrow (\iota \rightarrow t))$ terms.*

The operator *ext* enables the 'extensionalization' of the TY_2 translation, *temp*, of the noun temperature to the TY_2 term $\lambda i \lambda x \exists c. temp(i)(c) \wedge x = c(@)$. This term denotes a function from indices to the set of individuals whose members are identical to the result of applying some type-$(\sigma \rightarrow \iota)$ witness, c, of the property denoted by *temp* to the current index. To prevent an extensional interpretation of the second premise from $(\star)$ (cf. (ext-$\star$)), we restrict *ext* to the translations of *nouns*.

---

[17] Here, the type of the argument is underlined.

The possibility of interpreting intensional nouns in the type $\sigma \to (\iota \to t)$ enables the TY_1 translation of the first premise from $(\star)$:

1. $[_{NP}\text{ninety}] \rightsquigarrow ninety^\bullet = ninety$ $\qquad\qquad\qquad\qquad\qquad\qquad$ (4)

2. $[_{TV}\text{is}] \rightsquigarrow (\lambda y \lambda i \lambda x.x = y)^\bullet = (\lambda y \lambda x.x \doteq y)$

3. $[_{VP}[\text{is}][\text{ninety}]] \rightsquigarrow (\lambda i \lambda x.x = ninety)^\bullet$
   $= (\lambda y \lambda x.x \doteq y)[ninety] = (\lambda x.x \doteq ninety)$

4. $[_{N}\text{temperature}] \rightsquigarrow (ext(temp))^\bullet = (\lambda i \lambda x \exists c.\, temp\,(i)(c) \wedge x = c(@))^\bullet$
   $= (\lambda x \exists c.\, temp\,(@)(c) \wedge x = c(@))^\bullet = (\lambda x^\bullet \bigvee c^\bullet.\, temp^\bullet(c^\bullet) \wedge x^\bullet \doteq c^\bullet(@^\bullet))$
   $= \lambda x \bigvee c.\, \boldsymbol{temp}\,(c) \wedge x \doteq \boldsymbol{c}\,(@)$

5. $[_{DET}\text{the}] \rightsquigarrow (\lambda P_2 \lambda i \lambda P \exists x \forall y.(P_2(i)(y) \leftrightarrow x = y) \wedge P(i)(x))^\bullet$
   $= \lambda P_2 \lambda P \bigvee x \bigwedge y.(P_2(y) \leftrightarrow x \doteq y) \wedge P(x)$

6. $[_{NP}[_{DET}\text{the}][_{N}\text{temperature}]]$
   $\rightsquigarrow (\lambda i \lambda P \exists x \forall y.((\exists c.\, temp\,(i)(c) \wedge y = c(@)) \leftrightarrow x = y) \wedge P(i)(x))^\bullet$
   $= (\lambda P_2 \lambda P \bigvee x \bigwedge y.(P_2(y) \leftrightarrow x \doteq y) \wedge P(x))\,[\lambda z \bigvee c.\, \boldsymbol{temp}\,(c) \wedge z \doteq \boldsymbol{c}\,(@)]$
   $= \lambda P \bigvee x \bigwedge y.((\bigvee c.\, \boldsymbol{temp}\,(c) \wedge y \doteq \boldsymbol{c}\,(@)) \leftrightarrow x \doteq y) \wedge P(x)$

7. $[_{S}[_{NP}[_{DET}\text{the}][_{N}\text{temperature}]][_{VP}[\text{is}][\text{ninety}]]]$
   $\rightsquigarrow (\lambda i \exists x \forall y.((\exists c.\, temp\,(i)(c) \wedge y = c(@)) \leftrightarrow x = y) \wedge x = ninety)^\bullet$
   $= (\lambda P \bigvee x \bigwedge y.((\bigvee c.\, \boldsymbol{temp}\,(c) \wedge y \doteq \boldsymbol{c}\,(@)) \leftrightarrow x \doteq y) \wedge P(x))\,[\lambda z.z \doteq ninety]$
   $= \bigvee x \bigwedge y.((\bigvee c.\, \boldsymbol{temp}\,(c) \wedge y \doteq \boldsymbol{c}\,(@)) \leftrightarrow x \doteq y) \wedge x \doteq ninety$

Notably, the term from (4.7) does not result from the term in the first premise of $(PTQ\text{-}\star)$ by replacing 'c' and 'c_1' by '$\boldsymbol{c}$' and '$\boldsymbol{c_1}$', and by replacing '$temp$' and '$rise$' by '$\boldsymbol{temp}$', resp. '$\boldsymbol{rise}$'. In particular, while the term in the first premise of $(PTQ\text{-}\star)$ states the existence of a unique witness of the type-$((\sigma \to \iota) \to t)$ property of being a temperature, the term from (4.7) only states the existence of a unique witness of the TY_1 correlate of the type-$(\sigma \to (\iota \to t))$ property of being the temperature *at the current index*. Yet, since the occurrence of the temperature in the first premise of $(\star)$ receives an extensional interpretation (type ι), this weakening is unproblematic. We will see at the end of this section that (the TY_2 correlate of) our weaker TY_1 term still blocks Partee's temperature puzzle.

To enable an *intensional* (type-$(\sigma \to ((\sigma \to ((\sigma \to \iota) \to t)) \to t)))$ interpretation of the phrase the temperature, we introduce the *intensionalization* operator *int*. This operator sends the designators of individuals to the designators of individual concepts, and sends the designators of functions from parametrized sets of *individuals* to parametrized generalized quantifiers over *individuals* (type $(\sigma \to (\iota \to t)) \to (\sigma \to ((\sigma \to (\iota \to t)) \to t)))$ to the designators of functions from parametrized sets of *individual concepts* to generalized quantifiers over *individual concepts* (type $(\sigma \to ((\sigma \to \iota) \to t)) \to (\sigma \to ((\sigma \to ((\sigma \to \iota) \to t)) \to t))))$.[18]

---

[18] Since this operator is restricted to the types of proper names and determiners, it cannot be used to provide an intensional translation of the first premise from $(\star)$

Definition 12 (Intensionalization). *The operator 'int' then works as follows:*

$$int\,(ninety) := ninety$$
$$int\,(\lambda P_2 \lambda i \lambda P\, \exists x.\, P_2(i)(x) \wedge P(i)(x)) := \lambda T_2 \lambda i \lambda T\, \exists c.\, T_2(i)(c) \wedge T(i)(c)$$
$$int\,(\lambda P_2 \lambda i \lambda P\, \forall x.\, P_2(i)(x) \to P(i)(x)) := \lambda T_2 \lambda i \lambda T\, \exists c.\, T_2(i)(c) \to T(i)(c)$$
$$int\,(\lambda P_2 \lambda i \lambda P\, \exists x \forall y.\, (P_2(i)(y) \leftrightarrow x = y) \wedge P(i)(x))$$
$$:= \lambda T_2 \lambda i \lambda T\, \exists c \forall c_2.\, (T_2(i)(c_2) \leftrightarrow c = c_2) \wedge T(i)(c)$$

The operator *int* is an 'ι-to-$(\sigma \to \iota)$'-restricted partial variant of the intensionalization operator for extensional TY_2 terms from [6] (cf. [3, Chap. 8.4]). This operator systematically replaces each occurrence of ι in the type of a linguistic expression by the type $\sigma \to \iota$. As a result, the type for parametrized generalized quantifiers over individuals, $\sigma \to ((\sigma \to (\iota \to t)) \to t)$, will be replaced by the type $\sigma \to ((\sigma \to ((\sigma \to \iota) \to t)) \to t)$.

The interpretation of intensional noun phrases in the type $\sigma \to ((\sigma \to ((\sigma \to \iota) \to t)) \to t)$ enables the TY_1 translation of the second premise from $(\star)$:

1. $[_{IV}\text{rises}] \rightsquigarrow rise^{\bullet} = \boldsymbol{rise}$ \hfill (5)

2. $[_N\text{temperature}] \rightsquigarrow temp^{\bullet} = \boldsymbol{temp}$

3. $[_{DET}\text{the}] \rightsquigarrow (int\,(\lambda P_2 \lambda i \lambda P\, \exists x \forall y.\, (P_2(i)(y) \leftrightarrow x = y) \wedge P(i)(x)))^{\bullet}$
   $= (\lambda T_2 \lambda i \lambda T\, \exists c \forall c_2.\, (T_2(i)(c_2) \leftrightarrow c = c_2) \wedge T(i)(c))^{\bullet}$
   $= \lambda T_2^{\bullet} \lambda T^{\bullet} \bigvee c^{\bullet} \bigwedge c_2^{\bullet}.\, (T_2^{\bullet}(c_2^{\bullet}) \leftrightarrow c^{\bullet} \doteq c_2^{\bullet}) \wedge T^{\bullet}(c^{\bullet})$
   $= \lambda \boldsymbol{T_2} \lambda \boldsymbol{T} \bigvee \boldsymbol{c} \bigwedge \boldsymbol{c_2}.\, (\boldsymbol{T_2}(\boldsymbol{c_2}) \leftrightarrow \boldsymbol{c} \doteq \boldsymbol{c_2}) \wedge \boldsymbol{T}(\boldsymbol{c})$

4. $[_{NP}[_{DET}\text{the}][_N\text{temperature}]]$
   $\rightsquigarrow (\lambda i \lambda T\, \exists c \forall c_2.\, (temp\,(i)(c_2) \leftrightarrow c = c_2) \wedge T(i)(c))^{\bullet}$
   $= (\lambda \boldsymbol{T} \bigvee \boldsymbol{c} \bigwedge \boldsymbol{c_2}.\, (\boldsymbol{temp}(\boldsymbol{c_2}) \leftrightarrow \boldsymbol{c} \doteq \boldsymbol{c_2}) \wedge \boldsymbol{T}(\boldsymbol{c}))$

5. $[_S[_{NP}[_{DET}\text{the}][_N\text{temperature}]][_{IV}\text{rises}]]$
   $\rightsquigarrow (\lambda i\, \exists c \forall c_2.\, (temp\,(i)(c_2) \leftrightarrow c = c_2) \wedge rise\,(i)(c))^{\bullet}$
   $= \bigvee \boldsymbol{c} \bigwedge \boldsymbol{c_2}.\, (\boldsymbol{temp}(\boldsymbol{c_2}) \leftrightarrow \boldsymbol{c} \doteq \boldsymbol{c_2}) \wedge \boldsymbol{rise}(\boldsymbol{c})$

The possibility of interpreting proper names in the type for individual concepts enables us to translate the conclusion from $(\star)$ as follows:

1. $[_{NP}\text{ninety}] \rightsquigarrow (int\,(ninety))^{\bullet} = ninety^{\bullet} = \boldsymbol{ninety}$ \hfill (6)

2. $[_{IV}\text{rises}] \rightsquigarrow rise^{\bullet} = \boldsymbol{rise}$

3. $[_S[_{NP}\text{ninety}][_{IV}\text{rises}]] \rightsquigarrow (rise\,(ninety))^{\bullet} = \boldsymbol{rise}\,(\boldsymbol{ninety})$

This completes our translation of the 'ingredient sentences' for Partee's temperature puzzle. The invalid inference from the conjunction of (4.7) and (5.5) to (6.3) in the logic TY_1 is captured below:

(and, hence, to 'allow' Partee's temperature puzzle). I owe this observation to Ede Zimmermann.

$$\cfrac{\bigvee x \bigwedge y.((\bigvee c.\, \boldsymbol{temp}\,(c) \wedge y \doteq \boldsymbol{c}\,(\textbf{@})) \leftrightsquigarrow x \doteq y) \wedge x \doteq ninety \qquad \bigvee c \bigwedge c_2.(\boldsymbol{temp}\,(c_2) \leftrightsquigarrow c \doteq c_2) \wedge \boldsymbol{rise}(c)}{rise\,(ninety)} \quad (\text{EFL-}\star)$$

In particular, while the formula in the second premise attributes the property 'rise' to the type-$(o \to \iota)$ object which has the property of being a temperature, the formula in the first premise attributes the property 'is ninety' only to the result (type ι) of applying a temperature-object to the EFL-correlate of @. In virtue of this fact – and the resulting invalidity of substituting *ninety* for c in the second premise of (EFL-$\star$) –, the formula in the conclusion does not follow from the conjunction of the two premise-formulas by the (classical) rules of TY_1.

5   Conclusion

This paper has shown the possibility of interpreting Montague's PTQ-fragment in the class of EFL-models from [11], which only contain basic individuals and propositions. We have obtained this result by coding the interpretations of the PTQ-expressions from [12] into EFL-objects, and by translating the linguistically relevant sublanguage of a streamlined version, TY_2, of Montague's logic IL into the EFL-typed language TY_1 which respects this coding. Since this translation preserves the relation of logical consequence on the TY_2 translations of PTQ-sentences, it enables a new, *extensional*, solution to Partee's temperature puzzle.

The previously-assumed impossibility of such a solution can be attributed to the various challenges which emerge for any TY_2-to-TY_1 translation. These challenges include the different forms of the linguistically relevant TY_2 and TY_1 types, and the unavailability of truth-functional connectives or quantifiers in the language of TY_1. Our solutions to these challenges build on existing work on the relation between TY_2 and IL types [14], and on hyperintensional semantics [17].

Our TY_2-to-TY_1 translation enables a transfer of the interpretive success of PTQ-models to EFL-models (esp. w.r.t. the solvability of Partee's temperature puzzle) and a proof of the relative consistency of the two classes of models. At the same time, it identifies the minimal semantic requirements on formal models for the PTQ-fragment. Contrary to what is suggested by a comparison of [12] and [11], suitable PTQ-models need *not* contain a designated type for indices. We take these results to support the reduction view of formal natural language semantics.

References

1. Church, A.: A formulation of the simple theory of types. J. Symbolic Log. 5(2), 56–68 (1940)
2. Cresswell, M.J.: The semantics of degree. In: Partee, B. (ed.) Montague Grammar. Academic Press, New York (1976)
3. van Eijck, J., Unger, C.: Computational Semantics with Functional Programming. Cambridge University Press, Cambridge (2010)

4. Fox, C., Lappin, S.: An expressive first-order logic with flexible typing for natural language semantics. Log. J. IGPL **12**(2), 135–168 (2004)
5. Gallin, D.: Intensional and Higher-Order Modal Logic with Applications to Montague Semantics. North Holland, Amsterdam (1975)
6. de Groote, P., Kanazawa, M.: A note on intensionalization. J. Log. Lang. Inform. **22**(2), 173–194 (2013)
7. Henkin, L.: Completeness in the theory of types. J. of Symb. Log. **15**, 81–91 (1950)
8. Janssen, T.M.V.: Individual concepts are useful. In: Landman, F., Veltman, F. (eds.) Varieties of Formal Semantics: Proceedings of the 4th Amsterdam Colloquium (1984)
9. Klein, E., Sag, I.: Type-driven translation. Linguist. Philos. **8**, 163–201 (1985)
10. Loux, M.J.: Metaphysics: A Contemporary Introduction. Routledge, New York (2006)
11. Montague, R.: English as a formal language. In: Thomason, R.H. (ed.) Formal Philosophy: Selected papers of Richard Montague. Yale University Press, New Haven (1976)
12. Montague, R.: The proper treatment of quantification in ordinary English. In: Thomason, R.H. (ed.) Formal Philosophy: Selected papers of Richard Montague. Yale University Press, New Haven (1976)
13. Muskens, R.: Anaphora and the logic of change. Log. AI **478**, 412–427 (1991)
14. Muskens, R.: Meaning and Partiality. CSLI Lecture Notes. FoLLI, Stanford (1995)
15. Partee, B., Rooth, M.: Generalized conjunction and type ambiguity. In: Bauerle, R., Schwarz, C., von Stechow, A. (eds.) Meaning, Use and Interpretation of Language. Walter De Gruyter, Berlin (1983)
16. Schönfinkel, M.: Über die Bausteine der mathematischen Logik. Math. Ann. **92**, 305–316 (1924)
17. Thomason, R.H.: A model theory for the propositional attitudes. Linguist. Philos. **4**, 47–70 (1980)

CI via DTS

Daisuke Bekki[1,2,3](✉) and Elin McCready[4]

[1] Graduate School of Humanities and Sciences, Ochanomizu University,
2-1-1 Ohtsuka, Bunkyo-ku, Tokyo 112-8610, Japan
[2] National Institute of Informatics,
2-1-2 Hitotsubashi, Chiyoda-ku, Tokyo 101-8430, Japan
[3] CREST, Japan Science and Technology Agency,
4-1-8 Honcho, Kawaguchi, Saitama 332-0012, Japan
bekki@is.ocha.ac.jp
[4] Aoyama Gakuin University,
4-4-25 Shibuya, Shibuya-ku, Tokyo 150-8366, Japan

Abstract. It has been observed that conventionally implicated content interacts with at-issue content in a number of different ways. This paper focuses on the existence of anaphoric links between content of these two types, something disallowed by the system of Potts (2005), the original locus of work on these issues. The problem of characterizing this interaction has been considered by a number of authors. This paper proposes a new system for understanding it in the framework of Dependent Type Semantics. It is shown that the resulting system provides a good characterization of how "cross-dimensional" anaphoric links can be supported from a proof-theoretic perspective.

1 Conventional Implicatures

Conventional implicature (CI) is a kind of pragmatic content first discussed by Grice [7], which is taken to be (one part of the) nonasserted content conveyed by particular lexical items or linguistic constructions. Examples include appositives, non-restrictive relative clauses (NRRCs), expressive items, and speaker-oriented adverbs. Such content has been a focus of a great deal of research in linguistics and philosophy since the work of Potts [19]. According to Potts (who takes a position followed by much or most subsequent research), CIs have at least the following characteristics:

(1) a. CI content is independent from at-issue content (in the sense that the two are scopeless with respect to each other)

   b. CIs do not modify CIs.[1]

   c. Presupposition filters do not filter CIs

Our sincere thanks to the anonymous reviewers of LENLS11 who gave us insightful comments. Elin McCready and Daisuke Bekki are partially supported by a Grant-in-Aid for Scientific Research (C) (No. 25370441) from the Ministry of Education, Science, Sports and Culture. Daisuke Bekki is partially supported by JST, CREST.

[1] Though see footnote 3 for some necessary qualification.

T. Murata et al. (Eds.): JSAI-isAI 2014 Workshops, LNAI 9067, pp. 23–36, 2015.
DOI: 10.1007/978-3-662-48119-6_3

Potts models these features in a two-dimensional semantics for CIs in which CIs are associated with special semantic types. First, since CI content enters a dimension of meaning distinct from that of at-issue content, no scope relations are available, modeling (1a); characteristic (1b) follows from a lack of functional types with CI inputs in the type system; placing filters in the at-issue dimension also accounts for (1c). Although this system has been criticized for various reasons, it seems to be adequate for modeling the basic data associated with CIs.

2 Problem: Interaction Between At-Issue and CI Content

Potts's two-dimensional semantics, which utilizes distinct and dimensionally independent representations for at-issue and CI content, aims to capture their supposed mutual semantic independence. However, a fully separated multidimensional semantics is not fully satisfactory from an empirical perspective, as can be seen by focusing attention on the interaction between CIs and anaphora/presupposition (as also noted by Potts himself). In particular, the following facts are problematic for a treatment in which no interdimensional interaction is allowed:

1. CI content may serve as antecedent for later anaphoric items and presupposition triggers, meaning that discourse referents introduced in CI contexts are accessible to anaphora/presupposition triggers, as exemplified in the mini-discourse (2) ((3.15b) in [19], slightly simplified).
2. Anaphora/presupposition triggers introduced in CI environments may find their antecedents in the preceding discourse (their *local contexts*), i.e. anaphora/ presupposition triggers inside a conventional content require access to their left contexts, as exemplified in the mini-discourse (3) (see also [25].).

(2)   a.   Mary counseled John, who killed a coworker.
      b.   Unfortunately, Bill knows that he killed a coworker.

(3)   a.   John killed a coworker.
      b.   Mary, who knows that he killed a coworker, counseled him.

In both (2) and (3), the factive presupposition "he (=John) killed a coworker" can be bound by the antecedent in the first sentence. This behaviour of CIs fails to be explained by Potts's [19] analysis, where at-issue content and CI content are fully independent of each other, at least in their sentential representations. With regard to the cases such as (2), at minimum we require a mechanism to collect the CIs hanging in a sentential tree, and pass them to the succeeding discourse, where they can play the role of antecedents. In order to deal with the cases like (3), we also need a mechanism to pass the local context of a sentence to the collection of CIs which has been collected from it. (Here we assume an analysis which collects CIs from the (syntactic or semantic) tree in a Pottsian style, putting aside the arguments about compositionality raised by

Gutzmann [10] and others.) Neither extension, however, seems straightforward in Potts's [19] framework, nor in other frameworks that have been proposed for the analysis of CIs, but which have not attempted to account for the present set of phenomena (excluding theories using dynamic semantics, such as that of AnderBois et al. (2014) or Nouwen [18]).

3   Dependent Type Semantics

Dynamic solutions to these puzzles exist, such as the work of AnderBois et al. cited above; here, we take a different line, and propose a compositional analysis of conventional implicatures in the framework of Dependent Type Semantics (DTS; Bekki(2014)). DTS is based on dependent type theory (Martin-Löf [16]), Coquand and Huet [6]) which provides a proof-theoretic semantics in terms of the Curry-Howard correspondence between types and propositions, following the line of Sundholm [24]. This approach has been proved useful for linguistic analysis, especially in Ranta's [20] Type Theoretical Grammar and its successors. Krahmer and Piwek [14] found that anaphora resolution and presupposition binding/accommodation can be reduced to *proof search* (which is known as the "anaphora resolution as proof construction" paradigm). Bekki's [5] DTS inherits this paradigm and reformalizes the whole setting in a compositional manner; the resulting system can serve as the semantic component of any lexical grammar.

For example, the semantic representation of a classical relative donkey sentence as (4a) is calculated as (4b) in DTS.

(4)   a.   Every farmer who owns a donkey beats it$_1$.

$$
\text{b.} \quad \lambda c. \left(u: \begin{bmatrix} x:\textbf{entity} \\ \begin{bmatrix} \textbf{farmer}(x) \\ \begin{bmatrix} y:\textbf{entity} \\ \begin{bmatrix} \textbf{donkey}(y) \\ \textbf{own}(x,y) \end{bmatrix} \end{bmatrix} \end{bmatrix} \end{bmatrix} \right) \to \textbf{beat}(\pi_1(u), @_1(c,u))
$$

The semantic representation (4b) for the sentence (4a) contains an *underspecified term* $@_1$, which corresponds to the referent of the pronoun "it$_1$". Anaphora resolution in DTS then proceeds as follows: (1) the representation is given an *initial context* (which is () of type $\top$), (2) the resulting representation undergoes *type checking* (cf. Löh [15]) to check whether it has a type $\textsf{type}$ (i.e. the type of types (=propositions)), which in turn requires (3) that the underspecified term $@_1$ satisfies the following judgment:

$$
(5) \quad \Gamma, u: \begin{bmatrix} x:\textbf{entity} \\ \begin{bmatrix} \textbf{farmer}(x) \\ \begin{bmatrix} y:\textbf{entity} \\ \begin{bmatrix} \textbf{donkey}(y) \\ \textbf{own}(x,y) \end{bmatrix} \end{bmatrix} \end{bmatrix} \end{bmatrix} \vdash @_1 : \begin{bmatrix} \top \\ \begin{bmatrix} x:\textbf{entity} \\ \begin{bmatrix} \textbf{farmer}(x) \\ \begin{bmatrix} y:\textbf{entity} \\ \begin{bmatrix} \textbf{donkey}(y) \\ \textbf{own}(x,y) \end{bmatrix} \end{bmatrix} \end{bmatrix} \end{bmatrix} \end{bmatrix} \to \textbf{entity}
$$

Given the above, we arrive at a choice point. The first option: if the hearer chooses to *bind* $@_1$, he/she has to find a proof term of the specified type, to replace $@_1$. Here $\lambda c.\pi_1\pi_2\pi_2\pi_2(c)$ is a candidate for such a term, which corresponds to the intended donkey. Alternatively, the hearer may choose not to execute proof search and instead *accommodate* $@_1$, in which case he/she just assumes that there is such a term $@_1$ and uses it in the subsequent inferences. In either case, the semantic representation (4b) does not need drastic reconstruction, unlike van der Sandt's [22] DRT-based approach.

In intersentential composition cases, two sentential representations are merged into one by the *dynamic conjunction* operation defined below:

$$(6) \qquad M;N \stackrel{def}{\equiv} \lambda c. \begin{bmatrix} u{:}Mc \\ N(c,u) \end{bmatrix}$$

4   Representations of CIs in DTS

Our proposal is that a given bit of CI content A (again of type type) can be properly represented in terms of DTS in the following way:

4.1   The CI Operator

Definition 1 (The CI operator). *Let A be a type and $@_i$ be an underspecified term with an index i:*

$$\mathsf{CI}(@_i : A) \stackrel{def}{\equiv} \mathbf{eq}_A(@_i, @_i)$$

Let us call CI the *CI operator*, and a type of the form $\mathsf{CI}(@_i : A)$ a *CI type*. The CI operator is used with an underspecified term $@_i$ and a type A as its arguments (as will be demonstrated in the next section) to form a CI type. The CI type is defined in a rather technical way, but the content is simple: $\mathbf{eq}_A(M, N)$, with M, N any terms, is a type for equations between M and N in DTS, namely, it is the type of proofs of the proposition that M equals N ($@_i = @_i$, informally), both of which are of type A.

Thus, $\mathsf{CI}(@_i : A)$ is always true by the reflexivity law, under any context. In terms of DTS, $\mathsf{CI}(@_i : A)$ inhabits a canonical proof refl_A (i.e. $\vdash \mathsf{refl}_A :$ $\mathbf{eq}_A(@_i, @_i)$).

This means that the CI operator $\mathsf{CI}(@_i : A)$ does not contribute anything to at-issue content, since we know that it is always inhabited by the term refl_A. However, the type checking of a semantic representation which contains $\mathsf{CI}(@_i :$ $A)$ requires that the $\mathbf{eq}_A(@_i, @_i)$ has a type type, which in turn requires that the underspecified term $@_i$ has the type A. Therefore, the proposition A must have a proof term $@_i$ of type A (i.e. A must be true), which projects, regardless of the configuration in which it is embedded.

Moreover, unlike the cases of anaphora and presupposition, an underspecified term for a CI does not take any local context as its argument. This explains why CIs do not respect their left contexts.

Let us examine how our analysis how the CI operators are used to represent CIs and how they predict the set of benchmarks (1a)–(1c) for CIs.

4.2 Independence from At-Issue Content

The property (1a) is supported by the fact that both the sentences (7a) and (7b) entail the CI content *Lance Armstrong is an Arkansan*. Thus the CI content is not affected by, or projects through, logical operators such as negation that take scope over it.

(7) a. Lance Armstrong, an Arkansan, has won the 2003 Tour de France!

   b. It is not the case that Lance Armstrong, an Arkansan, has won the 2003 Tour de France!

The proposition *Lance Armstrong is an Arkansan* is represented in DTS as a type (=proposition) **arkansan**(*lance*). If it is embedded within the CI type as in $\mathsf{CI}(@_1 : \mathbf{arkansan}(lance))$, this proposition is a CI content, and $@_1$ is its proof term. This embedding for an indefinite appositive construction is done by applying the following Indefinite Appositive Rule.

Definition 2 (Indefinite Appositive Rule).

$$(IA_i)\ \dfrac{\begin{array}{c} S\backslash NP \\ : M \end{array}}{\begin{array}{c} S/(S\backslash NP)\backslash NP \\ : \lambda x.\lambda p.\lambda c.\begin{bmatrix} pxc \\ \mathsf{CI}(@_i : Mxc) \end{bmatrix} \end{array}}$$

This rule applies to an indefinite predicative noun phrase.[2] For example, the sentence (7a) is derived as follows,

(8)

$$>\ \dfrac{\begin{array}{cc} \dfrac{\text{an}}{\begin{array}{c} S\backslash NP/N \\ : id \end{array}} & \dfrac{\text{Arkansan}}{\begin{array}{c} N \\ : \lambda x.\lambda c.\mathbf{arkansan}(x) \end{array}} \end{array}}{\begin{array}{c} S\backslash NP \\ : \lambda x.\lambda c.\mathbf{arkansan}(x) \end{array}}$$

$$(IA_1)\ \dfrac{S\backslash NP\ :\ \lambda x.\lambda c.\mathbf{arkansan}(x)}{\begin{array}{c} S/(S\backslash NP)\backslash NP \\ : \lambda x.\lambda p.\lambda c.\begin{bmatrix} pxc \\ \mathsf{CI}(@_1 : \mathbf{arkansan}(x)) \end{bmatrix} \end{array}}$$

$$<\ \dfrac{\begin{array}{c} \text{Lance} \\ NP \\ : lance \end{array} \qquad S/(S\backslash NP)\backslash NP}{\begin{array}{c} S/(S\backslash NP) \\ : \lambda p.\lambda c.\begin{bmatrix} p(lance)c \\ \mathsf{CI}(@_1 : \mathbf{arkansan}(lance)) \end{bmatrix} \end{array}}$$

$$\begin{array}{c} \text{has won the 2003 Tour de France} \\ S\backslash NP \\ : \lambda x.\lambda c.\mathbf{won}(x) \end{array}$$

$$>\ \dfrac{}{\begin{array}{c} S \\ : \lambda c.\begin{bmatrix} \mathbf{won}(lance) \\ \mathsf{CI}(@_1 : \mathbf{arkansan}(lance)) \end{bmatrix} \end{array}}$$

The resulting SR entails that Lance is an Arkansan, because it contains the CI type $\mathsf{CI}(@_1 : \mathbf{arkansan}(lance))$ and type checking of this SR requires $\vdash @_1 : \mathbf{arkansan}(lance)$, namely, the underspecified term $@_1$ is

---

[2] We should specify some features of S both on the predicate side and the rule side, in order to prevent this rule to apply to other kinds of phrases of category $S\backslash NP$, such as verb phrases, which is a routine task we will not perform here.

of type **arkansan**(*lance*). In other words, the proposition that Lance is an Arkansan is inhabited. In contrast, a derivation of (7b) is shown in (9).

(9)

$$
\frac{
\underset{: \lambda p.\lambda c.\neg pc}{\underset{S/S}{\text{It is not the case that}}}
\quad
\underset{: \lambda c.\begin{bmatrix}\textbf{won}(lance)\\ \text{CI}(@_1 : \textbf{arkansan}(lance))\end{bmatrix}}{\underset{S}{\text{Lance, an Arkansan, has won the 2007 Tour de France}}}
}{
\underset{: \lambda c.\neg\begin{bmatrix}\textbf{won}(lance)\\ \text{CI}(@_1 : \textbf{arkansan}(lance))\end{bmatrix}}{S}
} >
$$

Here again, the resulting SR contains the CI type CI($@_1$: **arkansan**(*lance*)). Since type checking of this SR is not affected by the existence of the negation operator ¬ that encloses it, it also requires that the proposition that Lance is an Arkansan is inhabited. This way, the CI content is predicted to be independent from at-issue content, as expected.

4.3   Presupposition Filters Do Not Filter CIs

The contrast between (10a) and (10b) exemplifies (1b): in the sentence (10a), where the definite description *the cyclist* induces a presupposition that *Lance is a cyclist*, the presupposition is filtered by the antecedent of the conditional that entails the presupposition, so the whole sentence does not have any presupposition. On the other hand, in the sentence (10b), where the indefinite appositive *a cyclist* induces the CI that *Lance is a cyclist*, the CI is not filtered by the same antecedent thus projects over it, and moreover the whole sentence is infelicitous for Gricean reasons. There are various ways in which this infelicity could be viewed, but to us it is a violation of Quantity or Manner, in that the conditional clause is uninformative, as it is pre-satisfied by the appositive content.

(10)   a. If Lance is a cyclist, then the Boston Marathon was won by the cyclist.

   b. If Lance is a cyclist, then the Boston Marathon was won by Lance, a cyclist.

Let us explain how this contrast is predicted in DTS. First, the derivation of (10a) is as (11).

(11)

$$
\lambda c.\,(u{:}\textbf{cyclist}(lance)) \to \textbf{win}\left(\pi_1\left(@_1(c,u) : \begin{bmatrix}y{:}\textbf{entity}\\ \textbf{cyclist}(y)\end{bmatrix}, BM\right)\right)
$$

Then the type checking rules apply to the resulting SR under the initial context (), which require that the underspecified term $@_1$ satisfies the following judgment.

(12) $\quad \Gamma, u : \mathbf{cyclist}(lance) \vdash @_1 : \begin{bmatrix} \top \\ \mathbf{cyclist}(lance) \end{bmatrix} \to \begin{bmatrix} y\mathbf{:entity} \\ \mathbf{cyclist}(y) \end{bmatrix}$

In other words, the type checking launches a proof search, which tries to find a term of type:

(13) $\quad \begin{bmatrix} \top \\ \mathbf{cyclist}(lance) \end{bmatrix} \to \begin{bmatrix} y\mathbf{:entity} \\ \mathbf{cyclist}(y) \end{bmatrix}$

under a global context $\Gamma, u : \mathbf{cyclist}(lance)$. We assume that the hearer knows that Lance exists, i.e. we assume that the global context Γ includes the entry $lance : \mathbf{entity}$.

At a first glance, one may think that there are at least two different resolutions (14) and (15), and so that there are two different terms that satisfy (12):

(14)

$$\cfrac{\cfrac{lance : \mathbf{entity} \quad \cfrac{u : \mathbf{cyclist}(lance) \quad \cfrac{}{x : \begin{bmatrix} \top \\ \mathbf{cyclist}(lance) \end{bmatrix}}\,(1)}{u : \mathbf{cyclist}(lance)}\,(w)}{(lance, u) : \begin{bmatrix} y\mathbf{:entity} \\ \mathbf{cyclist}(y) \end{bmatrix}}\,(\Sigma I)}{\lambda x.(lance, u) : \begin{bmatrix} \top \\ \mathbf{cyclist}(lance) \end{bmatrix} \to \begin{bmatrix} y\mathbf{:entity} \\ \mathbf{cyclist}(y) \end{bmatrix}}\,(\Pi I)\,(1)$$

(15)

$$\cfrac{\cfrac{lance : \mathbf{entity} \quad \cfrac{\cfrac{}{x : \begin{bmatrix} \top \\ \mathbf{cyclist}(lance) \end{bmatrix}}\,(1)}{\pi_2 x : \mathbf{cyclist}(lance)}\,(\Sigma E)}{(lance, \pi_2 x) : \begin{bmatrix} y\mathbf{:entity} \\ \mathbf{cyclist}(y) \end{bmatrix}}\,(\Sigma I)}{\lambda x.(lance, \pi_2 x) : \begin{bmatrix} \top \\ \mathbf{cyclist}(lance) \end{bmatrix} \to \begin{bmatrix} y\mathbf{:entity} \\ \mathbf{cyclist}(y) \end{bmatrix}}\,(\Pi I)\,(1)$$

However, only (15) is licenced, because the underspecified term $@_1$ must not contain u as a free variable. The reason, which is a bit technical but empirically important, is that we implicitly assumed it in the derivation (11): more precisely, in the functional application between "If Lance is a cyclist" and "the BM was won by the cyclist", the following β-reduction took place.

(16)
$$(\lambda p.\lambda c.\,(u{:}\mathbf{cyclist}(lance)) \to p(c,u))\left(\lambda c.\mathbf{win}\left(\pi_1\left(@_1c : \begin{bmatrix} y{:}\mathbf{entity} \\ \mathbf{cyclist}(y) \end{bmatrix}\right], BM\right)\right)\right)$$
$$\longrightarrow_\beta \lambda c.\,(u{:}\mathbf{cyclist}(lance)) \to \mathbf{win}\left(\pi_1\left(@_1(c,u) : \begin{bmatrix} y{:}\mathbf{entity} \\ \mathbf{cyclist}(y) \end{bmatrix}\right], BM\right)$$

If the variable u occurs free in the underspecified term $@_1$, the variable u in the $\lambda p.\lambda c.\,(u{:}\mathbf{cyclist}(lance)) \to p(c,u)$ part should have been renamed. This means that $@_1$ does not contain u as a free variable, thus $\lambda x.(lance, u)$ is not a candidate to replace $@_1$.

Therefore, if one wants to bind $@_1$, then $@_1 = \lambda x.(lance, \pi_2 x)$. This shows that the presupposition triggered by "the cyclist" is bound by the local context, the information given by the antecedent of the conditional.

In the case of CI, the situation is different. The derivation of (10b) is as (17).

(17)

was won by : $S\backslash NP/NP$: $\lambda y.\lambda x.\lambda c.\mathbf{win}(y,x)$

Lance, a cyclist : $S/(S\backslash NP)$: $\lambda p.\lambda x.\lambda c.\begin{bmatrix} p(lance)xc \\ \mathbf{Cl}(@_2 : \mathbf{cyclist}(lance)) \end{bmatrix}$

the BM : NP : bm

$S\backslash NP$: $\lambda x.\lambda c.\begin{bmatrix} \mathbf{win}(lance, x) \\ \mathbf{Cl}(@_2 : \mathbf{cyclist}(lance)) \end{bmatrix}$

If Lance is a cyclist : S/S : $\lambda p.\lambda c.\,(u{:}\mathbf{cyclist}(lance)) \to p(c,u)$

S : $\lambda c.\begin{bmatrix} \mathbf{win}(lance, BM) \\ \mathbf{Cl}(@_2 : \mathbf{cyclist}(lance)) \end{bmatrix}$

$\lambda c.\,(u{:}\mathbf{cyclist}(lance)) \to \begin{bmatrix} \mathbf{win}(lance, BM) \\ \mathbf{Cl}(@_2 : \mathbf{cyclist}(lance)) \end{bmatrix}$

Type checking rules apply to the resulting SR under the initial context (), which requires that the underspecified term $@_2$ satisfies the following judgment.

(18) $\Gamma, u : \mathbf{cyclist}(lance) \vdash @_2 : \mathbf{cyclist}(lance)$

It seems that the variable u is an immediate candidate that can replace $@_2$, but this is not licenced, for the same reason as the case of definite descriptions: The underspecified term $@_2$ should not contain the free occurrence of u, since it is implicitly assumed in the β-reduction that took place at the bottom of (17).

Thus, there is no binding option for the CI in (10b) unless the global context Γ provides some knowledge that allows its inference. Otherwise, the hearer has to update Γ accordingly, i.e. accommodate it. The simplest way is to use the following updated global context Γ' (x is some variable chosen so that $x \notin \Gamma$).

(19) $\Gamma' \stackrel{def}{\equiv} \Gamma,\ x : \mathbf{cyclist}(lance)$

The difference between the two cases (10a) and (10b) is that in the formulation of CIs, the underspecified term does not take a local context as its argument, and so cannot refer to it, while in the formulation of presuppositions, the underspecified term is given a local context as its argument, and so is able to bind it by means of information deduced from the local context. This way, DTS predicts that antecedents of conditionals, which are of course presupposition filters, do not filter CI contents, thus deriving one of the empirical differences between these types of content.

It is also predicted in DTS that the sentence (10b) is pragmatically infelicitous. In order to *accept* (10b) as a felicitous sentence, one has to add the entry $x : $ **cyclist**(*lance*) to his/her global context in most cases. It is then inappropriate to assume that Lance is a cyclist, as in (10b), is redundant, since it is immediately derivable from the global context. This is one way to implement the idea of the infelicity of (10b) as a Gricean violation of the kind mentioned above.

4.4   CIs Do Not Modify CIs

Typical cases that exemplify (1c) are examples like (20), where the speaker-oriented adverb *surprisingly* does not modify the expressive content induced by *the bastard*, i.e. the bastardhood of Jerry is not surprising for the speaker.

(20)   Surprisingly, Jerry, the bastard, showed up with no money.

The derivation of (20) in DTS is as follows, assuming that the definite appositive is analyzed in the same way as the indefinite appositives, and the speaker-oriented adverb *surprisingly* takes a proof of the sentence it modifies.

(21)

$$
\cfrac{
  : \lambda p.\lambda c.\begin{bmatrix} u{:}pc \\ \mathsf{CI}(@_1 : \textbf{surprising}(u)) \end{bmatrix} \Big/ \cfrac{S/S}{\text{surprisingly}}
}{
  \cfrac{
    \cfrac{\text{Jerry, the bastard}}{S\backslash(S/NP)}\ : \lambda p.\lambda c.\begin{bmatrix} p(jerry)c \\ \mathsf{CI}(@_2 : \textbf{bastard}(jerry)) \end{bmatrix}
    \quad
    \cfrac{\text{showed up with no money}}{S\backslash NP}\ : \lambda x.\lambda c.\textbf{showedUpNoMoney}(x)
  }{
    \cfrac{S}{\ : \lambda c.\begin{bmatrix} \textbf{showedUpNoMoney}(jerry) \\ \mathsf{CI}(@_2 : \textbf{bastard}(jerry)) \end{bmatrix}}
  }
}
$$

$$
: \lambda c.\begin{bmatrix} u: \begin{bmatrix} \textbf{showedUpNoMoney}(jerry) \\ \mathsf{CI}(@_2 : \textbf{bastard}(jerry)) \end{bmatrix} \\ \mathsf{CI}(@_1 : \textbf{surprising}(u)) \end{bmatrix}
$$

Type checking of the resulting semantic representation in (20) under the initial context (), requires that the two underspecified terms are of the following types:

(22)   a. $\Gamma, u : \begin{bmatrix} \textbf{showedUpNoMoney}(jerry) \\ \mathsf{CI}(@_2 : \textbf{bastard}(jerry)) \end{bmatrix} \vdash @_1 : \textbf{surprising}(u)$

      b. $\Gamma \vdash @_2 : \textbf{bastard}(jerry)$

The judgment (22b) immediately requires the update of Γ, if it does not entail the bastardhood of Jerry. The case of judgment (22a), on the other hand, is more complex, since the **surprising** predicate is about a variable u, which is a proof term of type:

(23) $\begin{bmatrix} \textbf{showedUpNoMoney}(jerry) \\ \mathsf{CI}(@_2 : \textbf{bastard}(jerry)) \end{bmatrix}$

However, since the type $\mathsf{CI}(@_2 : \mathbf{bastard}(jerry))$ inhabits only one term $\mathsf{refl}_{\mathbf{bastard}(jerry)}$, the value of u only varies over the terms of type:

(24)   **showedUpNoMoney**(*jerry*)

and so states that Jerry showed up with no money. Thus, whether it is surprising only depends on how Jerry showed up with no money, and not on how the equality between two identical $@_2$ results in identity.

Thus DTS predicts that there is no interactions between different bits of CI content.[3]

5 Solution to the Puzzles

Let us now proceed to show how our analysis solves the puzzles regarding the interaction between CI contents and anaphora/presuppositions.

5.1 A CI can Serve as an Antecedent for the Subsequent Anaphora/presuppositions

The semantic representation for (2) is derived as (25). We assume a distinct lexical entry for "who" for NRRCs, which contains the CI operator for specifying their CI content.

The resulting discourse representation contains three underspecified terms: $@_1$ for the CI content, $@_2$ for the factive presupposition of "knows", and $@_3$ for the pronoun "he".

Type checking requires the term $@_1$ to be of type $\mathbf{KC}(john)$, which will be accommodated as new information to the hearer. The term $@_3$ can be independently resolved if it is intended to be coreferential to "John", namely, as $@_3 = \lambda c.john$. Then the term $@_2$, which is required to have type $\mathbf{KC}(john)$, can be bound just by being identified with $@_1$. In this way, what is introduced as a CI can bind the subsequent presuppositions, although it does not participate in the at-issue content.

---

[3] There is a possible problem with attributing the property (1c) to CIs. Gutzmann [9,10] argues that sentences such as (1) is a possible counter-example for (1c) in the sense that *fucking* in (1) serves to intensify the degree to which Jerry has the property of being an asshole, which is CI content that is induced by *asshole*; thus, the adjective works to strengthen not-at-issue content in cases of this kind.

(1) Jerry is a fucking asshole.

The current version of DTS, however, predicts that the target of the modification performed by *fucking* does not include the CI content of *asshole*, just as in the case of (20). We believe this issue relates to the sort of variance in what counts as "at-issue" discussed by Hom [12,13], and, as such, exhibits a level of complexity that requires a more detailed look at the pragmatics of these constructions (cf. Amaral et al. [1]). This difficult project is beyond the scope of the present paper.

(25)

$$\frac{\frac{,\text{who}_1}{\begin{array}{c}T\backslash(T/NP)\backslash NP/(S\backslash NP)\\: \lambda r.\lambda z.\lambda p.\lambda \boldsymbol{x}.\lambda c.\begin{bmatrix}pz\boldsymbol{x}c\\\textbf{CI}(@_1 : rzc)\end{bmatrix}\end{array}} \quad \frac{\text{killed a coworker}}{\begin{array}{c}S\backslash NP\\: \lambda x.\lambda c.\textbf{KC}(x)\end{array}}}{}$$

Full structure (25):

Top right combination:
$$\frac{\text{John}}{\begin{array}{c}NP\\:john\end{array}} \quad > \quad \frac{T\backslash(T/NP)\backslash NP}{: \lambda z.\lambda p.\lambda \boldsymbol{x}.\lambda c.\begin{bmatrix}pz\boldsymbol{x}c\\\textbf{CI}(@_1 : \textbf{KC}(z))\end{bmatrix}}$$

$$\frac{T\backslash(T/NP)}{: \lambda p.\lambda \boldsymbol{x}.\lambda c.\begin{bmatrix}pjohn\boldsymbol{x}c\\\textbf{CI}(@_1 : \textbf{KC}(john))\end{bmatrix}}$$

$$\frac{\text{counselled}}{\begin{array}{c}S\backslash NP/NP\\: \lambda y.\lambda x.\lambda c.\textbf{counsel}(x, y)\end{array}} \quad < \quad \frac{S\backslash NP}{: \lambda x.\lambda c.\begin{bmatrix}\textbf{counsel}(x, john)\\\textbf{CI}(@_1 : \textbf{KC}(john))\end{bmatrix}}$$

$$\frac{\text{Mary}}{\begin{array}{c}NP\\:mary\end{array}} \quad < \quad \frac{S}{: \lambda c.\begin{bmatrix}\textbf{counsel}(mary, john)\\\textbf{CI}(@_1 : \textbf{KC}(john))\end{bmatrix}} \quad ;$$

$$\frac{\text{Bill}}{\begin{array}{c}NP\\:bill\end{array}} \quad > \quad \frac{\frac{\text{knows}_2 \text{ that}}{\begin{array}{c}S\backslash NP/S\\: \lambda p.\lambda x.\lambda c.\textbf{know}(x, @_2 : pc)\end{array}} \quad \frac{\text{he}_3 \text{ killed a coworker}}{\begin{array}{c}S\\: \lambda c.\textbf{KC}(@_3c)\end{array}}}{\begin{array}{c}S\backslash NP\\: \lambda x.\lambda c.\textbf{know}(x, @_2 : \textbf{KC}(@_3c))\end{array}}$$

$$\frac{S}{: \lambda c.\textbf{know}(bill, @_2 : \textbf{KC}(@_3c))}$$

$$\xrightarrow{\text{Dynamic conjunction}} \lambda c.\begin{bmatrix}u:\begin{bmatrix}\textbf{counsel}(mary, john)\\\textbf{CI}(@_1 : \textbf{KC}(john))\end{bmatrix}\\\textbf{know}(bill, @_2 : \textbf{KC}(@_3(u, c)))\end{bmatrix}$$

5.2 Anaphora/Presuppositions Inside CIs Receive Their Left Contexts

The semantic representation for (3) is derived as follows.

(26)

$$\frac{\frac{\text{John}}{\begin{array}{c}NP\\:john\end{array}} \quad \frac{\text{killed a coworker}}{\begin{array}{c}S\backslash NP\\: \lambda x.\lambda c.\textbf{KC}(x)\end{array}}}{\begin{array}{c}S\\: \lambda c.\textbf{KC}(john)\end{array}} \quad > \quad ;$$

$$\frac{\frac{\text{knows}_2 \text{ that}}{\begin{array}{c}S\backslash NP/S\\: \lambda p.\lambda x.\lambda c.\textbf{know}(x, @_2 c : pc)\end{array}} \quad \frac{\text{John killed a coworker}}{\begin{array}{c}S\\: \lambda c.\textbf{KC}(john)\end{array}}}{\begin{array}{c}S\backslash NP\\: \lambda x.\lambda c.\textbf{know}(x, @_2 c : \textbf{KC}(john))\end{array}} \quad >$$

$$\frac{\frac{,\text{who}_1}{\begin{array}{c}T/(T\backslash NP)\backslash NP/(S\backslash NP)\\: \lambda r.\lambda z.\lambda p.\lambda \boldsymbol{x}.\lambda c.\begin{bmatrix}pz\boldsymbol{x}c\\\textbf{CI}(@_1 : rzc)\end{bmatrix}\end{array}}}{\begin{array}{c}T/(T\backslash NP)\backslash NP\\: \lambda z.\lambda p.\lambda c.\begin{bmatrix}pzc\\\textbf{CI}(@_1 : \textbf{know}(z, @_2c : \textbf{KC}(john)))\end{bmatrix}\end{array}} \quad >$$

$$\frac{\text{Mary}}{\begin{array}{c}NP\\:mary\end{array}} \quad < \quad \frac{T/(T\backslash NP)}{: \lambda p.\lambda c.\begin{bmatrix}pmaryc\\\textbf{CI}(@_1 : \textbf{know}(mary, @_2c : \textbf{KC}(john)))\end{bmatrix}} \quad \frac{\text{counselled him}_3}{\begin{array}{c}S\backslash NP\\: \lambda x.\lambda c.\textbf{counsel}(x, @_3c)\end{array}}$$

$$\frac{S}{: \lambda c.\begin{bmatrix}\textbf{counsel}(mary, @_3c)\\\textbf{CI}(@_1 : \textbf{know}(mary, @_2c : \textbf{KC}(john)))\end{bmatrix}} \quad >$$

$$\xrightarrow{\text{Dynamic conjunction}} \lambda c.\begin{bmatrix}u:\textbf{KC}(john)\\\begin{bmatrix}\textbf{CI}(@_1 : \textbf{know}(mary, @_2(c, u) : \textbf{KC}(john)))\end{bmatrix}\\\textbf{counsel}(mary, @_3(c, u))\end{bmatrix}$$

The resulting discourse representation contains three underspecified terms: $@_2$ for the factive presupposition triggered by "know" which states that John killed a coworker, $@_1$ for the NRRC that Mary knows it, and $@_3$ for the pronoun "him".

The factive presupposition $@_2$, which is embedded within the CI for NRRC, still receives its left context (c, u) that is a pair of the left context for this mini discourse and the proof of the first sentence. Obviously, the most salient resolution of this underspecification is $@_2 = \lambda c.\pi_2 c$, which returns the proof of the first sentence. What enables this solution is the flexibility of DTS in which the lexical entry of "who" can pass the left context it receives to the relative clause, while the CI content $@_1$ that it introduces does not receive it.

6   Conclusion

In this paper, we have given an analysis of conventional implicature in the framework of Dependent Type Semantics. In this framework, phenomena such as anaphora resolution and presupposition are viewed in terms of proof search; we have shown that this viewpoint, together with suitable constraints on conventional implicature, naturally derive certain observed behavior of conventional implicature with respect to semantic operators and interaction between at-issue and conventionally implicated content. We think the resulting picture is attractive, not least in that it is fully integrated with compositional, subsentential aspects of meaning derivation.

As we observed above, there are many other competing approaches to the derivation of conventional implicatures, and other analyses of their interaction with anaphora and presupposition. Analyses of the first type are generally based on type theory of the kind more standard in linguistic theory, as exemplified by [11,17,19]; the second sort of work tends to be set in dynamic semantics, in line with the majority of formal work on anaphora and presupposition in recent years. This paper removes the explicit focus on dynamics and works with a different view of type theory; as such, it can be placed directly within the recent movement to use continuations and other non-dynamic techniques to simultaneously model intersentential phenomena and to provide a compositional analysis of problems traditional views of composition have found difficult (e.g. [2–4,8]. We leave a full comparison of our theory here with existing views for future work.

This work exhibits many directions for future expansion. We would like to close with one that we believe is of general interest and that shows the power of the current approach. Roberts et al. [23] suggest that the projection behavior of not-at-issue content – i.e. that content which includes presupposition, conventional implicature, and possibly other types which do not play a direct role in the determination of truth conditions – depends on the relation of that content to the current Question Under Discussion, or QUD [21]. We are somewhat agnostic about the precise way in which this claim could or should be formalized, especially given the currently somewhat mysterious ontological status of QUDs; but we are highly sympathetic to the idea that projection behavior should be relativized in some manner to the discourse context, and possibly to the goals and desires of the participants (e.g. as realized by a QUD). But this view is clearly close to what we have set forward here. Plainly the discourse context makes various sorts of content available; if that content contains such things as goals and QUDs, then they ought to play a role in proof search as well, and so we might expect that different computations could be carried out in different contexts, yielding different projection behavior for not-at-issue content. The exact form by which this idea should be spelled out depends on a number of factors: the analysis of questions, probably the proper analysis of denial and other relational speech acts, the form of QUDs and the manner in which they are derived, and of course empirical facts about the projection behavior of not-at-issue content and its relation to contextual elements. We believe that exploring these issues is an exciting next step for the present project.

References

1. Amaral, P., Roberts, C., Smith, E.: Review of 'the logic of conventional implicatures' by Christopher Potts. Linguist. Philos. **30**, 707–749 (2008)
2. Asher, N., Pogodalla, S.: SDRT and continuation semantics. In: Onoda, T., Bekki, D., McCready, E. (eds.) JSAI-isAI 2010. LNCS, vol. 6797, pp. 3–15. Springer, Heidelberg (2011)
3. Barker, C., Bernardi, R., Shan, C.: Principles of interdemensional meaning interaction. In: Li, N., Lutz, D. (eds.) Semantics and Linguistic Theory (SALT) 20, pp. 109–127. eLanguage (2011)
4. Barker, C., Shan, C.C.: Donkey anaphora is in-scope binding. Semantics and Pragmatics **1**(1), 1–46 (2008)
5. Bekki, D.: Representing anaphora with dependent types. In: Asher, N., Soloviev, S. (eds.) LACL 2014. LNCS, vol. 8535, pp. 14–29. Springer, Heidelberg (2014)
6. Coquand, T., Huet, G.: The calculus of constructions. Inf. Comput. **76**(2–3), 95–120 (1988)
7. Grice, H.P.: Logic and conversation. In: Cole, P., Morgan, J.L. (eds.) Syntax and Semantics 3: Speech Acts, pp. 41–58. Academic Press, London (1975)
8. de Groote, P.: Towards a montagovian account of dynamics. In: Gibson, M., Howell, J. (eds.) 16th Semantics and Linguistic Theory Conference (SALT16), pp. 148–155. CLC Publications, University of Tokyo (2006)
9. Gutzmann, D.: Expressive modifiers & mixed expressives. In: Bonami, O., Cabredo Hofherr, P. (eds.) Empirical Issues in Syntax and Semantics 8, pp. 143–165 (2011)
10. Gutzmann, D.: Use-Conditional Meaning: studies in multidimensional semantics. Ph.D. thesis, Universität Frankfurt (2012)
11. Heim, I., Kratzer, A.: Semantics in Generative Grammar. Blackwell, Oxford (1998). No. 13 in Blackwell Textbooks in Linguistics
12. Hom, C.: The semantics of racial epithets. J. Philos. **105**, 416–440 (2008)
13. Hom, C.: A puzzle about pejoratives. Philos. Stud. **159**(3), 383–405 (2010)
14. Krahmer, E., Piwek, P.: Presupposition projection as proof construction. In: Bunt, H., Muskens, R. (eds.) Computing Meanings: Current Issues in Computational Semantics. Kluwer Academic Publishers, Dordrecht (1999). Studies in Linguistics Philosophy Series

15. Löh, A., McBride, C., Swierstra, W.: A tutorial implementation of a dependently typed lambda calculus. Fundamenta Informaticae - Dependently Typed Program. **102**(2), 177–207 (2010)
16. Martin-Löf, P.: Intuitionistic type theory. In: sambin, G. (ed.) vol. 17. Bibliopolis, Naples (1984)
17. Montague, R.: The proper treatment of quantification in ordinary english. In: Hintikka, J., Moravcsic, J., Suppes, P. (eds.) Approaches to Natural Language, pp. 221–242. Reidel, Dordrecht (1973)
18. Nouwen, R.: On appositives and dynamic binding. Res. Lang. Comput. **5**, 87–102 (2007)
19. Potts, C.: The Logic of Conventional Implicatures. Oxford University Press, New York (2005)
20. Ranta, A.: Type-Theoretical Grammar. Oxford University Press, New York (1994)
21. Roberts, C.: Information structure: towards an integrated formal theory of pragmatics. In: OSUWPL, Papers in Semantics, vol. 49. The Ohio State University Department of Linguistics (1996)
22. van der Sandt, R.: Presupposition projection as anaphora resolution. J. Seman. **9**, 333–377 (1992)
23. Simons, M., Tonhauser, J., Beaver, D., Roberts, C.: What projects and why. In: Proceedings of SALT 20, pp. 309–327. CLC Publications (2011)
24. Sundholm, G.: Proof theory and meaning. In: Gabbay, D., Guenthner, F. (eds.) Handbook of Philosophical Logic, vol. III, pp. 471–506. Kluwer, Reidel, Dordrecht (1986)
25. Wang, L., Reese, B., McCready, E.: The projection problem of nominal appositives. Snippets **11**, 13–14 (2005)

Formal Analysis of Epistemic Modalities and Conditionals Based on Logic of Belief Structures

Yasuo Nakayama(✉)

Graduate School of Human Sciences, Osaka University, Suita, Japan
nakayama@hus.osaka-u.ac.jp

Abstract. There is a strong context dependency in meaning of modalities in natural languages. Kratzer [9] demonstrates how to deal with this problem within possible world semantics. In this paper, we propose to interpret epistemic modalities in background of an epistemic state. Our analysis is a meta-linguistic one and we extensively use the proof-theoretic consequence relation. We define, then, a *belief structure* and introduce a *belief structure revision operator*. We call this framework *Logic of Belief Structures* (LBS). Then, we apply LBS to formalization of belief revision and interpretation of conditionals and investigate the relationship between belief revision and conditionals. Furthermore, we propose two types of conditionals, *epistemic* and *causal conditionals*.

Keywords: Conditionals · Epistemic modality · Belief structure · Belief revision · AGM theory · Sphere system

1 Logic for Epistemic Modalities

According to von Fintel [2], we can distinguish six kinds of modal meaning. They are *alethic, epistemic, deontic, bouletic, circumstantial*, and *teleological modality*. He characterized *epistemic modality* as the modality that is based on epistemic state:

(1a) [Epistemic modality] Epistemic modality concerns what is possible or necessary, given what is known and what the available evidence is.
(1b) [Example for epistemic modality] It has to be raining. [After observing people coming inside with wet umbrellas.]

Kratzer [9, pp. 4–6] proposes to explain the varieties of modalities in terms of the distinction of views. According to Kratzer, the core meaning of *must* can be interpreted as *must in view of*. This *must in view of* takes two arguments, namely *modal restriction* and *modal scope*. Then, we have the following schema for modal sentences:

must in view of (modal restriction, modal scope).

© Springer-Verlag Berlin Heidelberg 2015
T. Murata et al. (Eds.): JSAI-isAI 2014 Workshops, LNAI 9067, pp. 37–52, 2015.
DOI: 10.1007/978-3-662-48119-6_4

To demonstrate how to use this schema, let us take an example for epistemic modality:

(2a) [Example for epistemic modality] The ancestors of the Maoris must have arrived from Tahiti.
(2b) [*must-in-view-of* Interpretation] In view of what is known, the ancestors of the Maoris must have arrived from Tahiti.
(2c) [Application of Kratzer's schema] *must in view of* (what is known, the ancestors of the Maoris arrived from Tahiti).

Kratzer [9, pp. 10–11] defines a possible world semantics for *must in view of*; her definition is restricted to propositional logic.

Definition 1. (3a) *A proposition p is true in a world w in W iff $w \in p$.*
(3b) *The meaning of **must in view of** is a function ν that satisfies the following conditions:*
1. *The domain of ν is the set of all pairs $\langle p, f \rangle$ such that $p \in P(W)$ and f is a function from W to $P(P(W))$.*
2. *For any p and f such that $\langle p, f \rangle$ is in the domain of ν: $\nu(p, f) = \{w \in W : \bigcap f(w) \subseteq p\}$.*

The modal scope denotes a proposition p and the modal restriction denotes an individual concept f. The meaning of *must in view of* is a function that maps pairs consisting of a proposition and a function of the same type as f to another proposition. When we apply (3b) to (2a), (2a) is true in those worlds w such that it follows from what is known in w that the ancestors of the Maoris arrived from Tahiti.

Recently, I proposed a formal framework in which the epistemic and the deontic modality are relativized by an accepted epistemic and a deontic theory [13–15]. The framework is called *Logic for Normative Systems* (LNS). In this paper, we concentrate on the epistemic part of LNS and show that Krazer's view can be rewritten within our framework.

Logic for Epistemic Modalities (LEM) is a framework expressed in a meta-language of *First-order Logic* (FOL). We define LEM-sentences as follows:

Definition 2. (4a) *All FO-sentences (i.e., sentences in FOL) are LEM-sentences.*
(4b) *If p is a FO-sentence and T is a set of FO-sentences, then $MUST_T \, p$, $MIGHT_T \, p$, $KNOWN_T \, p$, and $BEL_T^{inf} p$ are LEM-sentences. In this paper, we use small letters p, q, ... to denote FO-sentences.*
(4c) *If ϕ and ψ are LEM-sentences, then not ϕ, $\phi \& \psi$, ϕ or ψ, $\phi \Rightarrow \psi$, and $\phi \Leftrightarrow \psi$ are LEM-sentences, where logical connectives, not, &, or, $\Rightarrow$, and $\Leftrightarrow$ belong to the meta-language.*
(4d) *If ϕ is a LEM-sentence, then ϕ satisfies (4a) or (4b) or (4c).*

Definition 2 indicates that no iteration of modal operators is allowed in LEM. The meaning of epistemic modalities is defined as follows.

Definition 3. *Let T be a set of FO-sentences and p be a FO-sentence. We use $cons(T)$ as an abbreviation of $\langle T$ is consistent$\rangle$. We call T in the following definitions $\langle$belief base$\rangle$. A belief base represents what is explicitly believed.*

(5a) $MUST_T\, p$ *iff* $(T \vdash p\ \&\ cons(T))$.
(5b) $MIGHT_T\, p$ *iff* $cons(T \cup \{p\})$.
(5c) *[Knowledge as Explicit Belief]* $KNOWN_T\, p$ *iff* $(p \in T\ \&\ cons(T))$.
(5d) *[Inferential Belief]* $BEL_T^{inf}\, p$ *iff* $(MUST_T\, p\ \&\ not\ KNOWN_T\, p)$.
(5e) $mod(T) = \{M : M \models T\}$.

Explicit belief and inferential belief play an important role for analysis of epistemic modalities (see Sect. 2). The semantics of LEM can be given in the same way as for FO-sentences.

To demonstrate the relationship to Krazer's approach, we introduce the following notations.

Definition 4. *Let T be a set of PL-formulas and p be a PL-formula. Let W be a set of possible worlds.*

(6a) v_W *is a function from PL-formulas to $P(W)$.*
(6b) $v_W^s(T) = \bigcap\{v_W(p) : p \in T\}$.
(6c) *W is a maximal set of worlds iff for any consistent set T of PL-formulas there is w such that $w \in W\ \&\ w \in v_W^s(T)$.*

Now, from Definitions 3 and 4, Propositions 5 and 6 immediately follow.

Proposition 5. *Let T be a set of FO-sentences and p be a FO-sentence.*

(7a) $MUST_T\, p \Rightarrow MIGHT_T\, p$.
(7b) $MUST_T\, (p \to q) \Rightarrow (MUST_T\, p \Rightarrow MUST_T\, q)$.
(7c) $(T_1 \subseteq T_2\ \&\ MIGHT_{T_2}(p \to p)) \Rightarrow (MUST_{T_1}\, p \Rightarrow MUST_{T_2}\, p)$.[1]
(7d) $(T_1 \subseteq T_2 \Rightarrow (MIGHT_{T_2}\, p \Rightarrow MIGHT_{T_1}\, p)$.
(7e) $KNOWN_T\, p \Rightarrow MUST_T\, p$.
(7f) $cons(T) \Rightarrow (KNOWN_T\, p \Leftrightarrow p \in T)$.
(7g) $BEL_T^{inf}\, p \Rightarrow p \notin T$.
(7h) $MUST_T\, p$ *iff* $(mod(T) \subseteq mod(\{p\})\ \&\ mod(T) \neq \emptyset)$.
(7i) $MIGHT_T\, p$ *iff* $mod(T \cup \{p\}) \neq \emptyset$.

Proposition 6. *Let T be a set of PL-formulas and p be a PL-formula. Let W be a maximal set of worlds.*

(8a) $MUST_T\, p$ *iff* $(v_W^s(T) \subseteq v_W(p)\ \&\ v_W^s(T) \neq \emptyset)$.
(8b) $MIGHT_T\, p$ *iff* $v_W^s(T \cup \{p\}) \neq \emptyset$.

In LEM, Krazer's *modal restriction* can be imitated by the *restriction given by a belief base*. We interpret, then, the modality not as a relation but as an operator restricted by a belief base: *[must in view of (T)]* *(proposition)*.

Now, let us reconsider Krazer's example (2a). We interpret it as (2d).

---

[1] Because $(p \to p)$ is a FOL-theorem, it holds: $MIGHT_{T_2}(p \to p)$ iff T_2 is consistent.

(2c) [Application of Kratzer's schema] *must in view of* (what is known, the ancestors of the Maoris arrived from Tahiti).

(2d) $MUST_T$ tr(the ancestors of the Maoris arrived from Tahiti).[2]

Thus, we are justified to say that theory T in $MUST_T$ expresses *the view of what is known*. In this context, $MUST_T$ can be understood as *must in view of what is known*.

2   Evidential Aspects of Epistemic Modalities

The interpretation of *must* as *must in view of what is known*, proposed in the previous section, is still inappropriate as an interpretation of *epistemic must*, because it ignores evidential aspects of *epistemic must*. According to von Fintel and Gilles [3, p. 357], *epistemic must* presupposes the presence of indirect inference rather than a direct observation. Karttunen [7] observed problems connected with the traditional interpretation of *epistemic must*. When one considers which of the answers to the question (9a) conveys more confidence, it is natural to feel that epistemic modal sentence (9c) is less forceful than simple sentence (9b).

(9a) Where are the keys?
(9b) They are in the kitchen drawer.
(9c) They must be in the kitchen drawer.

According to Karttunen [7], modal semantics predicts that (9c) is a stronger answer to the question than (9b), but our intuition goes the other way. To respect this intuition, we propose to analyze (9b) as (9d) and (9c) as (9e). Here, we presuppose that belief base T_{9b} represents *what is known* by the speaker of (9b) and that belief base T_{9c} represents *what is known* by the speaker of (9c). Let $p_{keys} = tr$(The keys are in the kitchen drawer).

(9d) Felicitous condition: $KNOWN_{T_{9b}}$ p_{keys}; Claim: p_{keys}.
(9e) Felicitous condition: $BEL_{T_{9c}}^{inf}$ p_{keys}; Claim: $MUST_{T_{9c}}$ p_{keys}.

It will be appropriate to interpret the situation described by (9a) $\sim$ (9c) as follows: Sentence (9b) is uttered by a person who is convinced that p_{keys}, while sentence (9c) is uttered by a person who has evidences for p_{keys} and accepts this proposition based on these evidences. According to our interpretation, (9b) is stronger than (9c) in the sense that the felicitous condition for (9b) implies $\langle must\ (9b)\rangle$.[3]

The bearer of T_{9b} knows that p_{keys}, while the bearer of T_{9c} does not know that p_{keys} and his belief of p_{keys} is supported by his inference based on his evidences.[4] Our interpretation fully supports the following observation of von Fintel and Gillies [3, p. 354]:

---

[2] Here, function tr is the translation function from English sentences to FO-sentences.

[3] Note that it holds: $KNOWN_{T_{9b}}$ $p_{keys} \Rightarrow MUST_{T_{9b}}$ p_{keys}. See (7e).

[4] Note that our interpretation agrees with Willet's taxonomy of evidential categories [16]. Willet interpret epistemic modalities as makers of indirect inference [3, p. 354].

epistemic modals are also evidential markers: they signal that the prejacent was reached through an inference rather than on the basis of direct
observation or trustworthy reports.

What does this signaling means? We propose to interpret it as a felicitous
condition (abbreviated as FC). This is described in Table 1.

S's utterance of p is felicitous iff S believes that S knows p.
S's utterance of $Must\ p$ is felicitous iff S accepts p based on an indirect
inference.

von Fintel and Gillies argue for the thesis that epistemic modalities signal not
weakness but indirect inference. This observation agrees with our interpretation
of epistemic modalities (See Table 1).

Table 1. Interpretation of simple sentences and epistemic modalities

	Claim	FC	Formal representation of FC
Simple sentence	p	$KNOWN_{T_1}\ p$	$p \in T_1\ \&\ cons(T_1)$
Epistemic *must*	$MUST_{T_2}\ p$	$BEL_{T_2}^{inf} p$	$p \notin T_2\ \&\ T_2 \vdash p\ \&\ cons(T_2)$

Let us consider some additional examples from von Fintel and Gillies [3,
p. 372]:

(9f) Seeing the pouring rain, Billy says: *It's raining.*
(9g) Seeing people coming inside with wet umbrellas, Billy says: *It must be
raining.*

We assume that $p_{rain} = tr$(it is raining) and $p_{umbrellas} = tr$(people coming
inside have wet umbrellas). Because of (9f) and (9g), it holds: $p_{rain} \in T_{9f}\ \&$
$p_{umbrellas} \in T_{9g}\ \&\ p_{rain} \notin T_{9g}$. In this case, the situation can be described as
Table 2.

Table 2. Examples for simple sentences and epistemic modalities

	Claim	FC
Simple sentence	p_{rain}	$KNOWN_{T_{9f}}\ p_{rain}$
Epistemic *must*	$MUST_{T_{9g}}\ p_{rain}$	$BEL_{T_{9g}}^{inf} p_{rain}$

We see that both (9f) and (9g) are appropriate, because felicitous conditions
for both cases are satisfied in these situations.

As von Fintel and Gillies [3] discuss, there are several semantic approaches for
epistemic modalities. Our approach is proof-theoretic and very straightforward.
It is directly based on the following observation: Epistemic modalities are used
in a situation in which the speaker has no direct but only indirect evidences for
the prejacent.

3   Logic of Belief Structures

To describe semantics for conditionals, we propose to represent an epistemic state by a belief structure, which is a linearly ordered set of consistent sets of FO-sentences. In this section, we define a logical framework for such belief structures and call it *Logic of Belief Structures* (LBS).

Definition 7. **(10a)** *[Belief structure BS] $BS = \langle ST, > \rangle$ is a belief structure, when the following three conditions are satisfied:*
  1. *$ST = \{T_i : 1 \le i \le n \ \& \ T_i$ is a consistent set of FO-sentences$\}$,*
  2. *$>$ is a total order on ST and $T_1 > ... > T_n$, and*
  3. *for all $T_i \in ST$ and $T_j \in ST$, $T_i \cap T_j = \emptyset$.*
(10b) *[k first fragment of BS] $top(BS, k) = \bigcup\{T_i : 1 \le i \le k$ and $T_i \in ST\}$. In other words, k first fragment of BS is the union of the first k elements of BS. We can also define $top(BS, k)$ recursively as follows:*
  1. *$top(BS, 1) = T_1$.*
  2. *$top(BS, k) = top(BS, k - 1) \cup T_k$.*
(10c) *[Consistent maximum of BS] $top(BS, k)$ is the consistent maximum of BS (abbreviated as cons-max(BS)) iff $(cons(top(BS, k)) \ \& \ not\ cons(top(BS, k + 1)))$. We call k the consistent maximum number of BS (abbreviated as cmn(BS)), when $top(BS, k) = cons\text{-}max(BS)$.*
(10d) *[Deductive closure] $Cn(T) = \{p : T \vdash p\}$.*
(10e) *[Belief set for BS] We call $Cn(cons\text{-}max(BS))$ the belief set for BS.*

Based on Definition 7, we can define some modal operators and some notions related to sphere systems.

Definition 8. *Let $BS = \langle ST, > \rangle$ be a belief structure with $T_1 > ... > T_n$. Let p and q be FO-sentences.*

(11a) *$MUST^*_{BS}\ p$ iff $MUST_{cons\text{-}max(BS)}\ p$.*
(11b) *$MIGHT^*_{BS}\ p$ iff $MIGHT_{cons\text{-}max(BS)}\ p$.*
(11c) *[Probability Order] $MORE\text{-}PROBABLE_{BS}(p, q)$ iff (there are $T_i \in ST$ and $T_j \in ST$ such that $(p \in T_i \ \& \ q \in T_j \ \& \ T_i > T_j)$).*
(11d) *$PROBABLY_{BS}\ p$ iff $(MIGHT^*_{BS}\ p \ \& \ not\ MUST^*_{BS}\ p \ \& \ p \in top(BS, n)$ $\& \ (\neg p \in top(BS, n) \Rightarrow MORE\text{-}PROBABLE_{BS}(p, \neg p)))$.*
(11e) *$MUST\text{-}min(BS, k, p)$ iff $(MUST_{top(BS,k)}\ p \ \& \ not\ MUST_{top(BS,k-1)}\ p)$.*
(11f) *$p \preccurlyeq_{BS} q$ iff there are k and m such that $(k \le m \le cmn(BS) \ \& \ MUST\text{-}min(BS, k, p) \ \& \ MUST\text{-}min(BS, m, q))$.*
(11g) *$p \approx_{BS} q$ iff $(p \preccurlyeq_{BS} q \ \& \ q \preccurlyeq_{BS} p)$.*
(11h) *$p \prec_{BS} q$ iff $(p \preccurlyeq_{BS} q \ \& \ not\ (p \approx_{BS} q))$.*
(11i) *[Sphere Model System]*
   SMS_{BS} is a sphere model system for BS iff
   1. *$SMS_{BS} = \{S_{cmn(BS)}, ..., S_1\}$, and*
   2. *$S_k = mod(top(BS, k))$ for k with $1 \le k \le cmn(BS)$.*
(11j) *[Sphere System] Let W be a maximal set of worlds. Let ST be a set of PL-formulas.*
   SS_{BS} is a sphere system for BS iff

1. $SS_{BS} = \{S_{cmn(BS)}, ..., S_1\}$, and
2. $S_k = v_W^s(top(BS, k))$ for k with $1 \leq k \leq cmn(BS)$.[5]

$p \preccurlyeq_{BS} q$ is read as ⟨Based on BS, it is at least as possible that p as it is that q⟩. $p \approx_{BS} q$ is read as ⟨Based on BS, it is equally possible that p and that q⟩. $p \prec_{BS} q$ is read as ⟨Based on BS, it is more possible that p than that q⟩.[6] From the view of belief change, we may read $p \prec_{BS} q$ as ⟨In BS, p is more entrenched than q⟩. Based on Definition 8, Propositions 9 and 10 can be easily shown.

Proposition 9. *Let BS be a belief structure with $T_1 > ... > T_n$. Let T_k ($1 \leq k \leq n$) be a set of FO-sentences.*

(12a) $k \leq m \leq n \Rightarrow top(BS, k) \subseteq top(BS, m)$.
(12b) $k \leq m \leq cmn(BS) \Rightarrow Cn(top(BS, k)) \subseteq Cn(top(BS, m))$.
(12c) $k \leq m \leq cmn(BS) \Rightarrow mod(top(BS, m)) \subseteq mod(top(BS, k))$.
(12d) *If SMS_{BS} is a sphere model system for BS, then SMS_{BS} satisfies the following four requirements:*
  1. *SMS_{BS} is centered on $S_{cmn(BS)}$, i.e., for all $S_k \in SMS_{BS}$, $S_{cmn(BS)} \subseteq S_k$.*
  2. *SMS_{BS} is nested, i.e., for all $Si, Sj \in SMS_{BS}$, $(S_i \subseteq Sj$ or $S_j \subseteq S_i)$.*
  3. *SMS_{BS} is closed under unions, i.e., $X \subseteq SMS_{BS} \Rightarrow \bigcup X \in SMS_{BS}$.*
  4. *SMS_{BS} is closed under (nonempty) intersections, i.e., $(X \subseteq SMS_{BS}$ & $X \neq \emptyset) \Rightarrow \bigcap X \in SMS_{BS}$.*

Proof. (12a) follows from (10b). (12b) follows from (10d) and (12a). (12c) follows from (5e) and (12a). (12d) 1, 2, 3, and 4 follow from (11i) and (12c). Q.E.D.

Proposition 10. *Let T_k ($1 \leq k$) be a set of PL-formulas and W be a maximal set of worlds. Let BS be a belief structure.*

(13a) $k \leq m \leq cmn(BS) \Rightarrow v_W^s(top(BS, m)) \subseteq v_W^s(top(BS, k))$.
(13b) *If SS_{BS} is a sphere system for BS, then SS_{BS} is centered on $S_{cmn(BS)}$, nested, closed under unions, and closed under (nonempty) intersections.*

Proof. (13a) follows from (6b) and (12a). (13b) follows from (11j) and (13a). Q.E.D.

Lewis defined a sphere system in [11, p. 14]. (12d) shows that only the first characterization is different from his definition. Lewis required that a sphere system is centered on a singleton $\{w_0\}$, where the intended reference of w_0 is the actual world. Our interpretation of the center of a sphere system is epistemic. The center, $S_{cmn(BS)}$, denotes the set of worlds (or the set of models) in which all of what are consistently believed are true.

---

[5] According to definition of v_W^s, $v_W^s(top(BS, k)) = \{w \in W: $ all formulas in $top(BS, k)$ are true in $w\}$.
[6] These orders are a modification of *comparative possibility* in Lewis [11, p. 52].

Proposition 11. *Let BS be a belief structure.*

(14a) $\preccurlyeq_{BS}$ *is transitive.*[7]
(14b) $\approx_{BS}$ *is symmetric and transitive.*[8]
(14c) $PROBABLY_{BS}\ p \Rightarrow (MIGHT^*_{BS}\ p\ \&\ not\ MUST^*_{BS}\ p).$

Proof. To show (14a), suppose that $p \preccurlyeq_{BS} q\ \&\ q \preccurlyeq_{BS} r$. Then from (11f), there are k, l, m such that $(k \leq l \leq m \leq cmn(BS)\ \&\ MUST\text{-}min(BS, k, p)\ \&\ MUST\text{-}min(BS, l, q)\ \&\ MUST\text{-}min(BS, m, r))$. Thus, from (11f), $p \preccurlyeq_{BS} r$. Therefore, transitivity holds for $\preccurlyeq_{BS}$. (14b) follows from (11g) and (14a). (14c) follows from (11d). Q.E.D.

4   Belief Revision Based on Logic of Belief Structures

We can divide a belief structure BS into two parts, namely the *consistent part*, $top(BS, k)$ with $k \leq cmn(BS)$, and the *inconsistent part*, $top(BS, k)$ with $cmn(BS) < k \leq n$. Now, let us define the *belief structure revision* and *expansion*.

Definition 12. *Let H be a consistent set of FO-sentences. Let BS be a belief structure with $T_1 > ... > T_n$.*

(15a) *We define $ext(H, BS)$ as the belief structure with $H > T_1 > ... > T_n$. In other words, the extended belief structure of BS by H is the belief structure that can be obtained from BS by adding H as the most reliable element.*
(15b) *[Belief structure revision] $bsR(BS, H) = Cn(cons\text{-}max(ext(H, BS)))$.*
(15c) *[Belief structure expansion] $bsEX(BS, H) = Cn(cons\text{-}max(BS) \cup H)$.*

We can show that our revision operator bsR satisfies all of postulates for the belief revision operator * in AGM-theory, if $H = \{p\}$ and p is a consistent FO-sentence.[9] Because the AGM-theory is a theory for propositional representation and our revision operator is defined for FO-sentences, our approach is broader than the AGM approach. The AGM postulates for belief revision can be defined as described in [6].

Definition 13. *Let p and q be PL-formulas and K be a set of PL-formulas. Let $K + p = Cn(K \cup p)$.*

(16a) *[Closure] $K^*p = Cn(K^*p)$.*
(16b) *[Success] $p \in K^*p$.*
(16c) *[Inclusion] $K^*p \subseteq K + p$.*
(16d) *[Vacuity] If $\neg p \notin K$, then $K^*p = K + p$.*
(16e) *[Consistency] K^*p is consistent if p is consistent.*
(16f) *[Extensionality] If p and q are logically equivalent, then $K^*p = K^*q$.*

---

[7] In domain $cons\text{-}max(BS)$, $\preccurlyeq_{BS}$ is also reflexive and connected.
[8] In domain $cons\text{-}max(BS)$, $\approx_{BS}$ is also reflexive. Thus, in $cons\text{-}max(BS)$, $\approx_{BS}$ is an equivalence relation.
[9] For AGM-theory, consult Gärdenfors [4, Sect. 3.3] and Hansson [6].

(16g) *[Superexpansion]* $K^*(p \wedge q) \subseteq (K^*p) + q$.
(16h) *[Subexpansion]* If $\neg q \notin K^*p$, then $(K^*p) + q \subseteq K^*(p \wedge q)$.

The following theorem shows that belief structure revision operator bsR satisfies all of the AGM postulates with the restriction that the revising FO-sentence is consistent.

Theorem 14. *Let p, q, and $p \wedge q$ be consistent FO-sentences.*

(17a) *[Closure]* $bsR(BS, \{p\})$ is a belief set.
(17b) *[Success]* $p \in bsR(BS, \{p\})$.
(17c) *[Inclusion]* $bsR(BS, \{p\}) \subseteq bsEX(BS, \{p\})$.
(17d) *[Vacuity]* $\neg p \notin Cn(cons\text{-}max(BS)) \Rightarrow bsR(BS, \{p\}) = bsEX(BS, \{p\})$.
(17e) *[Consistency]* $bsR(BS, \{p\})$ is consistent.
(17f) *[Extensionality]* If p and q are logically equivalent, then $bsR(BS, \{p\}) = bsR(BS, \{q\})$.
(17g) *[Superexpansion]* $bsR(BS, \{p \wedge q\}) \subseteq bsEX(bsR(BS, \{p\}), \{q\})$.
(17h) *[Subexpansion]* $\neg q \notin bsR(BS, \{p\}) \Rightarrow$
 $bsEX(bsR(BS, \{p\}), \{q\}) \subseteq bsR(BS, \{p \wedge q\})$.

Proof. We assume that p, q and $p \wedge q$ are consistent FO-sentences. Then, (17a) holds because of (15a), (15b), and Definition 7. Because $\{p\}$ is consistent, (17b) follows from Definitions 7 and 12. From Definitions 7 and 12 follows: $cons\text{-}max(ext(H, BS)) \subseteq cons\text{-}max(BS) \cup H$. Then, (17c) holds because of Definition 12. To show (17d), suppose $\neg p \notin Cn(cons\text{-}max(BS))$. Then, $cons\text{-}max(BS) \cup \{p\}$ is consistent. Thus, $cons\text{-}max(ext(\{p\}, BS)) = cons\text{-}max(BS) \cup \{p\}$. Hence, (17d) holds based on Definition 12. (17e) holds because of (15b). (17f) holds based on (15b) and inference rules of FOL. To show (17g), we assume: $k = cmn(ext(\{p \wedge q\}, BS)) - 1$ and $m = cmn(ext(\{p\}, BS)) - 1$. Then, from Definitions 7 and 12: $top(BS, k) \subseteq top(BS, m)$. In FOL, it holds: $T_1 \subseteq T_2 \Rightarrow Cn(Cn(T_1 \cup \{p \wedge q\}) \cup \{q\}) \subseteq Cn(Cn(T_2 \cup \{p\}) \cup \{q\})$. Because $Cn(Cn(T_1 \cup \{p \wedge q\}) \cup \{q\}) = Cn(T_1 \cup \{p \wedge q\})$, (17g) holds based on Definition 12. To show (17h), we assume $\neg q \notin bsR(BS, \{p\})$. In FOL, we can prove: If $T \cup \{p\} \nvdash \neg q$, then $[cons(T \cup \{p\})$ iff $cons(T \cup \{p \wedge q\})]$. Thus, $bsR(BS, \{p\}) = bsR(BS, \{p \wedge q\})$. Therefore, $(bsR(BS, \{p\}) \cup \{q\}) = (bsR(BS, \{p \wedge q\}) \cup \{q\})$. However, because q follows from $p \wedge q$, $Cn(bsR(BS, \{p \wedge q\}) \cup \{q\}) = bsR(BS, \{p \wedge q\})$. From these: $bsEX(bsR(BS, \{p\}), \{q\}) = bsR(BS, \{p \wedge q\})$. Thus, (17h) holds. Q.E.D.

AGM-theory is a standard framework for belief revision. Thus, Theorem 14 suggests the adequacy of our definition of belief structure revision. In fact, our approach provides a useful tool for belief revision, because it only requires a linearly order sets of FO-sentences. The original AGM requirements for the entrenchment relation are rather unnatural and difficult to use.[10]

---

[10] However, AGM-theory has a nice correspondence with the probability theory [4, Chap. 5]. Our approach is difficult to relate with a probability theory.

5 Conditionals and Belief Revision

Our analysis of conditionals in this paper is based on Ramsey Test [4, p. 147]:

[RT] Accept the sentence of the form ⟨If A, then C⟩ in a state of belief K if and only if the minimal change of K needed to accept A also requires accepting C.

This idea can be roughly expressed as follows: ⟨If A, then C⟩ is acceptable with respect to K iff $minimal\text{-}change(K, A)$ implies C.

This idea can be combined with Kratzer's approach to counterfactual conditionals. Kratzer [9, p. 64] suggests that there are (at least) three forms of conditionals: (*If ...*), (*necessarily/possibly/probably*). According to this observation, we have two types of operators in counterfactual conditionals (*If p, Modal q*). The operator *If* characterizes the considered situation, and the operator *Modal* makes a modal statement. The antecedent [*If p*] brings us to imagine a situation in which p is true, where the situation is described by T. Then, we consider whether the modal claim in the consequence [$MODAL_T q$] holds in the imagined situation. Based on this idea, we propose to interpret *If*-operator as a *belief structure revision operator* and p as the *revising consistent FO-sentence*.

Definition 15. *Let BS be a belief structure and H be a consistent set of FO-sentences. Let Modal $\in$ {Must, Might, Known} and MODAL $\in$ {MUST, MIGHT, KNOWN}.*

(18a) $IF_{BS}(H) = cons\text{-}max(ext(H, BS))$.
(18b) $[If^{BS}p](Modal\ q)$ *iff*
   $(not\ cons(top(BS, 1) \cup \{p\})$ *or* $(T = IF_{BS}(\{p\})$ & $MODAL_T\ q))$.
(18c) $[If^{BS}p](Probably\ q)$ *iff*
   $(not\ cons(top(BS, 1) \cup \{p\})$ *or* $PROBABLY_{ext(\{p\}, BS)}\ q)$.

From Definition (18b) follows: If $cons(top(BS, 1) \cup \{p\})$, then $[If^{BS}\ p](Must\ q)$ holds iff the minimal change of $cons\text{-}max(BS)$ needed to accept p also requires accepting q. This formulation roughly corresponds to [RT]. Based on Definition 15, we can prove Proposition 16.

Proposition 16. *Let BS be a belief structure and H be a consistent set of FO-sentences.*

(19a) $bsR(BS, H) = Cn(IF_{BS}(H))$.
(19b) $[If^{BS}\ p](Must\ q)$ *iff* $(IF_{BS}(\{p\}) = \{p\}$ *or* $q \in bsR(BS, \{p\}))$.
(19c) $MIGHT_{cons-max(BS)}\ p \Rightarrow$
   $([If^{BS}\ p](Must\ q) \Rightarrow MUST_{cons-max(BS)}\ (p \rightarrow q))$.
(19d) $[If^{BS}\ p](Must\ q) \Rightarrow$
   $(mod(top(BS, 1) \cup \{p\}) = \emptyset$ *or* $mod(IF_{BS}(\{p\})) \subseteq mod(\{q\}))$.

Proof. (19a) follows from (15b) and (18a). (19b) follows from (18a), (18b), and (19a). To show (19c), suppose that $MIGHT_{cons\text{-}max(BS)}$ p holds. Then, because of (5b), $cons\text{-}max(BS) \cup \{p\}$ is consistent. Thus, according to (15a) and (18a), $IF_{BS}(\{p\}) = cons\text{-}max(ext(\{p\}, BS)) = cons\text{-}max(BS) \cup \{p\}$. Now, suppose that $[If^{BS}\ p](Must\ q)$ holds. Then, from (18b), $MUST_{cons\text{-}max(ext(\{p\},BS))}$ q. Thus, $MUST_{cons\text{-}max(BS) \cup \{p\}}$ q. Then, because of (5a) and the deduction theorem of FOL, $MUST_{cons\text{-}max(BS)}$ $(p \to q)$ holds. Hence, (19c) holds. (19d) follows from (7h) and (18b). Q.E.D.

(19a) and (19b) show that our definition of counterfactual conditional is based on the belief structure revision. According to (19c), a material conditional follows from a counterfactual conditional, when no change is required to accept its antecedent. (19d) expresses the idea that the antecedent of a counterfactual conditional determines the range of models in which the consequent is evaluated.

Let us apply LBS to an example from Kratzer [9, p. 94].

(20a) If a wolf entered the house, he must have eaten grandma, since she was bedridden. He might have eaten the girl with the red cap, too. In fact, that's rather likely. The poor little thing wouldn't have been able to defend herself.

We assume that there are appropriate translations of sentences in (16a) into FO-sentences:

p: tr(a wolf entered the house).
q: tr(the wolf ate grandma).
r: tr(the grandma was bedridden).
s: tr(the wolf ate the girl with the red cap).
t: tr(the girl was not able to defend herself).

Now, we can express story (20a) within LBS as follows:

(20b) $[If^{BS}\ p](Must\ q)$ & $([If^{BS}\ p](Known\ r)$ & $not\ MUST_{\{p\}}\ r)$ & $[If^{BS}\ p]$ $((Might\ s)$ & $(Probably\ s)$ & $(Must\ t))$.

Here, $([If^{BS}\ p](Known\ r)$ & $not\ MUST_{\{p\}}\ r)$ expresses that r belongs to the part of BS that is kept in its consistent maximum after acceptance of p and that r is independent from p. When we assume $cons(top(BS, 1) \cup \{p\})$, from (20b) follows (20c).

(20c) $T = IF_{BS}(\{p\})$ & $MUST_T\ q$ & $(KNOWN_T\ r$ & $not\ MUST_{\{p\}}\ r)$ & $MIGHT_T\ s$ & $PROBABLY_{ext(\{p\},BS)}\ s$ & $MUST_T\ t$.

(20c) roughly means the following: (Suppose p. Then, (it must be q, because it is known that r & it might be s & it is probable that s & it must be t)). As Kratzer [9, p. 94] points out, the *if-clause* determines the evaluation range of modal operators in long stretches of subsequent discourse.[11].

---

[11] For interpretation of (20a), it would be more appropriate to deal with anaphoric relation. This can be done by using Skolem-symbols [12, 15].

6   Interpretation of Conditionals

In this section, we examine the relationship between our interpretation of conditionals and the standard interpretation. Lewis [11, p. 1] explains the standard interpretation as follows [8, p. 428]:

> A possible world in which the antecedent of a counterfactual is true is called an "antecedent-world." One can state the theory (in a somewhat simplified form) by saying that a counterfactual is true just in case its consequent is true in those antecedent-worlds that are most similar to the actual world.

Based on this idea, Lewis [11, p. 16] defines the truth condition for a counterfactual conditional as follows:[12]

> [LEWIS] $p \mapsto q$ is true at a world i (according to a system of spheres SS) iff either
>
> **1.** no p-world belongs to any sphere S in SS_i, or
> **2.** some sphere S in SS_i does contain at least one p-world, and $p \rightarrow q$ holds at every world in S.

Now, we examine the relationship between our interpretation and Lewis's standard interpretation. Actually, it turns out that our interpretation is very similar to [LEWIS]. The main difference lies in the notion of *center of a sphere system*, namely Lewis accepts only a singleton as the center. We can prove a proposition that is very close to [LEWIS].

Proposition 17. *Let BS be a belief structure with $T_1 > \ldots > T_n$.*

(21a) *Let $T_1, \ldots, T_n$ be sets of FO-sentences.*
    $[If^{BS} p](Must\, q)$ iff either
    1. *no p-model belongs to any sphere S in SMS_{BS}, or*
    2. *some sphere S in SMS_{BS} does contain at least one p-model, and $p \rightarrow q$ holds in every model in S.*
(21b) *Let $T_1, \ldots, T_n$ be sets of PL-formulas and W be a maximal set of worlds.*
    $[If^{BS} p](Must\, q)$ iff either
    1. *no p-model belongs to any sphere S in SS_{BS}, or*
    2. *some sphere S in SS_{BS} does contain at least one p-model, and $p \rightarrow q$ holds in every model in S.*

Proof. It can be easily shown: $mod(T_1 \cup \{p\}) = \emptyset$ iff (21a.1). Now, we consider cases in which $mod(T_1 \cup \{p\}) \neq \emptyset$. Let $k = cmn(ext(\{p\}, BS)) - 1$ and $S_k = mod(top(BS, k))$. Because $top(BS, k) \cup \{p\}$ is consistent, there is a model in S_k that makes p true. Furthermore, $IF_{BS}(\{p\}) = cons\text{-}max(ext(p, BS)) = top(BS, k) \cup \{p\}$. According to (5a) and (11i), $(T = IF_{BS}(\{p\})\ \&\ MUST_T\, q)$

---

[12] Here, we represent counterfactual conditional with $\mapsto$.

iff $mod(IF_{BS}(\{p\})) \subseteq mod(\{q\})$. Then, because of the deduction theorem in FOL: $mod(top(BS, k) \cup \{p\}) \subseteq mod(q)$ iff $mod(top(BS, k)) \subseteq mod(\{p \to q\}))$ iff $S_k \subseteq mod(\{p \to q\})$. Hence, $(T = IF_{BS}(\{p\})$ & $MUST_T\ q)$ iff (21a.2). Then, because of (18b), (21a) holds. (21b) can be proved in the same way as the proof of (21a). Q.E.D.

This result shows that our interpretation of conditionals is very similar to the standard one. In fact, with respect to the determination of *spheres*, our approach is more explicit than Lewis's approach (see (11i) and (11j)).

Grove [5] proposes sphere-semantics for theory change and shows that this semantics satisfies AGM postulates for the belief revision and that it is very similar to the sphere-semantics for counterfactual logic proposed by Lewis [11]. Thus, our results are similar to results in [5]. It is Grove's motivation for his investigation to connect the sphere semantics with the treatment of theory change.[13] His interest shares with ours. In fact, LBS is applicable to description of theory change in scientific activities.[14]

The main difference between two approaches lies in generality. Grove requires that the language is compact [5, p.157], while we deal with full FO-languages. Thus, our approach is broader than Grove's.

7   Two Types of Conditionals

Williams [17] points out a semantic difference between indicative and counterfactual conditionals.

(22a) [Indicative conditional] If Oswald didn't shoot Kennedy, someone else did.
(22b) [Counterfactual conditional] If Oswald hadn't shot Kennedy, someone else would have.

According to Williams, (22a) is true, while (22b) is false. This means that the meaning of indicative conditionals and that of counterfactual conditionals are different. We usually accept (22c) instead of (22b).

(22c) If Oswald hadn't shot Kennedy, Kennedy might not have been killed.

Williams explained this difference through a slight modification of the standard interpretation of conditionals proposed by Lewis [11]. Instead, we propose to distinguish both cases through a different relationship to causal dependencies.

---

[13] Gärdenfors [4, Sect. 4.5] gives an insightful description of Grove's system.

[14] Some parts of Lakatos' discussion on scientific research programs in [10] can be described within LBS. In belief structures of scientists, basic theories are more trusted than their auxiliary hypotheses $(BT > AH)$. Suppose that the set nO of observation data is consistent with BT but inconsistent with $BT \cup AH$. In such a case, scientists would try to find the set nAH of new auxiliary hypotheses such that $nO \cup BT \cup nAH$ is consistent. In this way, a basic theory can be protected against new anomalies.

It is usual to distinguish two types of conditionals [1, Sect. 1]. We propose that one type, like case (22a), is *epistemic* and the other type, like case (22c), is concerned with causal effects. The second type has the form "if-had A, then-would B", where the occurrence of B is causally dependent on the occurrence of A. In such a case, the shift of temporal perspective is often required. In (22a), our temporal view is fixed in the present and we assume that we know that Kennedy was killed. In this situation, we think about the possibility of Oswald's innocence. However, when we utter (22b) or (22c), our temporal viewpoint is shifted to the situation just before Kennedy was shot and we imagine what could happen after that situation. In this paper, we call the first type of conditionals *epistemic conditionals* and the second type *causal conditionals*.

Now, we define *casual dependency* as follows.

Definition 18. *Let BS be a belief structure with $T_1 > \ldots > T_n$. Let CT be a theory of causality that implies causal laws.*

A fact expressed by q is causally dependent on a fact expressed by p with respect to (wrt) BS iff there are i, j, and k such that ($i < j < k \leq cmn(BS)$ & $CT \subseteq T_i$ & $T_j = \{p\}$ & MUST-min(BS, j, p) & MUST-min(BS, k, q) & not $MUST_{top(BS,k)-CT}$ q & not $MUST_{top(BS,k)-T_j}$ q).

To explain this distinction of conditionals, let us consider examples (22a), (22b), and (22c). We use some abbreviations to improve readability:

killed: Kennedy was killed [$\exists t(killed(Kennedy, t) \land t <_t now)$];
someone: Someone shot Kennedy [$\exists t \exists x(shoot(x, Kennedy, t) \land t <_t now)$];
oswald: Oswald shot Kennedy [$\exists t(shoot(Oswald, Kennedy, t) \land t <_t now)$];
someone-else: Someone else shot Kennedy [$\exists t \exists x(shoot(x, Kennedy, t) \land x \neq Oswald \land t <_t now)$].

For the sake of simplicity, we assume that BS_1 is a belief structure with $CT > \{someone\} > \{killed\} > \{oswald\}$ and that BS_2 is a belief structure with $CT > \{oswald\} > \{killed\}$. We assume also that $CT \cup \{oswald\} \cup \{killed\}$ is consistent. Because ($oswald \rightarrow someone$) is a FOL-theorem, it holds $Cn(BS_1) = Cn(BS_2)$. Furthermore, according to Definition 18, the fact expressed by *killed* is causally dependent on the fact expressed by *oswald* only wrt BS_2. Because ($IF_{BS_1}(\{\neg oswald\}) = \{\neg oswald\} \cup CT \cup \{someone\} \cup \{killed\}$ & $IF_{BS2}(\{\neg oswald\}) = \{\neg oswald\} \cup CT$) and ($\neg oswald \land someone \rightarrow someone - else$) is a FOL-theorem, we obtain: $[If^{BS_1} \neg oswald](Must\ someone\text{-}else)$ & not $[If^{BS_2} \neg oswald](Must\ someone\text{-}else)$ & $[If^{BS_2} \neg oswald](Might\ \neg killed)$. This result can be summarized as follows:

BS_1: $CT > \{someone\} > \{killed\} > \{oswald\}$.
BS_2: $CT > \{oswald\} > \{killed\}$.
The fact expressed by *killed* is causally dependent on the fact expressed by *oswald* only wrt BS_2.
$[If^{BS_1} \neg oswald](Must\ someone\text{-}else)$.
not $[If^{BS_2} \neg oswald](Must\ someone\text{-}else)$.
$[If^{BS_2} \neg oswald](Might\ \neg killed)$.

To evaluate a causal conditional $\langle$If p, then $q\rangle$, we, at first, reformulate our belief structure, so that it reflects the causal dependency expressed by the conditional. Then, we imagine a situation in which p holds. Let us call this reformulated belief structure BS_c. The determination of the imagined situation can be achieved by calculating $[If^{BS_c} p]$. After that, we examine whether $MUST_T q$ holds, where $T = IF_{BS_c}(\{p\})$. In contrast, we do not need any reformulation of belief structures, when we evaluate epistemic conditionals.

8 Concluding Remarks

In the first part of this paper, we proposed *Logic for Epistemic Modalities* (LRM). LEM is based on the consequence relation of FOL. We have shown how to express in LEM some evidential features of epistemic modalities.

In the second part, we extended LEM to *Logic of Belief Structures* (LBS). Then, we defined a belief structure revision operator bsR based on LBS. A belief structure can be roughly understood as a linearly ordered set of sets of FO-sentences. We proved that bsR satisfies all postulates for belief revision in AGM-theory. Then, we defined the truth condition of counterfactual conditionals; we interpreted that the consequent of a conditional describes a modal state after a belief revision invoked by acceptance of the antecedent. We have also shown that *sphere semantics* can be defined for our treatment of conditionals. The characteristic feature of our approach lies in its explicitness. Instead of similarity relation among worlds, we use a reliability order among sets of FO-sentences. An example of causal interpretation of counterfactual conditionals demonstrated how LBS-approach can be used for describing truth conditions of modal statements in natural languages.[15]

References

1. Bennett, J.: A Philosophical Guide to Conditionals. Oxford University Press, New York (2003)
2. von Fintel, K.: Modality and language. In: Borchert, D.M. (ed.) Encyclopedia of Philosophy, vol. 10, 2nd edn, pp. 20–27. Macmillan Reference USA, Detroit (2006)
3. von Fintel, K., Gillies, A.S.: Must.. Stay.. Strong!. Nat. Lang. Semant. **18**, 351–383 (2010)
4. Gärdenfors, P.: Knowledge in Flux: Modeling the Dynamics of Epistemic States. MIT Press, Cambridge (1988)
5. Grove, A.: Two modellings for theory change. J. Philos. Logic **17**, 157–170 (1988)
6. Hansson, S.O: Logic of belief revision. In: Zalta, E.N. (ed.) The Stanford Encyclopedia of Philosophy (Winter 2014 edn.) (2014). http://plato.stanford.edu/archives/win2014/entries/logic-belief-revision/
7. Kartunen, L.: Possible and must. In: Kimball, J. (ed.) Syntax and Semantics, vol. 1, pp. 1–20. Academic Press, New York (1972)

[15] This research was supported by Grant-in-for Scientific Research, Scientific Research C (24520014): *The Construction of Philosophy of Science based on the Theory of Multiple Languages*. Finally, I would like to thank two reviewers for useful comments.

8. Kment, B.C.A.: Conditionals. In: Borchert, D.M. (ed.) Encyclopedia of Philosophy, vol. 2, 2nd edn, pp. 424–430. Macmillan Reference USA, Detroit (2006)
9. Kratzer, A.: Modals and Conditionals. Oxford University Press, Oxford (2012)
10. Lakatos, I.: The Methodology of Scientific Research Programmes: Philosophical Papers, vol. 1. Cambridge University Press, Cambridge (1978)
11. Lewis, D.K.: Counterfactuals. Harvard University Press, Cambridge (1973)
12. Nakayama, Y.: Dynamic interpretations and interpretation structures. In: Sakurai, A., Hasida, K., Nitta, K. (eds.) JSAI 2003. LNCS (LNAI), vol. 3609, pp. 394–404. Springer, Heidelberg (2007)
13. Nakayama, Y.: Logical framework for normative systems. In: SOCREAL 2010: Proceedings of the 2nd International Workshop on Philosophy and Ethics of Social Reality, pp. 19–24. Hokkaido University, Sapporo (2010)
14. Nakayama, Y.: Norms and Games: An Introduction to the Philosophy of Society, in Japanese. Keiso shobo, Tokyo (2011)
15. Nakayama, Y.: Analyzing speech acts based on dynamic normative logic. In: Nakano, Y., Satoh, K., Bekki, D. (eds.) JSAI-isAI 2013. LNCS, vol. 8417, pp. 98–114. Springer, Heidelberg (2014)
16. Willet, T.: A cross-linguistic survey of the grammarticalization of evidentiality. Stud. Lang. **12**(1), 51–97 (1988)
17. Williams, J.R.G.: Conversation and conditionals. Philos. Stud. **138**, 211–223 (2008)

A Type-Logical Account of Quantification in Event Semantics

Philippe de Groote[1]([✉]) and Yoad Winter[2]

[1] Inria Nancy - Grand Est, Villers-lés-Nancy, France
Philippe.deGroote@inria.fr
[2] UiL OTS, Utrecht University, Utrecht, The Netherlands

Abstract. It has been argued that Davidson's event semantics does not combine smoothly with Montague's compositional semantics. The difficulty, which we call the *event quantification problem*, comes from a possibly bad interaction between event existential closure, on the one hand, and quantification, negation, or conjunction, on the other hand. The recent literature provides two solutions to this problem. The first one is due to Champollion [2,3], and the second one to Winter and Zwarts [13]. The present paper elaborates on this second solution. In particular, it provides a treatment of quantified adverbial modifiers, which was absent from [13].

1 Introduction

It is well known that combining Davidsonian event semantics [5] with Montague's treatment of quantification [10] may give rise to unexpected semantic interpretations. To understand the potential problem, consider the standard interpretations of proper names and transitive verbs, which allow one to give a semantic interpretation to simple sentences like (2).

(1)   a. $[\![\text{John}]\!] = \lambda p.\, p\,\mathbf{j} : (\mathsf{e} \to \mathsf{t}) \to \mathsf{t}$
        b. $[\![\text{Mary}]\!] = \lambda p.\, p\,\mathbf{m} : (\mathsf{e} \to \mathsf{t}) \to \mathsf{t}$
        c. $[\![\text{kissed}]\!] = \lambda px.\, p\,(\lambda y.\, \mathbf{kissed}\, x\, y) : ((\mathsf{e} \to \mathsf{t}) \to \mathsf{t}) \to \mathsf{e} \to \mathsf{t}$

(2)   John kissed Mary.

Adapting lexical entry (1-c) to the Davidsonian approach consists in providing the binary relation **kissed** with an additional event argument of type v,[1] which results in lexical entry (3-a). This allows adverbial modifiers to parallel adnominal modifiers (see lexical entry (3-b)). Then, in order to interpret a sentence as a truth value rather than as a set of events, one has to apply an *existential closure* operator (3-c).

---

[1] We follow a Davidsonian approach as opposed to a neo-Davidsonian approach. We also distinguish the type of events (v) from the type of entities (e). These choices, which are rather arbitrary, will not affect our purpose.

© Springer-Verlag Berlin Heidelberg 2015
T. Murata et al. (Eds.): JSAI-isAI 2014 Workshops, LNAI 9067, pp. 53–65, 2015.
DOI: 10.1007/978-3-662-48119-6_5

(3)   a. $[\![\text{kissed}]\!] = \lambda pxe.\, p\,(\lambda y.\, \textbf{kissed}\, e\, x\, y)$:

$$((e \rightarrow t) \rightarrow t) \rightarrow e \rightarrow v \rightarrow t$$

      b. $[\![\text{passionately}]\!] = \lambda pe.\,(p\, e) \wedge (\textbf{passionate}\, e)$: $(v \rightarrow t) \rightarrow v \rightarrow t$

      c. $\text{E-CLOS} = \lambda p.\, \exists e.\, p\, e$: $(v \rightarrow t) \rightarrow t$

Using the above apparatus, we obtain the semantic interpretation of sentence (4) by computing the value of expression (5), which results in formula (6).

(4)   John kissed Mary passionately.

(5) $[\![\text{John}]\!]\,(\lambda x.\, \text{E-CLOS}\,([\![\text{passionately}]\!]\,([\![\text{kissed}]\!]\,[\![\text{Mary}]\!]\, x)))$

(6) $\exists e.\,(\textbf{kissed}\, e\, \textbf{j}\, \textbf{m}) \wedge (\textbf{passionate}\, e)$

Consider now sentence (7), which includes a quantified noun phrase.

(7)   John kissed every girl.

The standard interpretation of this quantified noun phrase is based on the following lexical interpretations:

(8)   a. $[\![\text{girl}]\!] = \lambda x.\, \textbf{girl}\, x$: $e \rightarrow t$

      b. $[\![\text{every}]\!] = \lambda pq.\, \forall x.\,(p\, x) \rightarrow (q\, x)$: $(e \rightarrow t) \rightarrow (e \rightarrow t) \rightarrow t$

Then, using expression (9) to compute the semantic interpretation of sentence (7) results in a counterintuitive interpretation (formula (10)).

(9) $[\![\text{John}]\!]\,(\lambda x.\, \text{E-CLOS}\,([\![\text{kissed}]\!]\,([\![\text{every}]\!]\,[\![\text{girl}]\!])\, x))$

(10) $\exists e.\, \forall x.\,(\textbf{girl}\, x) \rightarrow (\textbf{kissed}\, e\, \textbf{j}\, x)$

According to formula (10), there should be a single kissing event involving John and every girl. This requirement appears because the existential closure operator takes wide scope over the universally quantified noun phrase. The problem with this analysis becomes more apparent when we consider the interaction of events with quantifiers that are not upward-monotone, as in the following sentences.

(11)   a.   John kissed no girl.

      b.   John kissed less than five girls.

      c.   John kissed exactly one girl.

Consider the standard lexical interpretation of the *no* quantifier, together with a semantic analysis akin to expression (9).

(12)   a. $[\![\text{no}]\!] = \lambda pq.\, \forall x.\,(p\, x) \rightarrow \neg(q\, x)$: $(e \rightarrow t) \rightarrow (e \rightarrow t) \rightarrow t$

      b. $[\![\text{John}]\!]\,(\lambda x.\, \text{E-CLOS}\,([\![\text{kissed}]\!]\,([\![\text{no}]\!]\,[\![\text{girl}]\!])\, x))$

This leads to the following problematic interpretation:

(13) $\exists e.\, \forall x.\,(\textbf{girl}\, x) \rightarrow \neg(\textbf{kissed}\, e\, \textbf{j}\, x)$

According to formula (13), sentence (11-a) would be true if John kissed a girl, but there is another event where John did not kiss any girl. Indeed, since in the latter event John kissed no girl, the sentence might be incorrectly analyzed as true in this situation. The same kind of problem can also be shown with sentences (11-b) and (11-c).

A similar problem may arise with negation.

(14) a. John did not kiss Mary.
      b. $[\![\mathrm{not}]\!] = \lambda px.\,\neg(p\,x) : (\mathsf{e} \to \mathsf{t}) \to \mathsf{e} \to \mathsf{t}$
      c. $[\![\mathrm{John}]\!]\,([\![\mathrm{not}]\!]\,(\lambda x.\,\textsc{e-clos}\,([\![\mathrm{kissed}]\!]\,[\![\mathrm{Mary}]\!]\,x)))$
      d. $\neg(\exists e.\,\mathbf{kissed}\,e\,\mathbf{j}\,\mathbf{m})$
      e. $\textsc{e-clos}\,(\lambda e.\,[\![\mathrm{John}]\!]\,([\![\mathrm{not}]\!]\,(\lambda x.\,[\![\mathrm{kissed}]\!]\,[\![\mathrm{Mary}]\!]\,x\,e)))$
      f. $\exists e.\,\neg(\mathbf{kissed}\,e\,\mathbf{j}\,\mathbf{m})$

Using a standard interpretation of negation (lexical entry (14-b)), expression (14-c) leads to a correct semantic interpretation of sentence (14-a), namely, formula (14-d). There is, however, another possible analysis of sentence (14-a), which is expressed by expression (14-e). This leads to formula (14-f), where the existential quantifier over events takes scope over negation. According to this fallacious interpretation, sentence (14-a) might be true in a situation where John kissed Mary, provided there is another event where he did not kissed Mary.

Besides quantification and negation, conjunction is also known to interact badly with event existential closure. Consider the following sentence, whose interpretation is meant to be distributive:

(15)   John kissed Mary and [then] Sue.

The standard interpretation of distributive coordination is given by the following lexical entry:

(16) $[\![\mathrm{and}]\!] = \lambda pqr.\,(p\,r) \wedge (q\,r) :$
$$((\mathsf{e} \to \mathsf{t}) \to \mathsf{t}) \to ((\mathsf{e} \to \mathsf{t}) \to \mathsf{t}) \to (\mathsf{e} \to \mathsf{t}) \to \mathsf{t}$$

Nevertheless, the semantic analysis given by expression (17-a) yields an interpretation (formula (17-b)) akin to a collective reading. According to this problematic interpretation, Mary and Sue would be the patients of the same kissing event.

(17) a. $[\![\mathrm{John}]\!]\,(\lambda x.\,\textsc{e-clos}\,([\![\mathrm{kissed}]\!]\,([\![\mathrm{and}]\!]\,[\![\mathrm{Mary}]\!]\,[\![\mathrm{Sue}]\!])\,x))$
      b. $\exists e.\,(\mathbf{kissed}\,e\,\mathbf{j}\,\mathbf{m}) \wedge (\mathbf{kissed}\,e\,\mathbf{j}\,\mathbf{s})$

To obtain the right interpretations of sentences like (7), (11-a)-(11-c), (14-a), and (15), the event existential closure operator need to take narrow scope. For sentence (7), this is illustrated by expression (18), which is interpreted as in formula (19).

(18) $[\![\mathrm{John}]\!]\,(\lambda x.\,[\![\mathrm{every}]\!]\,[\![\mathrm{girl}]\!]\,(\lambda y.\,\textsc{e-clos}\,([\![\mathrm{kissed}]\!]\,(\lambda p.\,p\,y)\,x)))$
(19) $\forall x.\,(\mathbf{girl}\,x) \to (\exists e.\,\mathbf{kissed}\,e\,\mathbf{j}\,x)$

In many works, expressions such as (18) result from some covert movement operation. This explains why most approaches that combine event semantics and quantification either rely on syntactic devices that control the scope of the

quantifiers [9], or depart significantly from standard assumptions in compositional semantics [1,8]. Following Champollion [2,3] and Winter and Zwarts [13],
we see this enrichment of standard systems as problematic. Accordingly, we
refer to the problem of combining standard compositional semantics and event
semantics as the *event quantification problem*.

2 Two Solutions to the Event Quantification Problem

The recent literature provides two solutions to the event quantification problem.
The first one is due to Champollion [2,3]. It consists in interpreting sentences
as generalized quantifiers over events $((\mathsf{v} \to \mathsf{t}) \to \mathsf{t})$ rather than as sets of
events $(\mathsf{v} \to \mathsf{t})$. This allows the existential closure to occur at the lexical level.
Accordingly, the lexical entries in (3) are adapted as follows.

(20)   a. $[\![\text{kissed}]\!] = \lambda p x f. \, p \, (\lambda y. \, \exists e. \, (\textbf{kissed} \, e \, x \, y) \wedge (f \, e)) :$
$$((e \to t) \to t) \to e \to (\mathsf{v} \to \mathsf{t}) \to \mathsf{t}$$
        b. $[\![\text{passionately}]\!] = \lambda p f. \, p \, (\lambda e. \, (f \, e) \wedge (\textbf{passionate} \, e)) :$
$$((\mathsf{v} \to \mathsf{t}) \to \mathsf{t}) \to (\mathsf{v} \to \mathsf{t}) \to \mathsf{t}$$
        c. $\text{CLOS} = \lambda p. \, p \, (\lambda e. \, \textbf{true}) : ((\mathsf{v} \to \mathsf{t}) \to \mathsf{t}) \to \mathsf{t}$

The semantic interpretation of sentence (7) is then obtained by computing the
value of expression (21). This expression is akin to expression (9). Nevertheless,
this time, the computation yields the correct interpretation (formula (19)).

(21) $[\![\text{John}]\!] \, (\lambda x. \, \text{CLOS} \, ([\![\text{kissed}]\!] \, ([\![\text{every}]\!] \, [\![\text{girl}]\!]) \, x))$

The second solution, which is due to Winter and Zwarts [13], is expressed in
the framework of Abstract Categorial Grammar [7]. It takes advantage of the so-
called *tectogrammatic level* [4] for the treatment of scope interactions. Following
the type-logical tradition, Winter and Zwarts distinguish, at the abstract syntactic level, the category of noun phrases, NP, from the category of quantified
noun phrases, $(NP \to S) \to S$. They also distinguish the category S of sentences
interpreted as truth values (t) from the category V of sentences interpreted as
sets of events $(\mathsf{v} \to \mathsf{t})$.[2] This results in abstract syntactic structures specified by
a signature akin to the following one.

(22)   a.   JOHN : NP
        b.   GIRL : N
        c.   KISSED : $NP \to NP \to V$
        d.   EVERY : $N \to (NP \to S) \to S$
        e.   PASSIONATELY : $V \to V$
        f.   E-CLOS : $V \to S$

---

[2] Following Montague's homomorphism requirement, these two abstract categories
should indeed be distinguished since they correspond to different semantic types.
The fact that they share the same surface realizations may be considered as a mere
contingence. Alternatively, we may relax the homomorphism requirement, e.g. as in
[12], who treats indefinite NPs as ambiguous between predicates and quantifiers.

This signature comes with a lexicon that specifies the surface realization of the abstract syntactic structures:

(23) a. JOHN := *John*
      b. GIRL := *girl*
      c. KISSED := $\lambda xy.\, y + \textbf{\textit{kissed}} + x$
      d. EVERY := $\lambda xf.\, f\,(\textbf{\textit{every}} + x)$
      e. PASSIONATELY := $\lambda x.\, x + \textbf{\textit{passionately}}$
      f. E-CLOS := $\lambda x.\, x$

In this setting, the only abstract structure corresponding to sentence (7) is the following well-typed expression (whose derivation and surface realization are given in Appendix A).

(24) EVERY GIRL ($\lambda x.$ E-CLOS (KISSED x JOHN))

Then, using the semantic interpretation given here below, the evaluation of expression (24) yields the expected result, i.e., formula (19).

(25) a. JOHN := $\textbf{j} : \textsf{e}$
      b. GIRL := $\lambda x.\, \textbf{girl}\, x : \textsf{e} \to \textsf{t}$
      c. KISSED := $\lambda xye.\, \textbf{kissed}\, e\, x\, y : \textsf{e} \to \textsf{e} \to \textsf{v} \to \textsf{t}$
      d. EVERY := $\lambda pq.\, \forall x.\, (p\, x) \to (q\, x) : (\textsf{e} \to \textsf{t}) \to (\textsf{e} \to \textsf{t}) \to \textsf{t}$
      e. PASSIONATELY := $\lambda pe.\, (p\, e) \wedge (\textbf{passionate}\, e) :$
                                  $$(\textsf{v} \to \textsf{t}) \to (\textsf{v} \to \textsf{t})$$
      f. E-CLOS := $\lambda p.\, \exists e.\, p\, e : (\textsf{v} \to \textsf{t}) \to \textsf{t}$

Signature (22) compels the existential closure operator (E-CLOS) to take scope below the quantified noun phrase (EVERY GIRL). This is because the abstract syntactic category assigned to EVERY ($N \to (NP \to S) \to S$) is given in standard terms of the abstract category S (interpreted as truth-values) rather than V (interpreted as sets of events). Consequently, each of the derived sets of events (i.e., each expression of type V) must first be "closed" (i.e., turned into an expression of type S) before quantification can apply.

This solution is easily transferable to the cases of negation and conjunction. It suffices to express their abstract categories in term of S.[3]

(26) a. NOT : $(NP \to S_\circ) \to (NP \to S)$
      b. AND : $NP \to NP \to (NP \to S) \to S$

Thus, boolean negation and conjunction are still treated by using the boolean S-based types, rather than the Davidsonian V-based types.

Unlike Champollion, Winter and Zwarts do not consider the case of quantified adverbial modifiers such as *everyday* or *everywhere*. The present paper aims to fill this gap.

---

[3] The reason for distinguishing between type S and type $S_\circ$, which is merely syntactic, is explained in Appendix B.

3   Quantified Adverbial Modifiers

Consider sentence (27) together with a plausible semantic interpretation (formula (28)).

(27)   John kissed Mary everyday.

(28) $\forall x. (\mathbf{day}\, x) \rightarrow (\exists e. (\mathbf{kissed}\, e\, \mathbf{j}\, \mathbf{m}) \wedge (\mathbf{time}\, e\, x))$

In formula (28), the underlined subformulas are derived from the semantic interpretation of *everyday*. In a compositional setting, this makes it necessary that the lexical semantics of *everyday* acts both inside and outside of the scope of the existential closure.

The solution to this puzzle is in accordance with the type-logical tradition. It consists in distinguishing the category of adverbial modifiers from the category of quantified adverbial modifiers. Similarly to the treatment of noun phrases, the category of quantified adverbial modifiers is obtained from the category of non-quantified adverbial modifiers by type-shifting. This results in the following type assignment.

(29)   EVERYDAY : $((V \rightarrow V) \rightarrow S) \rightarrow S$

Then, the abstract structure corresponding to sentence (27) may be expressed as follows.

(30)   EVERYDAY $(\lambda q.\, \text{E-CLOS}\, (q\, (\text{KISSED MARY JOHN})))$

Finally, using lexical entry (31), one may compute compositionally the interpretation of sentence (27).

(31)   EVERYDAY $:= \lambda q. \forall x. (\mathbf{day}\, x) \rightarrow (q\, (\lambda pe.\, (p\, e) \wedge (\mathbf{time}\, e\, x)))$:
$$(((\mathsf{v} \rightarrow \mathsf{t}) \rightarrow (\mathsf{v} \rightarrow \mathsf{t})) \rightarrow \mathsf{t}) \rightarrow \mathsf{t}$$

It is to be noted that our approach is consistent with the treatment of quantified adverbial prepositional phrases such as *in every room*. Consider the category and the semantic interpretation assigned to the preposition *in*.

(32)   a.   IN : $NP \rightarrow V \rightarrow V$
          b.   IN $:= \lambda xpe.\, (p\, e) \wedge (\mathbf{location}\, e\, x)$

We may then derive an abstract structure of the appropriate type, corresponding to the prepositional phrase *in every room*.

(33) $\lambda q.\, \text{EVERY ROOM}\, (\lambda x.\, q\, (\text{IN}\, x)) : ((V \rightarrow V) \rightarrow S) \rightarrow S$

We give a toy-grammar that summarizes our approach in Appendix B. This grammar allows one to handle examples such as sentence (34).

(34)   John did not kiss Mary for one hour.

Sentences such as (34) are known to exhibit a scope ambiguity that yields two different semantic interpretations. In the present case, these two interpretations may be paraphrased as follows.

(35)   a.   For one hour, it was not the case that John kissed Mary.
       b.   It was not the case that John kissed Mary for one hour.

Using our toy-grammar, these two interpretations may be obtained by computing the semantic interpretations of the two following abstract structures whose surface realization is sentence (34).

(36)   a.   FOR$_o$ (ONE HOUR)
                 ($\lambda q.$ NOT ($\lambda x.$ E-CLOS$_o$ (q (KISSED$_o$ MARY x))) JOHN)
       b.   NOT ($\lambda x.$ FOR$_{oo}$ (ONE HOUR)
                 ($\lambda q.$ E-CLOS$_o$ (q (KISSED$_o$ MARY x)))))
             JOHN

The abstract constants marked with small circles (FOR$_o$, E-CLOS$_o$, etc.) involve a syntactic treatment of negation, which avoids ill-formed strings like *John did not kissed Mary. For more details see Appendix B.

   We conclude that the ambiguity of sentence (34), which is treated in Champollion's system as an ordinary case of scope ambiguity, is similarly treated here, by following an adaptation of the system proposed by Winter and Zwarts.

   A final remark has to do with sentences like the following.

(37)   John did not kiss Mary deliberately.

This sentence presents an ambiguity similar to the one of sentence (34):

(38)   a.   It was deliberate that John did not kiss Mary.
       b.   It was not deliberate that John kissed Mary.

Nevertheless, sentences like (37) cannot be treated like sentence (34) because the adverb is not quantificational. Following Davidson, we assume that adverbs like *deliberately*, unlike standard manner adverbs, have a modal element to their meaning, and accordingly modify full propositions rather than events. Like Champollion, we believe that the treatment of modal adverbs and other modal operators is orthogonal to the main tenets of event semantics.

Acknowledgement. The authors would like to thank Lucas Champollion for fruitful discussions, and the two anonymous referees. The work of the first author was supported by the French agency *Agence Nationale de la Recherche* (ANR-12-CORD-0004). The work of the second author was supported by a VICI grant 277-80-002 of the Netherlands Organisation for Scientific Research (NWO).

A   Appendix

Derivation of expression (24):

$$\frac{\vdash \text{EVERY} : N \rightarrow (NP \rightarrow S) \rightarrow S \qquad \vdash \text{GIRL} : N}{\vdash \text{EVERY GIRL} : (NP \rightarrow S) \rightarrow S} \quad (1)$$

$$\frac{\dfrac{\vdash \text{KISSED} : NP \rightarrow NP \rightarrow V \qquad x : NP \vdash x : NP}{x : NP \vdash \text{KISSED}\, x : NP \rightarrow V} \qquad \vdash \text{JOHN} : NP}{x : NP \vdash \text{KISSED}\, x\, \text{JOHN} : V} \quad (2)$$

$$\frac{\dfrac{\vdots \;\; (2)}{\vdash \text{E-CLOS} : V \rightarrow S \qquad x : NP \vdash \text{KISSED}\, x\, \text{JOHN} : V}{x : NP \vdash \text{E-CLOS}\,(\text{KISSED}\, x\, \text{JOHN}) : S}}{\vdash \lambda x.\, \text{E-CLOS}\,(\text{KISSED}\, x\, \text{JOHN}) : NP \rightarrow S} \quad (3)$$

$$\frac{\begin{array}{c}\vdots \;\; (1)\\ \vdash \text{EVERY GIRL} : (NP \rightarrow S) \rightarrow S\end{array} \qquad \begin{array}{c}\vdots \;\; (3)\\ \vdash \lambda x.\, \text{E-CLOS}\,(\text{KISSED}\, x\, \text{JOHN}) : NP \rightarrow S\end{array}}{\vdash \text{EVERY GIRL}\,(\lambda x.\, \text{E-CLOS}\,(\text{KISSED}\, x\, \text{JOHN})) : S}$$

Surface realization of expression (24):

$$
\begin{aligned}
&\text{EVERY GIRL}\,(\lambda x.\, \text{E-CLOS}\,(\text{KISSED}\, x\, \text{JOHN}))\\
&= (\lambda x f.\, f\,(\textit{every} + x))\,\text{GIRL}\,(\lambda x.\, \text{E-CLOS}\,(\text{KISSED}\, x\, \text{JOHN}))\\
&= (\lambda f.\, f\,(\textit{every} + \text{GIRL}))\,(\lambda x.\, \text{E-CLOS}\,(\text{KISSED}\, x\, \text{JOHN}))\\
&= (\lambda x.\, \text{E-CLOS}\,(\text{KISSED}\, x\, \text{JOHN}))\,(\textit{every} + \text{GIRL})\\
&= \text{E-CLOS}\,(\text{KISSED}\,(\textit{every} + \text{GIRL})\,\text{JOHN})\\
&= (\lambda x.\, x)\,(\text{KISSED}\,(\textit{every} + \text{GIRL})\,\text{JOHN})\\
&= \text{KISSED}\,(\textit{every} + \text{GIRL})\,\text{JOHN}\\
&= (\lambda x y.\, y + \textit{kiss} + x)\,(\textit{every} + \text{GIRL})\,\text{JOHN}\\
&= (\lambda y.\, y + \textit{kiss} + \textit{every} + \text{GIRL})\,\text{JOHN}\\
&= \text{JOHN} + \textit{kiss} + \textit{every} + \text{GIRL}\\
&= \textit{john} + \textit{kiss} + \textit{every} + \text{GIRL}\\
&= \textit{john} + \textit{kiss} + \textit{every} + \textit{girl}
\end{aligned}
$$

B Appendix

This appendix presents a toy grammar that covers the several examples that are under discussion in the course of the paper. It mainly consists of three parts:

- a set of abstract syntactic structures, specified by means of a higher-order signature;
- a surface realization of the abstract structures, specified by means of a homomorphic translation of the signature;
- a semantic interpretation of the abstract structures, specified by means of another homomorphic translation;

Table 1. Abstract syntax

Abstract Syntax
JOHN : NP
MARY : NP
SUE : NP
THE-KITCHEN : NP
ROOM : N
GIRL : N
HOUR : N_u
KISSED : $NP \to NP \to V$
KISSED$_o$: $NP \to NP \to V_o$
NOT : $(NP \to S_o) \to (NP \to S)$
ONE : $N_u \to NP_\tau$
A : $N \to (NP \to S) \to S$
A$_o$: $N \to (NP \to S_o) \to S_o$
EVERY : $N \to (NP \to S) \to S$
EVERY$_o$: $N \to (NP \to S_o) \to S_o$
NO : $N \to (NP \to S) \to S$
NO$_o$: $N \to (NP \to S_o) \to S_o$
AND : $NP \to NP \to (NP \to S) \to S$
AND$_o$: $NP \to NP \to (NP \to S_o) \to S_o$
PASSIONATELY : $V \to V$
PASSIONATELY$_o$: $V_o \to V_o$
EVERYDAY : $((V \to V) \to S) \to S$
EVERYDAY$_o$: $((V_o \to V_o) \to S) \to S$
IN : $NP \to V \to V$
IN$_o$: $NP \to V_o \to V_o$
FOR : $NP_\tau \to ((V \to V) \to S) \to S$
FOR$_o$: $NP_\tau \to ((V_o \to V_o) \to S) \to S$
FOR$_{oo}$: $NP_\tau \to ((V_o \to V_o) \to S_o) \to S_o$
E-CLOS : $V \to S$
E-CLOS$_o$: $V_o \to S_o$

B.1 Abstract Syntax

The signature specifying the abstract syntactic structures is given in Table 1. It uses a type system built upon the following set of atomic syntactic categories:

Table 2. Surface realization

Surface Realization
JOHN $:=$ *John*
MARY $:=$ *Mary*
SUE $:=$ *Sue*
THE-KITCHEN $:=$ *the* $+$ *kitchen*
ROOM $:=$ *room*
GIRL $:=$ *girl*
HOUR $:=$ *hour*
KISSED $:= \lambda xy.\, y + \textbf{kissed} + x$
KISSED$_\text{o} := \lambda xy.\, y + \textbf{kiss} + x$
NOT $:= \lambda fx.\, x + \textbf{did} + \textbf{not} + (f\,\epsilon)$
ONE $:= \lambda x.\, \textbf{one} + x$
A, A$_\text{o} := \lambda xf.\, f\,(\textbf{a} + x)$
EVERY, EVERY$_\text{o} := \lambda xf.\, f\,(\textbf{every} + x)$
NO, NO$_\text{o} := \lambda xf.\, f\,(\textbf{no} + x)$
AND, AND$_\text{o} := \lambda xyf.\, f\,(x + \textbf{and} + y)$
PASSIONATELY, PASSIONATELY$_\text{o} := \lambda x.\, x + \textbf{passionately}$
EVERYDAY, EVERYDAY$_\text{o} := \lambda f.\, f\,(\lambda x.\, x + \textbf{everyday})$
IN, IN$_\text{o} := \lambda xy.\, y + \textbf{in} + x$
FOR, FOR$_\text{o}$, FOR$_\text{oo} := \lambda xf.\, f\,(\lambda x.\, y + \textbf{for} + x)$
E-CLOS, E-CLOS$_\text{o} := \lambda x.\, x$

Table 3. Semantic types

Semantic Types
$N := \mathsf{e} \to \mathsf{t}$
$N_u := \mathsf{i} \to \mathsf{n} \to \mathsf{t}$
$NP := \mathsf{e}$
$NP_\tau := \mathsf{i} \to \mathsf{t}$
$V := \mathsf{v} \to \mathsf{t}$
$V_\text{o} := \mathsf{v} \to \mathsf{t}$
$S := \mathsf{t}$
$S_\text{o} := \mathsf{t}$

Table 4. Non-logical constants

Non-Logical Constants
$\mathbf{j}, \mathbf{m}, \mathbf{s}, \mathbf{k} : \mathsf{e}$
$\mathbf{room}, \mathbf{girl}, \mathbf{day} : \mathsf{e} \to \mathsf{t}$
$\mathbf{hour} : \mathsf{i} \to \mathsf{n} \to \mathsf{t}$
$\mathbf{kissed} : \mathsf{v} \to \mathsf{e} \to \mathsf{e} \to \mathsf{t}$
$\mathbf{1} : \mathsf{n}$
$\mathbf{passionate} : \mathsf{v} \to \mathsf{t}$
$\mathbf{time}, \mathbf{location} : \mathsf{v} \to \mathsf{e} \to \mathsf{t}$
$\mathbf{duration} : \mathsf{v} \to \mathsf{i} \to \mathsf{t}$

Table 5. Semantic Interpretation

Semantic Interpretation
JOHN $:=$ **j**
MARY $:=$ **m**
SUE $:=$ **s**
THE-KITCHEN $:=$ **k**
ROOM $:=$ $\lambda x.\,\textbf{room}\,x$
GIRL $:=$ $\lambda x.\,\textbf{girl}\,x$
HOUR $:=$ $\lambda xy.\,\textbf{hour}\,x\,y$
KISSED, KISSED$_\text{o}$ $:=$ $\lambda xye.\,\textbf{kissed}\,e\,x\,y$
NOT $:=$ $\lambda px.\,\neg(p\,x)$
ONE $:=$ $\lambda pt.\,p\,t\,1$
A, A$_\text{o}$ $:=$ $\lambda pq.\,\exists x.\,(p\,x) \wedge (q\,x)$
EVERY, EVERY$_\text{o}$ $:=$ $\lambda pq.\,\forall x.\,(p\,x) \rightarrow (q\,x)$
NO, NO$_\text{o}$ $:=$ $\lambda pq.\,\forall x.\,(p\,x) \rightarrow \neg(q\,x)$
AND, AND$_\text{o}$ $:=$ $\lambda xyp.\,(p\,x) \wedge (p\,y)$
PASSIONATELY,
PASSIONATELY$_\text{o}$ $:=$ $\lambda pe.\,(p\,e) \wedge (\textbf{passionate}\,e)$
EVERYDAY,
EVERYDAY$_\text{o}$ $:=$ $\lambda q.\,\forall x.\,(\textbf{day}\,x) \rightarrow (q\,(\lambda pe.\,(p\,e) \wedge (\textbf{time}\,e\,x)))$
IN, IN$_\text{o}$ $:=$ $\lambda xpe.\,(p\,e) \wedge (\textbf{location}\,e\,x)$
FOR, FOR$_\text{o}$, FOR$_\text{oo}$ $:=$ $\lambda pq.\,\exists t.\,(p\,t) \wedge (q\,(\lambda pe.\,(p\,e) \wedge (\textbf{duration}\,e\,t)))$
E-CLOS, E-CLOS$_\text{o}$ $:=$ $\lambda p.\,\exists e.\,p\,e$

- N, the category of *nouns*;
- N_u, the category of *nouns that name units of measurement*;
- NP, the category of *noun phrases*;
- NP_τ, the category of *noun phrases that denote time intervals*;
- S and S_o, the category of *sentences* (positive and negative);
- V and V_o, the category of *"open" sentences* (positive and negative).

The reason for distinguishing between the categories of positive and negative (open) sentences is merely syntactic. Without such a distinction, the surface realization of a negative expression such as:

NOT (KISSED MARY) JOHN

would be:

*John did not kissed Mary

Without this distinction, it would also be possible to iterate negation. This would allow the following ungrammatical sentences to be generated:

*John did not did not kiss Mary

*John did not did not did not kiss Mary

⋮

B.2   Surface Realization

The surface realization of the abstract syntactic structures is given in Table 2. This realization is such that every abstract term of an atomic type is interpreted as a string. Accordingly, abstract terms of a functional type are interpreted as functions acting on strings.

B.3   Semantic interpretation

The semantic interpretation of the abstract syntactic categories is given in Table 3. Besides the usual semantic types e and t, we also use v, i, and n. These stand for the semantic types of events, time intervals, and scalar quantities, respectively.

The semantic interpretation of the abstract constants is then given in Table 5. This interpretation makes use of the non-logical constants given in Table 4.

References

1. Beaver, D., Condoravdi, C.: On the logic of verbal modification. In: Aloni, M., Dekker, P., Roelofsen, F. (eds.) Proceedings of the Sixteenth Amsterdam Colloquium, pp. 3–9. University of Amsterdam (2007)
2. Champollion, L.: Quantification and negation in event semantics. The Baltic Int. Yearb. Cogn. Logic Commun. **6** (2010)
3. Champollion, L.: The interaction of compositional semantics and event semantics. Linguistic and Philosophy (To appear)
4. Curry, H.B.: Some logical aspects of grammatical structure. In: Jakobson, R. (ed.) Proceedings of the 12th Symposia on Applied Mathematics Studies of Language and its Mathematical Aspects, Providence, pp. 56–68 (1961)
5. Davidson, D.: The logical form of action sentences. In: Rescher, N. (ed.) The Logic of Decision and Action. University of Pittsburgh Press, Pittsburgh (1967) (Reprinted in [6])
6. Davidson, D.: Essays on Actions and Events. Clarendon Press, London (1980)
7. de Groote, Ph.: Towards abstract categorial grammars. In: Proceedings of the Conference on Association for Computational Linguistics, 39th Annual Meeting and 10th Conference of the European Chapter, pp. 148–155 (2001)
8. Krifka, M.: Nominal reference, temporal constitution and quantification in event semantics. In: Bartsch, R., van Benthem, J., van Emde Boas, P. (eds.) Semantics and Contextual Expression, pp. 75–115. Foris, Dordrecht (1989)
9. Landman, F.: Events and Plurality: The Jerusalem Lectures. Kluwer Academic Publishers, Boston (2000)

10. Montague, R.: The proper treatment of quantification in ordinary english. In; Hintikka, J., Moravcsik, J., Suppes, P. (eds.) Proceedings of the 1970 Stanford Workshop on Grammar and Semantics Approaches to Natural Language. Reidel, Dordrecht (1973) (Reprinted in [11])
11. Montague, R.: Formal Philosophy: Selected Papers of Richard Montague. Yale University Press, New Haven (1974). edited and with an introduction by Richmond Thomason
12. Partee, B.: Noun phrase interpretation and type shifting principles. In: Groenendijk, J., de Jong, D., Stokhof, M. (eds.) Studies in Discourse Representation Theories and the Theory of Generalized Quantifiers. Foris, Dordrecht (1987)
13. Winter, Y., Zwarts, J.: Event semantics and abstract categorial grammar. In: Kanazawa, M., Kornai, A., Kracht, M., Seki, H. (eds.) MOL 12. LNCS, vol. 6878, pp. 174–191. Springer, Heidelberg (2011)

Towards a Wide-Coverage Tableau Method
for Natural Logic

Lasha Abzianidze[✉]

Tilburg University - TiLPS, Tilburg, The Netherlands
L.Abzianidze@uvt.nl

Abstract. The first step towards a wide-coverage tableau prover for natural logic is presented. We describe an automatized method for obtaining Lambda Logical Forms from surface forms and use this method with an implemented prover to hunt for new tableau rules in textual entailment data sets. The collected tableau rules are presented and their usage is also exemplified in several tableau proofs. The performance of the prover is evaluated against the development data sets. The evaluation results show an extremely high precision above 97 % of the prover along with a decent recall around 40 %.

Keywords: Combinatory Categorial Grammar · Lambda Logical Form · Natural logic · Theorem prover · Tableau method · Textual entailment

1 Introduction

In this paper, we present a further development of the analytic tableau system for natural logic introduced by Muskens in [12]. The main goal of [12] was to initiate a novel formal method of modeling reasoning over linguistic expressions, namely, to model the reasoning in a signed analytic tableau system that is fed with Lambda Logical Forms (LLFs) of linguistic expressions. There are three straightforward advantages of this approach:

(i) since syntactic trees of LLFs roughly describe semantic composition of linguistic expressions, LLFs resemble surface forms (that is characteristic for natural logic); hence, obtaining LLFs is easier than translating linguistic expressions in some logical formula where problems of expressiveness of logic and proper translation come into play;

(ii) the approach captures an inventory of inference rules (where each rule is syntactically or semantically motivated and is applicable to particular linguistic phrases) in a modular way;

(iii) a model searching nature of a tableau method and freedom of choice in a rule application strategy seem to enable us to capture *quick* inferences that humans show over linguistic expressions.

© Springer-Verlag Berlin Heidelberg 2015
T. Murata et al. (Eds.): JSAI-isAI 2014 Workshops, LNAI 9067, pp. 66–82, 2015.
DOI: 10.1007/978-3-662-48119-6_6

The rest of the paper is organized as follows. First, we start with the syntax of LLFs and show how terms of syntactic and semantic types can be combined; then we briefly discuss a method of obtaining LLFs from surface forms as we aim to develop a wide-coverage natural tableau system (i.e. a tableau prover for natural logic). A combination of automatically generated LLFs and an implemented natural tableau prover makes it easy to extract a relevant set of inference rules from the data used in textual entailment challenges. In the end, we present the performance of the prover on several training data sets. The paper concludes with a discussion of further research plans.

Throughout the paper we assume the basic knowledge of a tableau method.

2 Lambda Logical Forms

The analytic tableau system of [12] uses LLFs as logical forms of linguistic expressions. They are simply typed λ-terms with semantic types built upon $\{e, s, t\}$ atomic types. For example, in [12] the LLF of no bird moved is (1) that is a term of type st.[1] As we aim to develop a wide-coverage tableau for natural logic, using only terms of semantic types does not seem to offer an efficient and elegant solution. Several reasons for this are given below.

$$(\text{no}_{(est)(est)st} \ \text{bird}_{est}) \ \text{moved}_{est} \tag{1}$$

$$(\text{no}_{n,(np,s),s} \ \text{bird}_n) \ \text{moved}_{np,s} \tag{2}$$

First, using only terms of the semantic types will violate the advantage (i) of the approach. This becomes clear when one tries to account for event semantics properly in LLFs as it needs an introduction of an event entity and *closure* or *existential closure* operators of [3,14] that do not always have a counterpart on a surface level.

Second, semantic types provide little syntactic information about the terms. For instance, bird_{est} and moved_{est} are both of type est in [12], hence there is no straightforward way to find out their syntactic categories. Furthermore, $M_{(est)est}H_{est}$ term can stand for adjective and noun, adverb and intransitive verb, or even noun and complement constructions. The lack of syntactic information about a term makes it impossible to find a correct tableau rule for the application to the term, i.e. it is difficult to meet property (ii). For example, for $A_{est}B_e$, it would be unclear whether to use a rule for an intransitive verb that introduces an event entity and a thematic relation between the event constant and B_e; or for $M_{(est)set}H_{est}$ whether to use a rule for adjective and noun or noun and complement constructions.[2]

---

[1] Hereafter we assume the following standard conventions while writing typed λ-terms: a type of a term is written in a subscript unless it is omitted, a term application is left-associative, and a type constructor comma is right-associative and is ignored if atomic types are single lettered.

[2] The latter two constructions have the same semantic types in the approach of [2], who also uses the C&C parser, like us, for obtaining logical forms but of first-order logic. The reason is that both PP and N categories for prepositional phrases and nouns, respectively, are mapped to et type.

Finally, a sentence generated from an open branch of a tableau proof can give us an explanation about failure of an entailment, but we will lose this option if we stay only with semantic types as it is not clear how to generate a grammatical sentence using only information about semantic types.[3]

In order to overcome the lack of syntactic information and remain LLFs similar to surface forms, we incorporate syntactic types and semantic types in the same type system. Let $\mathcal{A} = \{e, t, \mathbf{s}, \mathbf{np}, \mathbf{n}, \mathbf{pp}\}$ be a set of atomic types, where $\{e, t\}$ and $\{\mathbf{s}, \mathbf{np}, \mathbf{n}, \mathbf{pp}\}$ are sets of semantic and syntactic atomic types, respectively. Choosing these particular syntactic types is motivated by the syntactic categories of Combinatory Categorial Grammar (CCG) [13]. In contrast to the typing in [12], we drop s semantic type for states for simplicity reasons. Let $\mathcal{I}_{\mathcal{A}}$ be a set of all types, where complex types are constructed from atomic types in a usual way, e.g., $(\mathbf{np}, \mathbf{np}, \mathbf{s})$ is a type for a transitive verb. A type is called semantic or syntactic if it is constructed purely from semantic or syntactic atomic types, respectively; there are also types that are neither semantic nor syntactic, e.g., ees. After extending the type system with syntactic types, in addition to (1), (2) also becomes a well-typed term. For better readability, hereafter, we will use a boldface style for lexical constant terms with syntactic types.

The interaction between syntactic and semantic types is expressed by a subtyping relation ($\sqsubseteq$) that is a partial order, and for any $\alpha_1, \alpha_2, \beta_1, \beta_2 \in \mathcal{I}_{\mathcal{A}}$:

(a) $e \sqsubseteq \mathbf{np}$, $\mathbf{s} \sqsubseteq t$, $\mathbf{n} \sqsubseteq et$, $\mathbf{pp} \sqsubseteq et$;

(b) $(\alpha_1, \alpha_2) \sqsubseteq (\beta_1, \beta_2)$ iff $\beta_1 \sqsubseteq \alpha_1$ and $\alpha_2 \sqsubseteq \beta_2$

The introduction of subtyping requires a small change in typing rules, namely, if $\alpha \sqsubseteq \beta$ and A is of type α, then A is of type β too. From this new clause it follows that a term $A_\alpha B_\beta$ is of type γ if $\alpha \sqsubseteq (\beta, \gamma)$. Therefore, a term can have several types, which are partially ordered with respect to $\sqsubseteq$, with the least and greatest types. For example, a term $\mathbf{love}_{\mathbf{np},\mathbf{np},\mathbf{s}}$ is also of type eet (and of other five types too, where $(\mathbf{np},\mathbf{np},\mathbf{s})$ and eet are the least and greatest types, respectively). Note that all atomic syntactic types are subtypes of some semantic type except $e \sqsubseteq \mathbf{np}$. The latter relation, besides allowing relations like $(\mathbf{np}, \mathbf{s}) \sqsubseteq et$, also makes sense if we observe that any entity can be expressed in terms of a noun phrase (even if one considers event entities, e.g., singing$_{\mathrm{NP}}$ is difficult).

Now with the help of this *multiple typing* it is straightforward to apply $\mathbf{love}_{\mathbf{np},\mathbf{np},\mathbf{s}}$ $\mathbf{mary}_{\mathbf{np}}$ term to c_e constant, and there is no need to introduce new terms $\mathbf{love}_{eet}$ and $\mathbf{mary}_e$ just because $\mathbf{love}_{eet}\mathbf{mary}_e$ is applicable to c_e. For the same reason it is not necessary to introduce $\mathbf{man}_{et}$ for applying to c_e as $\mathbf{man}_{\mathbf{n}}c_e$ is already a well-formed term. From the latter examples, it is obvious that some syntactic terms (i.e. terms of syntactic type) can be used as semantic terms, hence minimize the number of terms in a tableau. Nevertheless, sometimes it will be inevitable to introduce a new term since its syntactic counterpart is not able to give a fine-grained semantics: if $\mathbf{red}_{\mathbf{n},\mathbf{n}}\mathbf{car}_{\mathbf{n}}c_e$ is evaluated as true, then

---

[3] An importance of the explanations is also shown by the fact that recently SemEval-2015 introduced a pilot task *interpretable STS* that requires systems to explain their decisions for semantic textual similarity.

one has to introduce $\mathbf{red}_{et}$ term in order to assert the redness of c_e by the term $\mathbf{red}_{et}c_e$ as $\mathbf{red}_{n,n}c_e$ is not typable. Finally, note that terms of type $\mathbf{s}$ can be evaluated either as true or false since they are also of type t.

Incorporating terms of syntactic and semantic types in one system can be seen as putting together two inference engines: one basically using syntactically-rich structures, and another one semantic properties of lexical entities. Yet another view from Abstract Categorial Grammars [7] or Lambda Grammars [11] would be to combine abstract and semantic levels, where terms of syntactic and semantic types can be seen as terms of abstract and semantic levels respectively, and the subtyping relation as a sort of simulation of the morphism between abstract and semantic types.[4]

3   Obtaining LLFs from CCG Trees

Automated generation of LLFs from unrestricted sentences is an important part in the development of the wide-coverage natural tableau prover. Combined with the implemented tableau prover, it facilitates exploring textual entailment data sets for extracting relevant tableau rules and allows us to evaluate the theory against these data sets.

We employ the C&C tools [4] as an initial step for obtaining LLFs. The C&C tools offer a pipeline of NLP systems like a POS-tagger, a chunker, a named entity recognizer, a super tagger, and a parser. The tools parse sentences in CCG framework with the help of a statistical parser. Altogether the tools are very efficient and this makes them suitable for wide-coverage applications [1]. In the current implementation we use the statistical parser that is trained on the *rebanked* version of CCGbank [8].

In order to get a semantically adequate LLF from a CCG parse tree (see Fig. 1), it requires much more effort than simply translating CCG trees to syntactic trees of typed lambda terms. There are two main reasons for this complication: (a) a trade-off that the parser makes while analyzing linguistic expressions in order to tackle unrestricted texts, and (b) accumulated wrong analyses in final parse trees introduced by the various C&C tools.

For instance, the parser uses combinatory rules that are not found in the CCG framework. One of such kind of rules is a lexical rule that simply changes a CCG category, for example, a category N into NP (see $\mathtt{lx}[np, n]$ combinatory rule in Fig. 1) or a category $S\backslash NP$ into $N\backslash N$. The pipeline of the tools can also introduce wrong analyses at any stage starting from the POS-tagger (e.g., assigning a wrong POS-tag) and finishing at the CCG parser (e.g., choosing a wrong combinatory rule). In order to overcome (at least partially) these problems, we use a pipeline consisting of several filters and transformation procedures. The general structure of the pipeline is the following:

---

[4] The connection between LLFs of [12] and the terms of an abstract level was already pointed out by Muskens in the project's description "Towards logics that model natural reasoning".

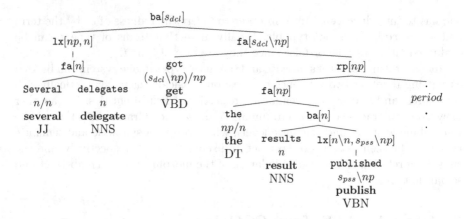

Fig. 1. A parse tree of several delegates got the results published. by the C&C parser

- *Transforming a CCG tree into a CCG term*: the procedure converts CCG categories in types by removing directionality from CCG categories (e.g., $S\backslash NP/NP \leadsto (\mathbf{np}, \mathbf{np}, \mathbf{s})$) and reordering tree nodes in a corresponding way.
- *Normalizing the CCG term*: since an obtained CCG term can be considered as a typed λ-term, it is possible to reduce it to $\beta\eta$-normal form.[5]
- *Identifying proper names*: if both function and argument terms are recognized as proper names by the C&C pipeline, then the terms are concatenated; for instance, **Leonardo**$_\mathbf{n,n}$(**da**$_\mathbf{n,n}$**Vinci**$_\mathbf{n}$) is changed in a constant term **Leonardo_da_Vinci**$_\mathbf{n}$ if all three terms are tagged as proper names.
- *Identifying multiword expressions (MWE)*: the CCG parser analyzes in a purely compositional way all phrases including MWEs like *a lot of, take part in, at least*, etc. To avoid these meaningless analyses, we replace them with constant terms (e.g., **a_lot_of** and **take_part_in**).
- *Correcting syntactic analyses*: this procedure is the most complex and extensive one as it corrects a CCG term by inserting, deleting or replacing terms. For example, the type shifts like $\mathbf{n} \leadsto \mathbf{np}$ are fixed by inserting corresponding determiners (e.g., $(\mathbf{oil_n})_{\mathbf{np}} \leadsto \mathbf{a_{n,np}oil_n}$) or by typing terms with adequate types (e.g., $(\mathbf{Leonardo\_da\_Vinci_n})_{\mathbf{np}} \leadsto \mathbf{Leonardo\_da\_Vinci_{np}}$ and $(\mathbf{several_{n,n}delegate_n})_{\mathbf{np}} \leadsto \mathbf{several_{n,np}delegate_n}$). More extensive corrections, like fixing a syntactically wrong analysis of a relative clause, like (3), are also performed in this procedure.
- *Type raising of quantifiers*: this is the final procedure, which takes a more or less fixed CCG term and returns terms where quantified noun phrases of type $\mathbf{np}$ have their types raised to $((\mathbf{np}, \mathbf{s}), \mathbf{s})$. As a result several LLFs are returned due to a scope ambiguity among quantifiers. The procedure makes sure that

---

[5] Actually the obtained CCG term is not completely a λ-term since it may contain type changes from lexical rules. For instance, in $(\mathbf{several_{n,n}delegate_n})_{\mathbf{np}}$ subterm, $(.)_{\mathbf{np}}$ operator changes a type of its argument into $\mathbf{np}$. Nevertheless, this kind of type changes are accommodated in the λ-term normalization calculus.

generalized quantifiers are applied to the clause they occur in if they do not take a scope over other quantifiers. For example, from a CCG term (4) only (5) is obtained and (6) is suppressed.

$$\text{old}_{n,n}(\text{who}_{(np,s),n,n}\text{cry}_{np,s}\text{man}_n) \rightsquigarrow \text{who}_{(np,s),n,n}\text{cry}_{np,s}(\text{old}_{n,n}\text{man}_n) \qquad (3)$$

$$\text{and}_{s,s,s}(\text{sleep}_{np,s}\text{john}_{np})(\text{snore}_{np,s}(\text{no}_{n,np}\text{man}_n)) \qquad (4)$$

$$\text{and}_{s,s,s}(\text{sleep}_{np,s}\text{john}_{np})(\text{no}_{n,(np,s),s}\text{man}_n\text{snore}_{np,s}) \qquad (5)$$

$$\text{no}_{n,(np,s),s}\text{man}_n(\lambda x.\,\text{and}_{s,s,s}(\text{sleep}_{np,s}\text{john}_{np})(\text{snore}_{np,s}x_{np})) \qquad (6)$$

The above described pipeline takes a single CCG tree generated from the C&C tools and returns a list of LLFs. For illustration purposes CCG term (7), which is obtained from the CCG tree of Fig. 1, and two LLFs, (8) and (9), generated from (7) are given below; here, vp abbreviates (np, s) and $s_{n,vp,s}$ term stands for the plural morpheme.

$$\text{got}_{np,vp}\Big(s_{n,np}(\text{who}_{vp,n,n}(\text{be}_{vp,vp}\text{publish}_{vp})\text{result}_n)\Big)\Big(\text{several}_{n,np}\text{delegate}_n\Big) \qquad (7)$$

$$\text{several}_{n,vp,s}\text{delegate}_n\Big(\lambda x.\,s_{n,vp,s}(\text{who}_{vp,n,n}(\text{be}_{vp,vp}\text{publish}_{vp})\text{result}_n)(\lambda y.\,\text{got}_{np,vp}\,y_{np}\,x_{np})\Big) \qquad (8)$$

$$s_{n,vp,s}(\text{who}_{vp,n,n}(\text{be}_{vp,vp}\text{publish}_{vp})\text{result}_n)(\lambda x.\,\text{several}_{n,vp,s}\text{delegate}_n(\text{got}_{np,vp}\,x_{np})) \qquad (9)$$

4 An Inventory of Natural Tableau Rules

The first collection of tableau rules for the natural tableau was offered in [12], where a wide range of rules are presented including Boolean rules, rules for algebraic properties (e.g., monotonicity), rules for determiners, etc. Despite this range of rules, they are insufficient for tackling problems found in textual entailment data sets. For instance, problems that only concentrate on quantifiers or Boolean operators are rare in the data sets. Syntactically motivated rules such as rules for passive and modifier-head constructions, structures with the copula, etc. are fruitful while dealing with wide-coverage sentences, and this is also confirmed by the problems found in entailment data sets. It would have been a quite difficult and time-consuming task to collect these syntactically motivated tableau rules without help of an implemented prover of natural tableau. For this reason the first thing we did was to implement a natural tableau prover, which could prove several toy entailment problems using a small inventory of rules mostly borrowed from [12].[6] With the help of the prover, then it was easier to explore manually the data sets and to introduce new rules in the prover that help it to further build tableaux and find proofs.

---

[6] Implementation of the prover, its computational model and functionality is a separate and extensive work, and it is out of scope of the current paper.

During collecting tableau rules we used a half portion of the FraCaS test suite [5] and the part of the SICK trial data [10] as a development set.[7] The reason behind opting for these data sets is that they do not contain long sentences, hence there is a higher chance that a CCG tree returned by the C&C tools will contain less number of wrong analyses, and it is more likely to obtain correct LLFs from the tree. Moreover, the FraCas test suite is considered to contain difficult entailment problems for textual entailment systems since its problems require more complex semantic reasoning than simply paraphrasing or relation extraction. We expect that interesting rules can be discovered from this set.

Hereafter, we will use several denotations while presenting the collected tableau rules. Uppercase letters $A, B, C, \ldots$ and lowercase letters $a, b, c, \ldots$ stand for meta variables over LLFs and constant LLFs, respectively. A variable letter with an arrow above it stands for a sequence of LLFs corresponding to the register of the variable (e.g. $\vec{C}$ is a sequence of LLFs). Let $[\,]$ denote an empty sequence. We assume that $e\!np$ is a variable type that can be either $\mathbf{np}$ or e and that $\mathbf{vp}$ abbreviates $(\mathbf{np}, \mathbf{s})$. Let $(-, \alpha) \in \mathcal{I}_{\mathcal{A}}$ for any $\alpha \in \mathcal{A}$ where the final (i.e. the most right) atomic type of $(-, \alpha)$ is α; for instance $(-, \mathbf{s})$ can be $\mathbf{s}$, $(\mathbf{np}, \mathbf{s})$, $(\mathbf{vp}, \mathbf{vp})$, etc. While writing terms we may omit their types if they are irrelevant for discussions, but often the omitted types can be inferred from the context the term occurs in. Tableau rules will be followed by the names that are the current names of the rules in the natural tableau prover. The same rule names with different subscripts mean that these rules are implemented in the prover by a single rule with this name. For instance, both mod_n_tr$_1$ and mod_n_tr$_2$ are implemented by a single rule mod_n_tr in the prover. Finally, we slightly change the format of nodes of [12]; namely, we place an argument list and a sign on the right side of a LLF – instead of $Tci : \mathbf{man}$ we write $\mathbf{man} : ci : \mathbb{T}$. We think that the latter order is more natural.

4.1   Rules from [12]

Most rules of [12] are introduced in the prover. Some of them were changed to more efficient versions. For example, the two rules deriving from format are modified and introduced in the prover as pull_arg, push_arg$_1$, and push_arg$_2$. These versions of the rules have more narrow application ranges. Hereafter, we assume that $\mathbb{X}$ can match both $\mathbb{T}$ and $\mathbb{F}$ signs.

$$\frac{\lambda x.\, A : c\vec{C} : \mathbb{X}}{(\lambda x.\, A)\, c : \vec{C} : \mathbb{X}}\ \text{pull\_arg} \qquad \frac{A\, c_e : \vec{C} : \mathbb{X}}{A : c_e\vec{C} : \mathbb{X}}\ \text{push\_arg}_1 \qquad \frac{A\, c_{\mathbf{np}} : \vec{C} : \mathbb{X}}{\begin{array}{c} A : c_{\mathbf{np}}\,\vec{C} : \mathbb{X} \\ A : c_e\,\vec{C} : \mathbb{X} \end{array}}\ \text{push\_arg}_2$$

The Boolean rules and rules for monotonic operators and determiners (namely, some, every, and no) are also implemented in the prover. It might be said that these rules are one of the crucial ones for almost any entailment problem.

---

[7] The Fracas test suite can be found at http://www-nlp.stanford.edu/~wcmac/ downloads, and the SICK trial data at http://alt.qcri.org/semeval2014/task1/index. php?id=data-and-tools.

4.2   Rules for Modifiers

One of the most frequently used set of rules is the rules for modifiers. These rules inspire us to slightly change the format of tableau nodes by adding an extra slot for *memory* on the left side of an LLF:

$$memorySet : LLF : argumentList : truthSign$$

An idea of using a memory set is to save modifiers that are not directly attached to a head of the phrase. Once a LLF becomes the head without any modifiers, the memory set is discharged and its elements are applied to the head. For example, if we want to entail beautiful car from beautiful red car, then there should be a way of obtaining (11) from (10) in a tableau. It is obvious how to produce (12) from (10) in the tableau settings, but this is not the case for producing (11) from (10), especially, when there are several modifiers for the head.

$$\mathbf{beautiful_{n,n}}(\mathbf{red_{n,n}car_n}) : c_e : \mathbb{T} \tag{10}$$

$$\mathbf{beautiful_{n,n}car_n} : c_e : \mathbb{T} \tag{11}$$

$$\mathbf{red_{n,n}car_n} : c_e : \mathbb{T} \tag{12}$$

With the help of a memory set, $\mathbf{beautiful_{n,n}}$ can be saved and retrieved back when the bare head is found. Saving *subsective* adjectives in a memory is done by mod_n_tr$_1$ rule while retrieval is processed by mods_noun$_1$ rule. In Fig. 2a, the closed tableau employs the latter rules in combination with int_mod_tr and proves that (10) entails (11).[8]

if b is subsective: $$\dfrac{\mathcal{M} : b_{n,n}A : c_e : \mathbb{T}}{\mathcal{M} \cup \{b_{n,n}\} : A : c_e : \mathbb{T}}\ \text{mod\_n\_tr}_1 \qquad \dfrac{\mathcal{M} \cup \{m_{n,n}\} : a_n : c_e : \mathbb{T}}{m_{n,n}a_n : c_e : \mathbb{T}}\ \text{mods\_noun}_1$$

if b is intersective: $$\dfrac{\mathcal{M} : b_{n,n}A : c_e : \mathbb{T}}{\begin{array}{c}\mathcal{M} : A : c_e : \mathbb{T} \\ b_{et} : c_e : \mathbb{T}\end{array}}\ \text{int\_mod\_tr} \qquad \dfrac{b_{n,n}A : c_e : \mathbb{F}}{\begin{array}{c}A : c_e : \mathbb{F} \quad A : c_e : \mathbb{T} \\ b_{et} : c_e : \mathbb{F}\end{array}}\ \text{int\_mod\_fl}$$

Hereafter, if a rule do not employ memory sets of antecedent nodes, then we simply ignore the slots by omitting them from nodes. The same applies to precedent nodes that contain an empty memory set. In rule int_mod_tr, a memory of a premise node is copied to one of conclusion nodes while rule int_mod_fl attaches empty memories to conclusion nodes, hence, they are omitted. The convention about omitting memory sets is compatible with rules found in [12].

---

[8] It is not true that mod_n_tr$_1$ always gives correct conclusions for the constructions similar to (10). In case of small beer glass the rule entails small glass that is not always the case, but this can be avoided in the future by having more fine-grained analysis of phrases (that beer glass is a compound noun), richer semantic knowledge about concepts and more restricted version of the rule; currently rule mod_n_tr$_1$ can be considered as a default rule for analyzing this kind of constructions.

1 $\textbf{beautiful}_{n,n}(\textbf{red}_{n,n}\textbf{car}_n) : c_e : \mathbb{T}$
2 $\textbf{beautiful}_{n,n}\textbf{car}_n : c_e : \mathbb{F}$

|

3 $\{\textbf{beautiful}_{n,n}\} : \textbf{red}_{n,n}\textbf{car}_n : c_e : \mathbb{T}$

|

4 $\{\textbf{beautiful}_{n,n}\} : \textbf{car}_n : c_e : \mathbb{T}$
5 $\textbf{red}_{et} : c_e : \mathbb{T}$

|

6 $\textbf{beautiful}_{n,n}\textbf{car}_n : c_e : \mathbb{T}$
7 $\times$

(a)

1 $\textbf{today}_{vp,vp}(\textbf{slowly}_{vp,vp}\textbf{ran}_{vp}) : \textbf{john}_{np} : \mathbb{T}$
2 $\textbf{today}_{vp,vp}\textbf{ran}_{vp} : \textbf{john}_{np} : \mathbb{F}$

|

3 $\{\textbf{today}_{vp,vp}\} : \textbf{slowly}_{vp,vp}\textbf{ran}_{vp} : \textbf{john}_{np} : \mathbb{T}$

|

4 $\{\textbf{today}_{vp,vp}, \textbf{slowly}_{vp,vp}\} : \textbf{ran}_{vp} : \textbf{john}_{np} : \mathbb{T}$

|

5 $\{\textbf{slowly}_{vp,vp}\} : \textbf{today}_{vp,vp}\textbf{ran}_{vp} : \textbf{john}_{np} : \mathbb{T}$
7 $\times$

(b)

Fig. 2. Tableaux that use rules for pulling and pushing modifiers in a memory: (a) beautiful red car $\Rightarrow$ beautiful car; (b) john ran slowly today $\Rightarrow$ john ran today

$$\frac{\mathcal{M} : B_{vp,vp}A : \vec{C} : \mathbb{T}}{\mathcal{M}\cup\{B_{vp,vp}\} : A : \vec{C} : \mathbb{T}} \text{ mod\_push} \qquad \frac{\mathcal{M}\cup\{B_{vp,vp}\} : A : \vec{C} : \mathbb{T}}{\mathcal{M} : B_{vp,vp}A : \vec{C} : \mathbb{T}} \text{ mod\_pull}$$

if p is a preposition:

$$\frac{\mathcal{M}\cup\{p_{np,vp,vp}\,d_{evp}\} : A_n : c_e : \mathbb{T}}{\begin{array}{c} p_{eet}\,d_e : c_e : \mathbb{T} \\ p_{np,n,n}\,d_{evp}A_n : c_e : \mathbb{T} \end{array}} \text{ mods\_noun}_2 \qquad \frac{\mathcal{M} : p_{np,n,n}\,d_{evp}\,A_n : c_e : \mathbb{T}}{\begin{array}{c} \mathcal{M} : A_n : c_e : \mathbb{T} \\ p_{eet}\,d_e : c_e : \mathbb{T} \end{array}} \text{ pp\_mod\_n}$$

The other rules that save a modifier or discharge it are mod_push, mod_pull and mods_noun$_2$. They do this job for any LLF of type (vp, vp). For instance, using these rules (in conjunction with other rules) it is possible to prove that (13) entails (14); moreover, the tableau in Fig. 2b employs push_mod and pull_mod rules and demonstrates how to capture an entailment about events with the help of a memory set without introducing an event entity.

Yet another rules for modifiers are pp_mod_n, n_pp_mod and aux_verb. If a modifier of a noun is headed by a preposition like in the premise of (15), then pp_mod_n rule can treat a modifier as an intersective one, and hence capture entailment (15). In the case when a propositional phrase is a complement of a noun, rule n_pp_mod treats the complement as an intersective property and attaches the memory to the noun head. This rule with mod_n_tr$_1$ and mods_noun$_1$ allows the entailment in (16).[9]

$$\textbf{in}_{np,vp,vp}\textbf{paris}_{np}\big(\lambda x.\, \textbf{a}_{n,vp,s}\textbf{tourist}_n(\lambda y.\, \textbf{is}_{np,vp}\,y_{np}\,x_{np})\big)\textbf{john}_{np} \qquad (13)$$

$$\textbf{in}_{np,n,n}\textbf{paris}_{np}\textbf{tourist}_n\textbf{john}_e \qquad (14)$$

---

[9] Note that the phrase in (16) is wrongly analyzed by the CCG parser; the correct analysis is $\textbf{for}_{np,n,n}C_{np}(\textbf{nobel}_{n,n}\textbf{prize}_n)$. Moreover, entailments similar to (16) are not always valid (e.g. $\textbf{short}_{n,n}(\textbf{man}_{pp,n}(\textbf{in}_{np,pp}\textbf{netherlands}_{np})) \not\Rightarrow \textbf{short}_{n,n}\textbf{man}_n)$. Since the parser and our implemented filters, at this stage, are not able to give correct analysis of noun complementation and post-nominal modification, we adopt n_pp_mod as a default rule for these constructions.

$$\text{in}_{\text{np,n,n}}\text{paris}_{\text{np}}\text{tourist}_{\text{n}}\text{john}_{e} \quad \Rightarrow \quad \text{in}_{eet}\text{paris}_{e}\text{john}_{e} \qquad (15)$$

$$\text{nobel}_{\text{n,n}}\big(\text{prize}_{\text{pp,n}}(\text{for}_{\text{np,pp}}C_{\text{np}})\big) \quad \Rightarrow \quad \text{nobel}_{\text{n,n}}\text{prize}_{\text{n}} \qquad (16)$$

Problems in data sets rarely contain entailments involving the tense, and hence aux_verb is a rule that ignores auxiliary verbs and an infinitive particle to. In Fig. 4, it is shown how aux_verb applies to 4 and yields 5. The rule also *accidentally* accounts for predicative adjectives since they are analyzed as $\text{be}_{\text{vp,vp}}P_{\text{vp}}$, and when aux_verb is applied to a copula-adjective construction, it discards the copula. The rule can be modified in the future to account for tense and aspect.

$$\frac{\mathcal{M} : d_{\text{pp,n}}A_{\text{pp}} : c_e : \mathbb{T}}{\begin{array}{c}\mathcal{M} : d_{\text{n}} : c_e : \mathbb{T} \\ A_{\text{pp}} : c_e : \mathbb{T}\end{array}} \text{ n\_pp\_mod} \qquad \frac{\mathcal{M} : b_{(-,\text{s}),(-,\text{s})}A : \vec{C} : \mathbb{X}}{\mathcal{M} : A : \vec{C} : \mathbb{X}} \text{ aux\_verb}$$

$$\text{where } b \in \{\textbf{do}, \textbf{will}, \textbf{be}, \textbf{to}\}$$

4.3 Rules for the Copula *be*

The copula be is often considered as a semantically vacuous word and, at the same time, it is sometimes a source of introduction of the equality relation in logical forms. Taking into account how the equality complicates tableau systems (e.g., a first-order logic tableau with the equality) and makes them inefficient, we want to get rid of be in LLFs whenever it is possible. The first rule that ignores the copula was already introduced in the previous subsection.

$$\text{If } p_{\text{np,pp}} \text{ is a preposition:} \qquad \frac{\mathcal{M} : \text{be}_{\text{pp,np,s}}\,(p_{\text{np,pp}}\,c_{\text{enp}}) : d_{\text{enp}} : \mathbb{X}}{\begin{array}{c}\mathcal{M} : p_{\text{np,pp}}\,c_{\text{enp}} : d_e : \mathbb{X} \\ p_{eet} : c_e\,d_e : \mathbb{X}\end{array}} \text{ be\_pp}$$

The second rule that does the removal of the copula is be_pp. It treats a propositional phrase following the copula as a predicate, and, for example, allows to capture the entailment in (17). Note that the rule is applicable with both truth signs, and the constants c and d are of type e or np.

$$\text{be}_{\text{pp,np,s}}(\text{in}_{\text{np,pp}}\,\text{paris}_{\text{np}})\,\text{john}_{\text{np}} \quad \Rightarrow \quad \text{in}_{\text{np,pp}}\,\text{paris}_{\text{np}}\,\text{john}_{e} \qquad (17)$$

The other two rules a_subj_be and be_a_obj apply to NP-be-NP constructions and introduce LLFs with a simpler structure. If we recall that quantifier terms like $\text{a}_{\text{n,vp,s}}$ and $\text{s}_{\text{n,vp,s}}$ are inserted in a CCG term as described in Sect. 3, then it is clear that there are many quantifiers that can introduce a fresh constant; more fresh constants usually mean a larger tableau and a greater choice in rule application strategies, which as a result decrease chances of finding proofs. Therefore, these two rules prevent tableaux from getting larger as they avoid introduction of a fresh constant. In Fig. 3, the tableau uses be_a_obj rule as the first rule application. This rule is also used for entailing (14) from (13).

If $a \in \{\mathbf{a}, \mathbf{the}\}$ and $c \neq \mathbf{there}$
$$\frac{\mathcal{M} : a_{\mathrm{n,vp,s}} \, N_{\mathrm{n}}(\mathbf{be} \, c_{e\eta p}) : [\,] : \mathbb{X}}{\mathcal{M} : N_{\mathrm{n}} : c_e : \mathbb{X}} \; \text{a\_subj\_be}$$

If $a \in \{\mathbf{a}, \mathbf{s}, \mathbf{the}\}$ and $c \neq \mathbf{there}$
$$\frac{\mathcal{M} : a_{\mathrm{n,vp,s}} \, N_{\mathrm{n}}(\lambda x.\, \mathbf{be} \, x_{e\eta p} \, c_{e\eta p}) : [\,] : \mathbb{X}}{\mathcal{M} : N_{\mathrm{n}} : c_e : \mathbb{X}} \; \text{be\_a\_obj}$$

4.4   Rules for the Definite Determiner *the*

We have already presented several new rules in the previous section that apply to certain constructions with the copula and the determiner the. Here we give two more rules that are applicable to a wider range of LLFs containing the.

Since the definite determiner presupposes a unique referent inside a context, rule the_c requires two nodes to be in a tableau branch: the node with the definite description and the node with the head noun of this definite description. In case these nodes are found, the constant becomes the referent of the definite description, and the verb phrase is applied to it. The rule avoids introduction of a fresh constant. The same idea is behind rule the but it introduces a fresh constant when no referent is found on the branch. The rule is similar to the one for existential quantifier some of [12], except that the is applicable to false nodes as well due to the presupposition attached to the semantics of the.

$$\frac{\mathcal{M} : \mathbf{the}_{\mathrm{n,vp,s}} \, N \, V : [\,] : \mathbb{X}}{\mathcal{M} : V : d_e : \mathbb{X}} \; \text{the\_c} \qquad \frac{\mathcal{M} : \mathbf{the}_{\mathrm{n,vp,s}} \, N \, V : [\,] : \mathbb{X}}{N : c_e : \mathbb{T} \atop \mathcal{M} : V : c_e : \mathbb{X}} \; \text{the where } c_e \text{ is fresh}$$

4.5   Rules for Passives

Textual entailment problems often contain passive paraphrases, therefore, from the practical point of view it is important to have rules for passives too. Two rules for passives correspond to two types the CCG parser can assign to a by-phrase: either pp while being a complement of VP, or (vp, vp) – being a VP modifier. Since these rules model paraphrasing, they are applicable to nodes with both signs. In Fig. 4, nodes 6 and 7 are obtained by applying vp_pass$_2$ to 5.

$$\frac{\mathcal{M} : V_{\mathrm{pp,vp}} \, (\mathbf{by}_{\mathrm{np,pp}} C_{e\eta p}) : D_{e\eta p} : \mathbb{X}}{\mathcal{M} : V_{\mathrm{np,vp}} : D_{e\eta p} C_{e\eta p} : \mathbb{X}} \; \text{vp\_pass}_1 \qquad \frac{\mathcal{M} : \mathbf{by}_{\mathrm{np,vp,vp}} C_{e\eta p} \, V_{\mathrm{vp}} : D_{e\eta p} : \mathbb{X}}{\mathcal{M} : V_{\mathrm{np,vp}} : D_{e\eta p} C_{e\eta p} : \mathbb{X}} \; \text{vp\_pass}_2$$

4.6   Closure Rules

In general, closure rules identify or introduce an inconsistency in a tableau branch, and they are sometimes considered as closure conditions on tableau branches. Besides the revised version of the closure rule $\bot\leq$ found in [12], we add three new closure rules to the inventory of rules.

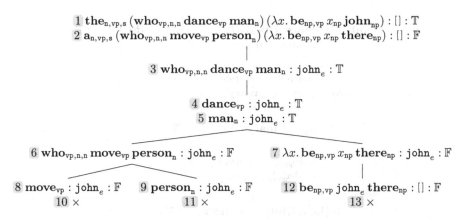

Fig. 3. A tableau for the man who dances is John ⇒ there is a person who moves. The tableau employs be_a_obj rule to introduce 3 from 1. The first two branches are closed by ⊥≤ taking into account that **man** ≤ **person** and **dance** ≤ **move**. The last branch is closed by applying ⊥there to 12.

$$\frac{\{\} : \textbf{be}_{\text{np,np,s}}\ c\ \textbf{there}_{\text{np}} : [] : \mathbb{F}}{\times}\ \bot\text{there} \qquad \frac{\mathcal{M}_1 : A : \vec{C} : \mathbb{T} \quad \mathcal{M}_2 : B : \vec{C} : \mathbb{F}}{\times}\ \bot\leq \text{ where } A \leq B,\ \mathcal{M}_2 \subseteq \mathcal{M}_1$$

$$\frac{\textbf{do}_{\text{np,vp}} : c_e\ D_{\text{evp}} : \mathbb{T} \quad A_\text{n} : c_e : \mathbb{T} \quad \{\} : A_{\text{vp}} : D_{\text{evp}} : \mathbb{F}}{\times}\ \bot\text{do\_vp}_1 \qquad \frac{\{\} : \textbf{do}_{\text{np,vp}} : c_e\ D_{\text{evp}} : \mathbb{F} \quad A_\text{n} : c_e : \mathbb{T} \quad A_{\text{vp}} : D_{\text{evp}} : \mathbb{T}}{\times}\ \bot\text{do\_vp}_2$$

Rule ⊥there considers a predicate corresponding to there is as a universal one. For example, in Fig. 3, you can find the rule in action (where be_a_obj rule is also used). The rules ⊥do_vp$_1$ and ⊥do_vp$_2$ model a light verb construction. See Fig. 4, where the tableau is closed by applying ⊥do_vp$_1$ to 6, 8 and 2.

The tableau rules presented in this section are the rules that are found necessary for proving certain entailment problems in the development set. For the complete picture, some rules are missing their counterparts for the false sign. This is justified by two reasons: those missing rules are inefficient from the computational point of view, and furthermore, they do not contribute to the prover's accuracy with respect to the development set. Although some rules are too specific and several ones seem too general (and in few cases unsound), at the moment our main goal is to list the fast and, at the same time, useful rules for textual entailments. The final analysis and organization of the inventory of rules will be carried out later when most of these rules will be collected. It is worth mentioning that the current tableau prover employs more computationally efficient versions of the rules of [12] and, in addition to it, admissible rules (unnecessary from the completeness viewpoint) are also used since they significantly decrease the size of tableau proofs.

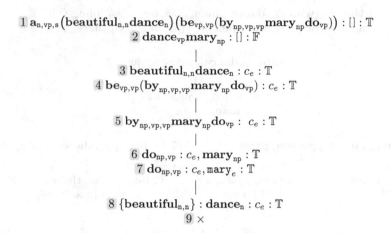

$$1 \ a_{n,vp,s}\big(beautiful_{n,n}dance_n\big)\big(be_{vp,vp}(by_{np,vp,vp}mary_{np}do_{vp})\big) : [] : \mathbb{T}$$
$$2 \ dance_{vp}mary_{np} : [] : \mathbb{F}$$

$$3 \ beautiful_{n,n}dance_n : c_e : \mathbb{T}$$
$$4 \ be_{vp,vp}(by_{np,vp,vp}mary_{np}do_{vp}) : c_e : \mathbb{T}$$

$$5 \ by_{np,vp,vp}mary_{np}do_{vp} : c_e : \mathbb{T}$$

$$6 \ do_{np,vp} : c_e, mary_{np} : \mathbb{T}$$
$$7 \ do_{np,vp} : c_e, mary_e : \mathbb{T}$$

$$8 \ \{beautiful_{n,n}\} : dance_n : c_e : \mathbb{T}$$
$$9 \times$$

Fig. 4. A tableau proof for a beautiful dance was done by Mary $\Rightarrow$ Mary danced

5   Evaluation

In order to demonstrate the productivity of the current inventory of tableau rules, we present the performance of the prover on the development set. As it was already mentioned in Sect. 4, we employ the part of the SICK trial data (100 problems) and a half of the FraCaS data (173 problems) as the development set. In these data sets, problems have one of three answers: entailment, contradiction, and neutral. Many entailment problems contain sentences that are long but have significant overlap in terms of constituents with other sentences. To prevent the prover from analyzing the common chunks (that is often unnecessary for finding the proof), we combine the prover with an optional simple aligner that aligns LLFs before a proof procedure. The prover also considers only a single LLF (i.e. semantic reading) for each sentence in a problem. Entailment relations between lexical words are modeled by the hyponymy and hypernymy relations of WordNet-3.0 [6]: $term_1 \leq term_2$ holds if there is a sense of $term_1$ that is a hyponym of some sense of $term_2$.

We evaluate the prover against the FraCaS development set (173 problems) and the whole SICK trial data (500 problems). The confusion matrix of the performance over the FraCaS development set is given in white columns of Table 1. As it is shown the prover was able to prove 31 true entailment and 2 contradiction problems. From all 173 problems, 18 problems are categorizes as defected since LLFs of those problems were not obtained properly – either the parser could not parse a sentence in a problem or it parsed the sentences but they were not comparable as their CCG categories were different. If we ignore the obvious errors from the parser by excluding the defected problems, the prover

Table 1. The confusion matrix of the prover's performance on the FraCaS dev-set and on the SICK trial data (with a gray background). Numbers in parentheses are the cases when the prover was force terminated (after 200 rule applications).

Problem answer	Prover status	Proof		No proof		Defected input	
Entailment		31	48	58 (6)	95 (18)	10	1
Contradiction		2	42	13 (0)	31 (1)	3	1
Neutral		1	2	50 (6)	277 (29)	5	3

1 $\text{every}_{n,vp,s}(\text{apcom}_{n,n}\text{manager}_n)(\lambda y.\, s_{n,vp,s}(\text{company}_{n,n}\text{car}_n)(\lambda x.\, \text{have}_{np,vp}\, x_{np}\, y_{np})) : [\,] : \mathbb{T}$

2 $a_{n,vp,s}(\text{apcom}_{n,n}\text{manager}_n)(\lambda x.\, \text{be}_{np,vp}\, x_{np}\, \text{jones}_{np}) : [\,] : \mathbb{T}$

3 $a_{n,vp,s}(\text{company}_{n,n}\text{car}_n)(\lambda x.\, \text{have}_{np,vp}\, x_{np}\, \text{jones}_{np}) : [\,] : \mathbb{F}$

|

4 $\text{apcom}_{n,n}\text{manager}_n : \text{jones}_e : \mathbb{T}$

|

5 $\lambda y.\, s_{n,vp,s}(\text{company}_{n,n}\text{car}_n)(\lambda x.\, \text{have}_{np,vp}\, x_{np}\, y_{np}) : \text{jones}_e : \mathbb{T}$

6 $\lambda y.\, s_{n,vp,s}(\text{company}_{n,n}\text{car}_n)(\lambda x.\, \text{have}_{np,vp}\, x_{np}\, y_{np}) : \text{jones}_{np} : \mathbb{T}$

|

7 $s_{n,vp,s}(\text{company}_{n,n}\text{car}_n)(\lambda x.\, \text{have}_{np,vp}\, x_{np}\, \text{jones}_{np}) : [\,] : \mathbb{T}$

|

8 $s_{n,vp,s}(\text{company}_{n,n}\text{car}_n) : \lambda x.\, \text{have}_{np,vp}\, x_{np}\, \text{jones}_{np} : \mathbb{T}$

9 $a_{n,vp,s}(\text{company}_{n,n}\text{car}_n) : \lambda x.\, \text{have}_{np,vp}\, x_{np}\, \text{jones}_{np} : \mathbb{F}$

|

10 $s_{n,vp,s} : \text{company}_{n,n}\text{car}_n,\ \lambda x.\, \text{have}_{np,vp}\, x_{np}\, \text{jones}_{np} : \mathbb{T}$

11 $a_{n,vp,s} : \text{company}_{n,n}\text{car}_n,\ \lambda x.\, \text{have}_{np,vp}\, x_{np}\, \text{jones}_{np} : \mathbb{F}$

12 $\times$

Fig. 5. A tableau proof for FraCaS-103 problem: all APCOM managers have company cars. Jones is an APCOM manager $\Rightarrow$ Jones has a company car (Note that 4 was obtained from 2 by be_a_obj, 5 and 6 from 1 and 4 by the efficient version of the rule for every of [12], 7 from 6, 8 and 9 from 3 and 7 by the efficient version of the monotone rule of [12], 10 and 11 from 8 and 9 in the same way as the previous nodes, and 12 from 10 and 11 by $\bot\leq$ and the fact that $a \leq s$).

get the precision of .97 and the recall of .32. For this evaluation the prover is limited by the number of rule applications; if it is not possible to close a tableau after 200 rule applications, then the tableau is considered as open. For instance, the prover reached the rule application limit and was forcibly terminated for 12 problems (see the numbers in parentheses in 'No proof' column). In Fig. 5, it is shown a closed tableau found by the prover for the FraCaS problem with multiple premises. The first three entries in the tableau exhibit the LLFs of the sentences that were obtained by the LLF generator.

The results over the FraCaS data set seem promising taking into account that the set contains sentences with linguistic phenomena (such as anaphora, ellipsis, comparatives, attitudes, etc.) that were not modeled by the tableau rules.[10]

The evaluation over the SICK trial data is given in gray columns of Table 1. Despite exploring only a fifth portion of the SICK trial data, the prover showed decent results on the data (see them in gray columns of Table 1). The evaluation again shows the extremely high precision of .98 and the more improved recall of .42 than in case of the FraCaS data. The alignment preprocessing drastically decreases complexity of proof search for the problems of the SICK data since usually there is a significant overlap between a premise and a conclusion. The tableau proof in Fig. 6 demonstrates this fact, where treating shared complex LLFs as a constant results in closing the tableau in three rule applications.

1 $two_{n,vp,s}$ $person_n$ be_and_watch_the_sunset_stand_in_the_ocean$_{vp}$: [] : $\mathbb{T}$
2 $no_{n,vp,s}$ $person_n$ be_and_watch_the_sunset_stand_in_the_ocean$_{vp}$: [] : $\mathbb{T}$

3 $person_n$: c_e : $\mathbb{T}$
4 be_and_watch_the_sunset_stand_in_the_ocean$_{vp}$: c_e : $\mathbb{T}$

5 $person_n$: c_e : $\mathbb{F}$
6 ×

Fig. 6. A tableau proof for SICK-6146 problem:two people are standing in the ocean and watching the sunset ⊥ nobody is standing in the ocean and watching the sunset. The tableau starts with $\mathbb{T}$ sign assigned to initial LLFs for proving the contradiction. The proof introduces 5 from 2 and 4 using the efficient version of the rule for no of [12] and, in this way, avoids branching of the tableau.

Table 2. The false positive problems

ID	Answer	Prover	Problem (premises ⇒ hypothesis)
FraCaS-287	Neutral	Entailment	Smith wrote a report in two hours ⇒ Smith wrote a report in one hour
SICK-1400	Neutral	Entailment	A sad man is crying ⇒ A man is screaming
SICK-8461	Neutral	Contradiction	A man with no hat is sitting on the ground ⇒ A man with backwards hat is sitting on the ground

The problems that were classified as neutral but proved by the prover represent also the subject of interest (see Table 2). The first problem was proved

---

[10] The FraCaS data contains entailment problems requiring deep semantic analysis and it is rarely used for system evaluation. We are aware of a single case of evaluating the system against this data; namely, the NatLog system [9] achieves quite high accuracy on the data but only on problems with a single premise. The comparison of our prover to it must await future research.

due to the poor treatment of cardinals by the prover – there is no distinction between them. The second problem was identified as entailment since cry may also have a meaning of shout. The last one was proved because the prover used LLFs, where no hat and a backwards hat had the widest scopes.

6 Future Work

Our future plan is to continue enriching the inventory of tableau rules. Namely, the SICK training data is not yet explored entirely, and we expect to collect several (mainly syntax-driven) rules that are necessary for *unfolding* certain LLFs. We also aim to further explore the FraCaS data and find the ways to accommodate in natural tableau settings semantic phenomena contained in plurals, comparatives, anaphora, temporal adverbials, events and attitude verbs.

At this stage it is early to compare the current performance of the prover to those of other entailment systems. The reason is that entailment systems usually do not depend on a single classifier but the combination of several shallow (e.g., word overlap) or deep (e.g., using semantically relevant rules) classifiers. In the future, we plan to make the prover more robust by combining it with a shallow classifier or by adding default rules (not necessarily sound) that are relevant for a development data set, and then compare it with other entailment systems. We also intend to employ other parsers for improving the LLF generator and knowledge bases for enriching the lexical knowledge of the prover.

Acknowledgements. I would like to thank Reinhard Muskens for his discussions and continuous feedback on this work. I also thank Matthew Honnibal, James R. Curran and Johan Bos for sharing the retrained CCG parser and anonymous reviewers of LENLS11 for their valuable comments. The research is part of the project "Towards Logics that Model Natural Reasoning" and supported by the NWO grant (project number 360-80-050).

References

1. Bos, J., Clark, S., Steedman, M., Curran, J.R., Hockenmaier, J.: Wide-coverage semantic representations from a CCG parser. In: Proceedings of the 20th International Conference on Computational Linguistics (COLING 2004), pp. 1240–1246 (2004)
2. Bos, J.: Towards a large-scale formal semantic lexicon for text processing. from form to meaning: processing texts automatically. In: Proceedings of the Biennal GSCL Conference, pp. 3–14 (2009)
3. Champollioni, L.: Quantification and negation in event semantics. In: Baltic International Yearbook of Cognition, Logic and Communication, vol. 6 (2010)
4. Clark, S., Curran, J.R.: Wide-coverage efficient statistical parsing with CCG and log-linear models. Comput. Linguist. **33**(4), 493–552 (2007)
5. Cooper, R., Crouch, D., van Eijck, J., Fox, C., van Genabith, J., Jaspars, J., Kamp, H., Milward, D., Pinkal, M., Poesio, M., Pulman, S.: Using the framework. Technical Report LRE 62–051 D-16. The FraCaS Consortium (1996)

6. Fellbaum, Ch. (ed.): WordNet: An Electronic Lexical Database. MIT press, Cambridge (1998)
7. de Groote, Ph.: Towards abstract categorial grammars. In: Proceedings of the Conference on ACL 39th Annual Meeting and 10th Conference of the European Chapter, pp. 148–155 (2001)
8. Honnibal, M., Curran, J.R., Bos, J.: Rebanking CCGbank for improved NP interpretation. In: Proceedings of the 48th Meeting of the Association for Computational Linguistics (ACL), pp. 207–215 (2010)
9. MacCartney, B., Manning, C.D.: Modeling semantic containment and exclusion in natural language inference. In: Proceedings of Coling-2008, Manchester, UK (2008)
10. Marelli, M., et al.: A sick cure for the evaluation of compositional distributional semantic models. In Proceedings of LREC, Reykjavik (2014)
11. Muskens, R.: Language, lambdas, and logic. In: Kruijff, G., Oehrle, R. (eds.) Resource-Sensitivity, Binding and Anaphora. Studies in Linguistics and Philosophy, vol. 80, pp. 23–54. Springer, Heidelberg (2003)
12. Muskens, R.: An analytic tableau system for natural logic. In: Aloni, M., Bastiaanse, H., de Jager, T., Schulz, K. (eds.) Logic, Language and Meaning. LNCS, vol. 6042, pp. 104–113. Springer, Heidelberg (2010)
13. Steedman, M., Baldridge, J.: Combinatory Categorial Grammar. In: Borsley, R.D., Borjars, K. (eds.) pp. 181–224. Blackwell Publishing (2011)
14. Winter, Y., Zwarts, J.: Event semantics and abstract categorial grammar. In: Kanazawa, M., Kornai, A., Kracht, M., Seki, H. (eds.) MOL 12. LNCS, vol. 6878, pp. 174–191. Springer, Heidelberg (2011)

Resolving Modal Anaphora in Dependent Type Semantics

Ribeka Tanaka[1(✉)], Koji Mineshima[1,3], and Daisuke Bekki[1,2,3]

[1] Ochanomizu University, Tokyo, Japan
{tanaka.ribeka,bekki}@is.ocha.ac.jp
[2] National Institute of Informatics, Tokyo, Japan
[3] CREST, Japan Science and Technology Agency, Tokyo, Japan
mineshima.koji@ocha.ac.jp

Abstract. This paper presents an analysis of modal subordination in the framework of Dependent Type Semantics, a framework of natural language semantics based on dependent type theory. Dependent types provide powerful type structures that have been applied to various discourse phenomena in natural language, yet there has been little attempt to produce an account of modality and its interaction with anaphora from the perspective of dependent type theory. We extend the framework of Dependent Type Semantics with a mechanism of handling explicit quantification over possible worlds, and show how modal anaphora and subordination can be handled within this framework.

1 Introduction

Modal anaphora and subordination have been extensively studied within *model-theoretic* approaches to discourse semantics, including Discourse Representation Theory (DRT) and Dynamic Semantics (Roberts [13], Frank and Kamp [7], van Rooij [15], Asher and McCready [1]). In contrast, *proof-theoretic* approaches to natural language semantics have been developed within a framework of dependent type theory and have been applied to dynamic discourse phenomena (Sundholm [17], Ranta [12]). The proof-theoretic framework is attractive in that entailment relations can be directly computed without referring to models; it provides a foundation of computational semantics that can be applied to the problems of natural language inference and of recognizing textual entailment using modern proof assistants (Chatzikyriakidis and Luo [5]). However, there has been little attempt to produce an account of modality and its interaction with anaphora from the perspective of dependent type theory, or more generally, from a proof-theoretic perspective on natural language semantics. Here we provide such an account: we present an analysis of *modal subordination* (MS) within a framework of proof-theoretic natural language semantics called *Dependent Type Semantics* (DTS).

There are at least three possible approaches to treating modality in natural language from a proof-theoretic perspective. One is to construct a proof system for natural language inference that contains modal expressions as primitive; the program of *natural logic* (e.g. Muskens [11]) can be regarded as an instance of

© Springer-Verlag Berlin Heidelberg 2015
T. Murata et al. (Eds.): JSAI-isAI 2014 Workshops, LNAI 9067, pp. 83–98, 2015.
DOI: 10.1007/978-3-662-48119-6_7

this approach. As far as we can see, however, the treatment of discourse phenomena such as anaphora and presupposition in this approach is underdeveloped; in particular, at the current stage it is not clear how to handle various discourse phenomena in a proof system based on the surface structure of natural language sentences, or, more generally, a variable-free proof system. Another approach is the one proposed by Ranta [12], according to which the notion of contexts in dependent type theory plays a central role in explaining modal constructions. A problem with this approach is that the single notion of context seems to be insufficient to account for various kinds of modal expressions in natural language, including epistemic and deontic modality as well as a variety of propositional attitudes. The third approach is to analyze modality in terms of explicit quantification over possible worlds using dependent type theory. This approach enables us to make use of the findings that have been accumulated in formal semantics of natural language over past half a century. We adopt this explicit approach to modal semantics and attempt to integrate these findings with the framework of DTS. This will enable us to handle modals, conditionals and attitude verbs in a unified framework and thereby to broaden the empirical coverage of DTS.

This paper is structured as follows. Section 2 motivates our proof-theoretic approach to the phenomena of MS. Section 3 provides an overview of the framework of DTS, and then, in Sect. 4, we extend it with modal semantics and present an analysis of MS in terms of dependent types. Section 5 extends our approach to the analysis of conditionals. Section 6 provides a dynamic lexicon and compositional analysis in a setting of categorial grammar.

2   Modal Subordination

The phenomena known as MS were first investigated by Roberts [13] in the framework of DRT. A characteristic of MS is that, as is exemplified in the contrast shown in (1), modal expressions like *might* introduce a proposition that is passed to a subsequent modal discourse but not to a discourse with factual mood.

(1)   a. *A wolf* might enter. *It* would growl.

b. *A wolf* might enter. #*It* growls.

Note that even if the indefinite *a wolf* in (1a) is interpreted as taking scope under the modal *might*, the modal *would* enables the pronoun *it* to pick up its antecedent. The intended reading of the second sentence in (1a) can be paraphrased as *If a wolf entered, it would growl.* The problem can be formulated as follows: how to explain the following valid pattern of inference under the intended reading?

(2) $\dfrac{A \text{ } wolf \text{ might enter. } It \text{ would growl.}}{\text{If } a \text{ } wolf \text{ entered, } it \text{ would growl.}}$

Schematically, the problem can also be formulated as: how to derive if φ, $\text{modal}_2 \psi$ from the discourse $\text{modal}_1 \varphi$. $\text{modal}_2 \psi$ in terms of some reasoning mechanism, where modal$_1$ and modal$_2$ are suitable modal expressions. A desirable account has to be powerful enough to provide such a derivation, while it must be suitably constrained so as to block the anaphoric link as shown in (1b).

MS also arises with presuppositions. Consider the following sentence with a classical example of presupposition trigger (van Rooij [15]).

(3)   It is possible that John used to smoke and possible that he just *stopped* doing so.

Here the presupposition trigger *stopped* occurs in a modal environment, and carries a presupposition which is successfully satisfied by the proposition introduced by the antecedent clause having a modal force. Though there is a difference in the ways presupposition and anaphora are resolved (see e.g., Geurts [8]), henceforth we use "anaphora" as a cover term for both pronominal anaphora and presupposition.

Roberts [13,14] developed an account based on *accommodation* of the antecedent clause If φ in the schematic representation mentioned above. Subsequent authors criticized this approach mainly on the grounds that the process of accommodation is too unconstrained and hence over-generates; since then, various theories of MS have been developed in a model-theoretic tradition, in particular, in the framework of DRT (Frank and Kamp [7]; Geurts [8]) and Dynamic Semantics (van Rooij [15]; Asher and McCready [1]).

In addition to the general attractiveness of a proof-theoretic approach to natural language inference, let us mention an advantage of handling MS from the perspective emphasizing the role of inference in resolving anaphora. A problem with the treatment of anaphora in model-theoretic approaches including DRT and Dynamic Semantics is that they do not do justice to the fact that anaphora resolution often requires the hearer to perform *inference*. A typical example of such a case is one involving the so-called *bridging* inference (Clark [6]). The following is an example of the interaction of MS and bridging.

(4)   John might have a new house. The front door would be blue.

The definite description *the front door* in the second sentence does not have an overt antecedent, but a suitable antecedent is easily *inferred* using the commonplace assumption that a house has a front door. According to the standard account in DRT and Dynamic Semantics, the presupposed information is identified with some element present in the previous discourse or copied in a suitable place via accommodation. However, examples such as (4) suggest that resolving anaphora is not simply a matter of matching or adding information; rather, it crucially involves inferences with assumptions that are not directly provided in a discourse.[1] The proof-theoretic approach provides a well-developed proof system that accounts for the fact that inferences with implicit assumptions play a crucial role in identifying the antecedent of anaphora.

---

[1] Geurts [8] (pages 72–79) admits the importance of inferences with world knowledge in resolving presuppositions, but provides no clues on how to incorporate additional inferential architectures into the framework of DRT.

3 Dependent Type Semantics

DTS (Bekki [2]) is a framework of natural language semantics that extends dependent type theory with a mechanism of context passing to account for anaphora resolution processes and with a component to derive semantic representations in a compositional way.

The syntax is similar to that of dependent type theory [10], except it is extended with an @-term that can be annotated with some type Λ, written as $@_i^\Lambda$. The syntax for *raw terms* in DTS is specified as follows.[2]

$$
\begin{array}{llll}
\Lambda := & x & & \text{variable} \\
& |\quad c & & \text{constant} \\
& |\quad @_i^\Lambda & & \text{underspecified term annotated with type } \Lambda \\
& |\quad (\Pi x : \Lambda)\Lambda & & \text{dependent function type } (\Pi\text{-type}) \\
& |\quad (\Sigma x : \Lambda)\Lambda & & \text{dependent sum type } (\Sigma\text{-type}) \\
& |\quad (\lambda x : \Lambda)\Lambda & & \text{lambda abstraction} \\
& |\quad \Lambda\Lambda & & \text{function application} \\
& |\quad (\Lambda, \Lambda) & & \text{pair} \\
& |\quad \pi_1 & & \text{the first projection function} \\
& |\quad \pi_2 & & \text{the second projection function} \\
& |\quad \Lambda =_\Lambda \Lambda & & \text{propositional equality}
\end{array}
$$

We will often omit type τ in $(\lambda x : \tau)M$ and abbreviate $(\lambda x_1)\ldots(\lambda x_n)M$ as $(\lambda x_1 \ldots x_n)M$.

The type constructor Σ is a generalized form of the product type and serves as an existential quantifier. An object of type $(\Sigma x : A)B(x)$ is a pair (m, n) such that m is of type A and n is of type $B(m)$. Conjunction (or, product type) $A \wedge B$ is a degenerate form of $(\Sigma x : A)B$ if x does not occur free in B. Σ-types are associated with projection functions π_1 and π_2 that are computed with the rules $\pi_1(m, n) \equiv m$ and $\pi_2(m, n) \equiv n$, respectively.

The type constructor Π is a generalized form of functional type and serves as a universal quantifier. Implication $A \to B$ is a degenerate form of $(\Pi x : A)B$ if x does not occur free in B. An object of type $(\Pi x : A)B(x)$ is a function f such that for any object a of type A, fa is an object of type $B(a)$. See e.g., Martin-Löf [10] and Ranta [12] for more details and inference rules for Π-types and Σ-types.

DTS is based on the paradigm of the Curry-Howard correspondence, according to which propositions are identified with *types*; the truth of a proposition is then defined as the existence of a proof (i.e., proof-term) of the proposition. In other words, for any (static) proposition P, we can say that P is true if and only if P is *inhabited*, that is, there exists a proof-term t such that $t : P$. In this paper, we will denote the type of (static) proposition by prop.

A *dynamic* proposition in DTS is a function mapping a proof c of a static proposition γ, a proposition representing the preceding discourse, to a static

---

[2] In dependent type theory, terms and types can be mutually dependent; thus, the terms defined here can serve as types as well.

proposition; hence it has type $\gamma \to \mathsf{prop}$. Such a proof c is called a *context*. For instance, a sentence *a man entered* is represented as

(5) $(\lambda c)(\Sigma u : (\Sigma x : \mathsf{E})\,\mathsf{man}\,x)\,\mathsf{enter}\,\pi_1 u,$

where E is a type of entities and c is a variable for a given context. In this case, the sentence does not have any anaphora or presupposition trigger; accordingly, the variable c does not appear in the body of the representation. A sentence containing an anaphoric element is represented using an @-term. For instance, the sentence *he whistled* is represented as $(\lambda c)\,\mathsf{whistle}(@_0^{\gamma_0 \to \mathsf{E}} c)$, where the annotated term $@_0^{\gamma_0 \to \mathsf{E}}$ corresponds to the pronoun *he*. Here γ_0 is the type of the context variable c. The term $@_0^{\gamma_0 \to \mathsf{E}}$ will eventually be replaced by some term A having the annotated type $\gamma_0 \to \mathsf{E}$, in which case, we say that the @-term is *bound* to A.

Two dynamic propositions are conjoined by *dynamic conjunction*, defined as:

(6) $M; N \equiv (\lambda c)(\Sigma u : Mc)N(c, u).$

Here the information from the left context, represented as a proof term c, is passed to the first conjunct M. Then the second conjunct N receives the pair (c, u), where the proof term u represents the information from M. As a result, an anaphoric element in N can refer to an object introduced in the left context as well as that introduced in M.

As an illustration, let us consider how to derive the following simple inference:

(7) $\dfrac{\text{A } man \text{ entered. } He \text{ whistled.}}{\text{There is a man who entered and whistled.}}$

By dynamic conjunction, the semantic representations for *a man entered* and *he whistled* are merged into the following:

(8) $(\lambda c)(\Sigma v : (\Sigma u : (\Sigma x : \mathsf{E})\,\mathsf{man}\,x)\,\mathsf{enter}\,\pi_1 u))\,\mathsf{whistle}(@_0^{\gamma_0 \to \mathsf{E}}(c, v))$

How to resolve the type γ_0 and the term $@_0^{\gamma_0 \to \mathsf{E}}$ can be inferred based on a type checking algorithm (see Bekki [2]). In the present case, given that $@_0^{\gamma_0 \to \mathsf{E}}$ takes the pair (c, v) as an argument, one can infer that γ_0 is set to

(9) $\gamma \wedge (\Sigma u : (\Sigma x : \mathsf{E})\,\mathsf{man}\,x)\,\mathsf{enter}\,\pi_1 u,$

and that a term that can be substituted for $@_0^{\gamma_0 \to \mathsf{E}}$ is $\pi_1 \pi_1 \pi_2$. The resulting representation reduces to the following:

(10) $(\lambda c)(\Sigma v : (\Sigma u : (\Sigma x : \mathsf{E})\,\mathsf{man}\,x)\,\mathsf{enter}\,\pi_1 u))\,\mathsf{whistle}\,\pi_1 \pi_1 v.$

This gives the semantic representation after anaphora resolution (in terms of the substitution of $@_0$ for $\pi_1 \pi_1 \pi_2$) for the discourse which appears as a premise in (7). Let us assume that the conclusion of (7) is represented as:

(11) $(\Sigma u : (\Sigma x : \mathsf{E})\,\mathsf{man}\,x)(\mathsf{enter}\,\pi_1 u \wedge \mathsf{whistle}\,\pi_1 u).$

Then, it is easily checked that given an initial context c, if the body of the representation in (10) is true, the proposition in (11) is true as well; in other words, from the given assumption, one can construct a proof-term for (11). In this way, we can derive the inference in (7).

To see how anaphora resolution interacts with inferences involving implicit assumptions, consider a simple example of bridging:

(12)   John has a house. The door is blue.

The second sentence can be represented as $(\lambda c)\mathsf{blue}\,\pi_1(@_0^{\gamma_0 \to (\Sigma x:\mathsf{E})\,\mathsf{door}\,x}c)$, where the definite description *the door* is represented by the first projection of the annotated @-term applied to a given context c. The annotated type

$$\gamma_0 \to (\Sigma x : \mathsf{E})\,\mathsf{door}\,x$$

means that the definite article *the* selects a pair having the type $(\Sigma x : \mathsf{E})\,\mathsf{door}\,x$ from a context of type γ_0. Such a pair consists of some entity x and a proof that x is a door, and its first projection, i.e., an entity x, is applied to the predicate blue. This means that for the whole term to be typable, one needs to give a proof of the existence of a door. Intuitively, this captures the existence presupposition triggered by the definite description *the door*.

In the same way as (8) above, the two sentences in (12) are conjoined by dynamic conjunction and reduced to the following, with an initial context c:

(13) $(\Sigma v: (\Sigma u: (\Sigma x: \mathsf{E})\mathsf{house}\ x)\ \mathsf{have}(j, \pi_1 u))\ \mathsf{blue}(\pi_1(@_0^{\gamma_0 \to (\Sigma x:\mathsf{E})\mathsf{door}x}(c, v)))$.

Given that the annotated @-term takes a pair (c, v) as an argument, one can infer that γ_0 is $\gamma \wedge (\Sigma u: (\Sigma x: \mathsf{E})\ \mathsf{house}\,x)\ \mathsf{have}\,(j, \pi_1 u)$. Thus given a term (c, v) of this type, the @-term requires to construct an object of type $(\Sigma x : \mathsf{E})\,\mathsf{door}\,x$. Let us assume that judgement $f : (\Pi x: \mathsf{E})(\mathsf{house}\,x \to (\Sigma y: \mathsf{E})(\mathsf{door}\,y \wedge \mathsf{have}(x, y)))$ is taken as an axiom in the global context that represents our commonplace knowledge. Let t be a term $f(\pi_1\pi_1\pi_2(c, v))(\pi_2\pi_1\pi_2(c, v))$. Then, it can be easily verified that t is of type $(\Sigma y: \mathsf{E})(\mathsf{door}\,y \wedge \mathsf{have}(\pi_1\pi_1 v, y))$, and hence, $(\pi_1 t, \pi_1\pi_2 t)$ has the required type $(\Sigma x : \mathsf{E})\,\mathsf{door}\,x$. By taking the first projection of this pair, one can eventually obtain the following proposition:

$$(\Sigma v: (\Sigma u: (\Sigma x: \mathsf{E})\ \mathsf{house}\,x)\ \mathsf{have}\,(j, \pi_1 u))\ \mathsf{blue}(\pi_1 f(\pi_1\pi_1 v)(\pi_2\pi_1 v)).$$

This can be read as *A door of John's house is blue*, which captures correct information derivable from the discourse in (12).

4   Modality and Modal Subordination in DTS

To represent modal propositions in DTS, we parameterize propositions over worlds and contexts. Let W be a type of worlds and γ a type of contexts. Then dynamic propositions have type $\mathsf{W} \to \gamma \to \mathsf{prop}$, abbreviated henceforth as κ. Let M, N be of type κ. We define $\Diamond$ (epistemic possibility), $\square$ (epistemic necessity),

; (dynamic conjunction), and $\triangleright$ (dynamic implication) as follows:

$$\Diamond M \equiv (\lambda wc)(\Sigma w' : \mathsf{W})(\mathsf{R_{epi}}\, ww' \wedge Mw'c)$$

$$\Box M \equiv (\lambda wc)(\Pi w' : \mathsf{W})(\mathsf{R_{epi}}\, ww' \to Mw'c)$$

$$M \,;\, N \equiv (\lambda wc)(\Sigma u : Mwc)Nw(c, u)$$

$$M \triangleright N \equiv (\lambda wc)(\Pi u : Mwc)Nw(c, u)$$

Since our focus is on the phenomena of MS, we take epistemic accessibility relation $\mathsf{R_{epi}}$ as primitive and remain neutral with respect to the particular analysis of it.[3]

Let rprop be a subtype of propositions with the axiom $p : \mathsf{rprop} \to \mathsf{prop}$. Intuitively, rprop denotes a class of *root* propositions, i.e., propositions embedded under modal operators and introduced as hypothetical ones by modal sentences. Type $\mathsf{W} \to \gamma \to \mathsf{rprop}$ of parameterized root proposition will be abbreviated as $\widehat{\kappa}$. Then we have $(\lambda gwc)p(gwc) : \widehat{\kappa} \to \kappa$. The function $(\lambda gwc)p(gwc)$, which maps a parameterized root proposition to a parameterized proposition, will be abbreviated as $^{\downarrow}(\cdot)$. Now *might A* and *would A*, where A is of type κ, are defined as follows:[4]

$$[\![might]\!](A) = (\lambda wc)(\Diamond(^{\downarrow}(@_ic)\,;\,A)wc \wedge (\Sigma P : \widehat{\kappa})(^{\downarrow}P =_\kappa {}^{\downarrow}(@_ic)\,;\,A))$$

$$[\![would]\!](A) = (\lambda wc)(\Box(^{\downarrow}(@_ic) \triangleright A)wc \wedge (\Sigma P : \widehat{\kappa})(^{\downarrow}P =_\kappa {}^{\downarrow}(@_ic)\,;\,A))$$

For brevity, here and henceforth we usually omit the annotated type ending with $\widehat{\kappa}$ and write $@_i$ for $@_i^{\gamma \to \widehat{\kappa}}$.

As usual, *might* and *would* are analyzed as involving existential and universal quantification over worlds, respectively. One difference from the standard account is that modal operators involve an @-term that triggers anaphoric reference to an antecedent parameterized root proposition of type $\widehat{\kappa}$. This is because we have to take into account discourse meaning: if there is a root proposition of type $\widehat{\kappa}$ introduced in the previous modal context, it can be anaphorically picked up by the @-term and embedded in the restrictor of the modal operator, i.e., in the position before dynamic conjunction or dynamic implication. The right conjuncts of the definitions introduce such a root proposition of type $\widehat{\kappa}$ in terms of Σ types. Thus, modal operators can both receive and introduce a hypothetical proposition. Together with the context-passing mechanism of DTS, this enables us to handle cross-sentential anaphora resolution.

To represent the empty modal context, we let $\mathbb{T} : \mathsf{rprop}$ and $f(\mathbb{T}) = \top$, where $\top$ is a unit type with the unique element $\star : \top$. Then we have $^{\downarrow}(\lambda wc)\,\mathbb{T} =_\kappa (\lambda wc)\top$, where $(\lambda wc)\top$ is used to represent the empty non-modal dynamic context and abbreviated as ε. If there is no appropriate antecedent for $@_i$, for example, if a sentence is uttered in a null context, $@_i$ can be bound to $(\lambda xwc)\,\mathbb{T}$ of type $\gamma \to \widehat{\kappa}$, and we can obtain $^{\downarrow}(@_ic) = \varepsilon$.

---

[3] Kratzer (2012) derives accessibility relation from a *modal base* and an *ordering source*. Our analysis would be compatible with such a decomposition.

[4] In this section, *might* and *would* will be treated as propositional operators. A fully compositional analysis will be given in Sect. 6.

As an illustration, consider how to derive the basic inference in (2). The two sentences in the premise are conjoined as

$$[\![might]\!](A)\,;\,[\![would]\!](B),$$

where A is short for

$$(\lambda wc)(\varSigma x : \mathsf{E}_w)(\mathsf{wolf}_w\, x \wedge \mathsf{enter}_w\, x)$$

and B for

$$(\lambda wc)\,\mathsf{growl}_w(@_1^{\gamma_1 \to \mathsf{E}_w} c),$$

both being of type κ. Note that the type of entities, E, is parameterized over worlds. Thus, a one-place predicate, say wolf, has the dependent function type $(\varPi w : \mathsf{W})(\mathsf{E}_w \to \gamma \to \mathsf{prop})$, instead of the function type $\mathsf{E} \to \gamma \to \mathsf{prop}$.

By binding the @-term occurring in $[\![might]\!]$ to the empty informational context, the representation can be reduced as follows:

$$
\begin{aligned}
&[\![might]\!](A)\,;\,[\![would]\!](B)\\
&\equiv\ (\lambda wc)(\varSigma u : (\Diamond(\varepsilon\,;A)wc \wedge (\varSigma P : \widehat{\kappa})({}^{\downarrow}P =_\kappa \varepsilon\,;A)))\\
&\quad (\Box({}^{\downarrow}(@_0(c,u)) \rhd B)w(c,u) \wedge (\varSigma Q : \widehat{\kappa})({}^{\downarrow}Q =_\kappa {}^{\downarrow}(@_0(c,u))\,;B))
\end{aligned}
$$

Here $@_0$ can be bound to $\pi_1\pi_2\pi_2$, resulting in the following (parameterized) proposition:

$$
\begin{aligned}
&(\lambda wc)(\varSigma u : (\Diamond(\varepsilon\,;A)wc \wedge (\varSigma P : \widehat{\kappa})({}^{\downarrow}P =_\kappa \varepsilon\,;A)))\\
&\quad (\Box({}^{\downarrow}(\pi_1\pi_2 u) \rhd B)w(c,u) \wedge (\varSigma Q : \widehat{\kappa})({}^{\downarrow}Q =_\kappa {}^{\downarrow}(\pi_1\pi_2 u)\,;B)).
\end{aligned}
$$

This gives the semantic representation for the premise in (2) after anaphora resolution. Given a world w and an initial context c, suppose that the proposition in the premise is true, i.e., there is a term t such that

$$
\begin{aligned}
t :\ &(\varSigma u : (\Diamond(\varepsilon\,;A)wc \wedge (\varSigma P : \widehat{\kappa})({}^{\downarrow}P =_\kappa \varepsilon\,;A)))\\
&(\Box({}^{\downarrow}(\pi_1\pi_2 u) \rhd B)w(c,u) \wedge (\varSigma Q : \widehat{\kappa})({}^{\downarrow}Q =_\kappa {}^{\downarrow}(\pi_1\pi_2 u)\,;B)).
\end{aligned}
$$

Then we have

$$\pi_2\pi_2\pi_1\, t :\ {}^{\downarrow}(\pi_1\pi_2\pi_1 t) =_\kappa \varepsilon\,;A$$

and

$$\pi_1\pi_2\, t : \Box({}^{\downarrow}(\pi_1\pi_2\pi_1 t) \rhd B)w(c, \pi_1 t).$$

Thus we obtain $\pi_1\pi_2\, t : \Box((\varepsilon\,;A) \rhd B)w(c, \pi_1 t)$. By unfolding $\Box, ;, \rhd, A$, and B, we obtain:

$$
\begin{aligned}
\pi_1\pi_2\, t :\ &(\varPi w' : \mathsf{W})(\mathsf{R}_{\mathsf{epi}}\, ww' \to (\varPi u : (\top \wedge (\varSigma x : \mathsf{E}_{w'})(\mathsf{wolf}_{w'}\, x \wedge \mathsf{enter}_{w'}\, x)))\\
&\mathsf{growl}_{w'}(@_1^{\gamma_1 \to \mathsf{E}_{w'}}((c, \pi_1 t), u))).
\end{aligned}
$$

Here $@_1$ can be bound to $\pi_1\pi_2\pi_2$, thus we have

$$\pi_1\pi_2\,t : (\Pi w' : \mathsf{W})(\mathsf{R}_{\mathsf{epi}}\,ww' \to$$
$$(\Pi u : (\top \wedge (\Sigma x : \mathsf{E}_{w'})(\mathsf{wolf}_{w'}\,x \wedge \mathsf{enter}_{w'}\,x)))\,\mathsf{growl}_{w'}(\pi_1\pi_2 u)).$$

The resulting proposition can be read as *If a wolf entered, it would growl.* In this way we can derive the inference in (2).

An advantage of the present analysis is that no extension is needed to block anaphoric link as shown in (1b). In the discourse in (1b), the first sentence introduces an entity of type $\mathsf{E}_{w'}$, where w' is a world accessible from the current world w. However, the pronoun in the second sentence has the annotated term $@_1^{\gamma_1 \to \mathsf{E}_w}$ that requires an entity of type E_w as an antecedent, and hence, it fails to be bound.

Another advantage is that the present analysis can be applied to modal subordination phenomena involving presupposition. For instance, in the case of (3), the object argument of *stopped* can be analyzed as involving the @-term annotated with the type that specifies the relevant presupposition, say, $\mathsf{used\_to}_w(\mathsf{smoke}_w\,x)$. Nested presuppositions and "quantifying in to presuppositions" (i.e., presuppositions containing a free variable) can also be dealt with in this approach.[5] We leave a detailed analysis of presuppositional inferences for another occasion.

It is not difficult to see that the interaction of MS and bridging inferences as exemplified in (4) can be dealt with by combining the analysis given to the simple case in (12) and the mechanism to handle MS presented in this section. In the case of (4), the representation like (13) is embedded in the scope of modal operator *would*; then the @-term in its restrictor can find an antecedent root proposition introduced in the previous modal sentence. This ensures that the whole discourse implies that the proposition *If John had a new house, the front door would be blue* is true.

The analysis so far has been confined to epistemic modality, but it can be readily extended to other kinds of modal expressions, including attitude verbs, by giving suitable accessibility relations. For instance, using the deontic accessibility relation $\mathsf{R}_{\mathsf{deon}}$, deontic modals can be analyzed along the following lines:

$[\![should]\!](A)$

$= (\lambda wc)(\Pi w' : \mathsf{W})((\mathsf{R}_{\mathsf{deon}}ww' \to (^{\downarrow}(@_i c) \rhd A)w'c) \wedge (\Sigma P : \widehat{\kappa})(^{\downarrow}P =_\kappa {}^{\downarrow}(@_i c)\,;A))$

$[\![may]\!](A)$

$= (\lambda wc)(\Sigma w' : \mathsf{W})((\mathsf{R}_{\mathsf{deon}}ww' \wedge (^{\downarrow}(@_i c)\,;A)w'c) \wedge (\Sigma P : \widehat{\kappa})(^{\downarrow}P =_\kappa {}^{\downarrow}(@_i c)\,;A))$

---

[5] Presuppositional contents can be independent from asserted contents. A classical example is *too*; for example, *John$_i$ is leaving, too$_i$* is said to be presupposing that some (particular) person other than John is leaving. Such cases can be treated within the present framework by incorporating the mechanism developed in Bekki and McCready [3] to handle semantic contents independent of the asserted meaning. The aim of Bekki and McCready [3] is to analyze conventional implicature in the framework of DTS, but their analysis can be applied, with a suitable modification, to the analysis of presuppositions that are independent of asserted contents.

Note that the present analysis does not prevent anaphoric dependencies (in terms of @-terms) from being made between different kinds of modalities. For example, a hypothetical proposition of type $\widehat{\kappa}$ introduced by a deontic modal can be picked up by the @-term in a subsequent sentence with an epistemic modal. Although the issues surrounding what kinds of modality support modal subordination are complicated, modal subordination phenomena can occur between different kinds of modality, as witnessed by the following example (Roberts [14]).

(14)   You should buy *a lottery ticket*. *It* might be worth a million dollars.

Here, an anaphoric dependency is made between deontic and epistemic modalities. The analysis presented above can capture this kind of dependency.

   We agree with Roberts [14] that the infelicity of the example like (15b) is accounted for, not directly by entailment relations induced by attitude verbs, but by pragmatic considerations pertaining to anaphora resolution.

(15)   a. John tries to find *a unicorn* and wishes to eat *it*.

        b. #John wishes to find *a unicorn* and tries to eat *it*.

As is indicated by the treatment of bridging inferences, the proof-theoretic framework presented here is flexible enough to handle the interaction of entailment and anaphora resolution. We leave a detailed analysis of the interaction of attitude verbs and MS for another occasion.

5   Conditionals

The present analysis can be naturally extended to handle examples involving conditionals like (16):

(16)   a. If a farmer owns a donkey, he beats it. # He doesn't like it.

        b. If a farmer owns a donkey, he beats it. It might kick back.

Following the standard assumption in the literature (cf. Kratzer [9]), we assume:

(i) A modal expression is a binary propositional operator having the structure modal (φ, ψ), where φ is a restrictor and ψ is a scope.

(ii) *if-clause* contributes to a restrictor of a modal expression, i.e., If φ, modal ψ is represented as modal (φ, ψ);

(iii) If a modal expression is left implicit as in the first sentence in (16a), it is assumed by default that it has universal modal force: If φ, ψ is represented as $\Box(\varphi, \psi)$.

Binary modal operators then are analyzed as follows.

$$[\![might]\!](A, B)$$
$$= (\lambda wc)(\Diamond((^{\downarrow}(@_i c)\,;A)\,;B)\,wc \wedge (\Sigma P : \widehat{\kappa})(^{\downarrow}P =_\kappa (^{\downarrow}(@_i c)\,;A);B))$$
$$[\![would]\!](A, B)$$
$$= (\lambda wc)(\Box((^{\downarrow}(@_i c)\,;A) \triangleright B)\,wc \wedge (\Sigma P : \widehat{\kappa})(^{\downarrow}P =_\kappa (^{\downarrow}(@_i c)\,;A);B))$$

Both *would* and *might* introduce a (parameterized) propositional object P of type $\widehat{\kappa}$, which inherits the content of the antecedent A as well as the consequent B. This object is identified with $(^{\downarrow}(@_i c)\,;A)\,;B$. Now it is not difficult to derive the following pattern of inference under the current analysis.

(17) $\dfrac{\text{If } \varphi_1, \text{ would } \varphi_2. \text{ might } \varphi_3.}{\text{might } (\varphi_1 \text{ and } \varphi_2 \text{ and } \varphi_3).}$

A compositional analysis of conditionals will be provided in the next section.

According to the present analysis, the antecedent φ_1 in If φ_1, would φ_2 is passed to the first argument of the binary modal operator $[\![would]\!]$. Here it is worth pointing out an alternative analysis that attempts to establish the relationship between an *if*-clause and a modal expression in terms of @-operators. According to the alternative analysis, the semantic role of *if*-clause is to introduce a propositional object in terms of Σ-type:

$$[\![if]\!](A) = (\lambda wc)(\Sigma P : \widehat{\kappa})(^{\downarrow}P =_{\kappa} {}^{\downarrow}(@_i c) \, ; A)$$

Modal expressions are taken as unary operators: the definition is repeated here.

$$[\![might]\!](A) = (\lambda wc)(\Diamond(^{\downarrow}(@_i c) \, ; A) \, wc \wedge (\Sigma P : \widehat{\kappa})(^{\downarrow}P =_{\kappa} {}^{\downarrow}(@_i c) \, ; A))$$

$$[\![would]\!](A) = (\lambda wc)(\Box(^{\downarrow}(@_i c) \triangleright A) \, wc \wedge (\Sigma P : \widehat{\kappa})(^{\downarrow}P =_{\kappa} {}^{\downarrow}(@_i c) \, ; A))$$

Then the *if*-clause and the main clause are combined by dynamic conjunction:

$$[\![if \, A, \, would \, B]\!] = [\![if \, A]\!] \, ; [\![would \, B]\!]$$

The @-term in $[\![would \, B]\!]$ can be bound to the root proposition introduced in $[\![if \, A]\!]$, hence we can obtain the same result as the first approach. An advantage of this alternative approach is that it simplifies the semantics of modal expressions *might* and *would* by taking them as unary operators and reducing the role of restrictor arguments to @-operators. However, one drawback is that it allows the @-operator associated with a modal expression to be bound by a proposition other than the one introduced by the *if*-clause: the @-operator can in principle be bound by any proposition of type $\widehat{\kappa}$ appearing in a suitable antecedent context. According to the first approach, in contrast, the binary *would* has the representation $((^{\downarrow}(@_i c) \, ; A) \triangleright B)$, where $@_i c$ is responsible for capturing the information given in a context and A for the information given in the *if*-clause. In this way, we can distinguish two aspects of the meaning of a conditional, i.e., grammatically determined meaning and contextually inferred meaning. For this reason, we adopt the first approach in this paper.

6   Compositional Analysis

In this section, we give a compositional analysis of constructions involving modal anaphora and subordination we discussed so far. To be concrete, we will adopt Combinatory Categorial Grammar (CCG) as our syntactic framework (see Steedman [16] for an overview). Generally speaking, categorial grammar can be seen as a framework based on the idea of *direct compositionality*, i.e., the idea of providing a compositional derivation of semantic representations based on surface structures of sentences. To provide a compositional analysis of modal constructions in such a setting is not a trivial task, since modal auxiliaries tend to take a scope that is unexpected from their surface position.

Consider again the initial example in (1a), repeated here as (18).

(18)   a. *A wolf* might enter.

        b. *It* would growl.

We are concerned with the reading of (18a) in which *a wolf* is interpreted as *de dicto*, i.e., as taking narrow scope with respect to the modal *might*. The issue of how to analyze the *de re* reading in which the subject NP takes scope over the modal seems to be orthogonal to the issue of how to handle modal subordination phenomena, so we leave it for another occasion.

A lexicon for the compositional analysis of (18) and related constructions is given in Table 1. Here we will write VP for $S\backslash NP$. In CCG, function categories of the form X/Y expect their argument Y to its right, while those of the form $X\backslash Y$ expect Y to their left. The forward slash / and the backward slash $\backslash$ are left-associative: for example, $S/VP/N$ means $(S/VP)/N$.

The lexical entries provided here yield the following derivation tree for (18a).

$$
\begin{array}{c}
\dfrac{
\begin{array}{cc}
a_{nom} & wolf \\
S/VP/N & N
\end{array}
}{
\dfrac{S/VP}{\qquad}
} >
\quad
\dfrac{
\begin{array}{cc}
might^1 & enter \\
S\backslash(S/VP)/VP & VP
\end{array}
}{
S\backslash(S/VP)
} >
\\[2ex]
\dfrac{\qquad\qquad\qquad\qquad S \qquad\qquad\qquad\qquad}{} <
\end{array}
$$

Given this derivation tree, the semantic representation for (18a) is derived in the following way.

Table 1. Dynamic lexicon of DTS for basic modal semantics

Expression	Syntactic category	Semantic representation
wolf	N	$(\lambda wxc)(\mathsf{wolf}_w x)$
enter	VP	$(\lambda wxc)(\mathsf{enter}_w x)$
growl	VP	$(\lambda wxc)(\mathsf{growl}_w x)$
beat	VP/NP	$(\lambda wyxc)(\mathsf{beat}_w(x,y))$
John	S/VP	$(\lambda vwc)(vw\,\mathsf{john}\,c)$
a_{nom}	$S/VP/N$	$(\lambda nvwc)(\Sigma u:(\Sigma x:\mathsf{E}_w)(nwxc))(vw(\pi_1 u)(c,u))$
a_{acc}	$VP\backslash(VP/NP)/N$	$(\lambda nvwxc)(\Sigma u:(\Sigma y:\mathsf{E}_w)(nwyc))(vw(\pi_1 u)x(c,u))$
it^i_{nom}	S/VP	$(\lambda vwc)(vw(@_i^{\gamma\to\mathsf{E}_w}c)c)$
it^i_{acc}	$VP\backslash(VP/NP)$	$(\lambda vwxc)(vw(@_i^{\gamma\to\mathsf{E}_w}c)xc)$
the^i_{nom}	$S/VP/N$	$(\lambda nvwc)(vw(\pi_1(@_i^{\gamma\to(\Sigma x:\mathsf{E}_w)nwxc}c))c)$
the^i_{acc}	$VP\backslash(VP/NP)/N$	$(\lambda nvwxc)(vw(\pi_1(@_i^{\gamma\to(\Sigma y:\mathsf{E}_w)nwyc}c))xc)$
$might^i$	$S\backslash(S/VP)/VP$	$(\lambda vqwc)((\Sigma w':\mathsf{W})(\mathsf{R}_{\mathsf{epi}}\,ww'\wedge(\,^\downarrow(@_i c)\,;\,qv)w'c)$ $\wedge(\Sigma P:\widehat{\kappa})(\,^\downarrow P =_\kappa\,^\downarrow(@_i c)\,;\,qv))$
$would^i$	$S\backslash(S/VP)/VP$	$(\lambda vqwc)((\Pi w':\mathsf{W})(\mathsf{R}_{\mathsf{epi}}\,ww'\to(\,^\downarrow(@_i c)\rhd qv)w'c)$ $\wedge(\Sigma P:\widehat{\kappa})(\,^\downarrow P =_\kappa\,^\downarrow(@_i c)\,;\,qv))$

$\llbracket a_{nom} \rrbracket (\llbracket wolf \rrbracket)$

$\equiv_\beta (\lambda vwc)(\Sigma u : (\Sigma x : \mathsf{E}_w)(\mathsf{wolf}_w x))(vw(\pi_1 u)(c, u))$

$\llbracket might^1 \rrbracket (\llbracket enter \rrbracket)$

$\equiv_\beta (\lambda qwc)((\Sigma w' : \mathsf{W})(\mathsf{R}_{\mathsf{epi}} ww' \wedge (^\downarrow(@_1 c) \, ; q(\lambda wxc)(\mathsf{enter}_w x))w'c)$

$\qquad \wedge (\Sigma P : \widehat{\kappa})(^\downarrow P =_\kappa {}^\downarrow(@_1 c) \, ; q(\lambda wxc)(\mathsf{enter}_w x)))$

Let $@_1$ be bound to the empty context, i.e., $^\downarrow(@_1 c) = \varepsilon$. For simplicity, henceforth we will omit ε and $\top$ throughout this section.

$\llbracket might^1 \rrbracket (\llbracket enter \rrbracket)(\llbracket a_{nom} \rrbracket (\llbracket wolf \rrbracket))$

$\equiv_\beta (\lambda wc)((\Sigma w' : \mathsf{W})(\mathsf{R}_{\mathsf{epi}} ww' \wedge (\Sigma u : (\Sigma x : \mathsf{E}_{w'})(\mathsf{wolf}_{w'} x))(\mathsf{enter}_{w'}(\pi_1 u)))$

$\qquad \wedge (\Sigma P : \widehat{\kappa})(^\downarrow P =_\kappa (\lambda wc)(\Sigma u : (\Sigma x : \mathsf{E}_w)(\mathsf{wolf}_w x))(\mathsf{enter}_w (\pi_1 u))))$

The derivation tree of (18b) is given as follows.

$$
\cfrac{it_{nom}^3 \quad \cfrac{\cfrac{would^2}{S\backslash(S/VP)/VP} \quad \cfrac{growl}{VP}}{S\backslash(S/VP)} >}{\cfrac{S/VP}{S}} <
$$

The semantic representation of (18b) is derived in a similar way. Note that the pronoun *it* here is interpreted, in a sense, as *de dicto*, taking scope under the modal *would*.

$\llbracket would^2 \rrbracket (\llbracket growl \rrbracket)(\llbracket it_{nom}^3 \rrbracket)$

$\equiv_\beta (\lambda wc)((\Pi w' : \mathsf{W})(\mathsf{R}_{\mathsf{epi}} ww' \rightarrow (^\downarrow(@_2 c) \, ; (\lambda wc)(\mathsf{growl}_w(@_3^{\gamma \rightarrow \mathsf{E}_w} c)))w'c)$

$\qquad \wedge (\Sigma P : \widehat{\kappa})(^\downarrow P =_\kappa {}^\downarrow(@_2 c) \, ; (\lambda wc)(\mathsf{growl}_w(@_3^{\gamma \rightarrow \mathsf{E}_w} c)))))$

As is easily checked, combining the semantic representations for (18a) and (18b) by dynamic conjunction yields the same semantic representation as the one for (1a) presented in Sect. 4.

According to the lexicon given in Table 1, the object NP in a modal sentence is interpreted as taking scope under the modal. This enables us to handle the anaphoric dependency in the following discourse.

(19)   a. *A wolf might enter.*

   b. *John would beat it.*

For the modal subordination reading to be derivable, the pronoun *it* in object position of (19b) has to be interpreted as taking scope under *would*. Our lexical entries yield the following derivation tree for (19b).

$$
\cfrac{John \quad \cfrac{\cfrac{would^1}{S\backslash(S/VP)/VP} \quad \cfrac{\cfrac{beat}{VP/NP} \quad \cfrac{it_{acc}^2}{VP\backslash(VP/NP)}}{VP} <}{S\backslash(S/VP)} >}{\cfrac{S/VP}{S}} <
$$

The semantic representation is derived as follows:

$$[\![would^1]\!]([\![it^2_{acc}]\!]([\![beat]\!]))([\![John]\!])$$

$$\equiv_\beta (\lambda wc)((\Pi w' : \mathsf{W})(\mathsf{R}_{\mathsf{epi}}\, ww' \to (^\downarrow(@_1c)\,;(\lambda wc)(\mathsf{beat}_w(\mathsf{john}, @_2^{\gamma \to \mathsf{E}_w}c)))w'c)$$

$$\wedge (\Sigma P : \widehat{\kappa})(^\downarrow P =_\kappa {}^\downarrow(@_1c)\,;(\lambda wc)(\mathsf{beat}_w(\mathsf{john}, @_2^{\gamma \to \mathsf{E}_w}c))))).$$

In the same way as the derivation for (18) shown above, this yields the desired reading of the discourse in (19).

For the analysis of the interaction of modals and conditionals, we introduce lexical entries for *if* and binary modal operators *would* and *might*, which are shown in Table 2. As an illustration, consider the following example.

(20)   If a farmer owns a donkey, he would beat it.

The derivation tree of (20) is given as follows:

$$
\begin{array}{c}
\dfrac{\dfrac{\dfrac{If}{S/S/S} \quad \dfrac{a\ farmer\ owns\ a\ donkey}{S}}{S/S} > \quad \dfrac{he^1_{nom}}{S/VP} \quad \dfrac{\dfrac{would^2}{S\backslash(S/S)\backslash(S/VP)/VP} \quad \dfrac{beat\ it^3_{acc}}{VP}}{\dfrac{S\backslash(S/S)\backslash(S/VP)}{S\backslash(S/S)} <}{}>}{S}
\end{array}
$$

The semantic representation for *a farmer owns a donkey* is computed as follows:

$$[\![a_{nom}]\!]([\![farmer]\!])([\![owns]\!]([\![a_{acc}]\!]([\![donkey]\!])))$$

$$\equiv_\beta (\lambda wc)(\Sigma v : (\Sigma x : \mathsf{E}_w)(\mathsf{farmer}_w x))(\Sigma u : (\Sigma y : \mathsf{E}_w)(\mathsf{donkey}_w y))\,\mathsf{own}_w(\pi_1 v, \pi_1 u).$$

Let us abbreviate this representation as A. Then the derivation tree above generates the following semantic representation:

$$[\![would^2]\!]([\![it^3_{acc}]\!]([\![beat]\!]))([\![he^1_{nom}]\!])([\![if]\!](A))$$

$$\equiv_\beta (\lambda wc)((\Pi w' : \mathsf{W})(\mathsf{R}_{\mathsf{epi}}\, ww'$$

$$\to ((^\downarrow(@_2c)\,;A) \rhd (\lambda wc)(\mathsf{beat}_w(@_1^{\gamma \to \mathsf{E}_w}c, @_3^{\gamma \to \mathsf{E}_w}c)))w'c)$$

$$\wedge (\Sigma P : \widehat{\kappa})(^\downarrow P =_\kappa (^\downarrow(@_2c)\,;A)\,;(\lambda wc)(\mathsf{beat}_w(@_1^{\gamma \to \mathsf{E}_w}c, @_3^{\gamma \to \mathsf{E}_w}c))))$$

Table 2. Dynamic lexicon of DTS for conditionals

Expression	Syntactic category	Semantic representation
if	$S/S/S$	$(\lambda spwc)pswc$
might^i	$S\backslash(S/S)\backslash(S/VP)/VP$	$(\lambda vqpwc)(p((\lambda swc)((\Sigma w' : \mathsf{W})(\mathsf{R}_{\mathsf{epi}}\, ww' \wedge$
		$((^\downarrow(@_i c)\,;s)\,;qv)w'c)$
		$\wedge (\Sigma P : \widehat{\kappa})(^\downarrow P =_\kappa\ (^\downarrow(@_i c)\,;s)\,;qv))))$
would^i	$S\backslash(S/S)\backslash(S/VP)/VP$	$(\lambda vqpwc)(p((\lambda swc)((\Pi w' : \mathsf{W})(\mathsf{R}_{\mathsf{epi}}\, ww' \to$
		$((^\downarrow(@_i c)\,;s) \rhd qv)w'c)$
		$\wedge (\Sigma P : \widehat{\kappa})(^\downarrow P =_\kappa\ (^\downarrow(@_i c)\,;s)\,;qv))))$

The resulting semantic representation corresponds to the one presented in Sect. 5. Here, the dynamic proposition corresponding to the root sentence *he beats it* appears in the nuclear scope of the binary modal operator *would*. The dynamic proposition expressed by *a farmer owns a donkey* in the *if*-clause fills in the restrictor of *would*. It can be easily seen that this representation enables the pronouns *he* and *it* to establish the intended anaphoric relation to their antecedents.

The unary modal operators *might* and *would* shown in Table 1 can be regarded as a special case of the binary modal operators introduced here. We can assume that when modal expressions *might* and *would* appear without *if*-clauses, the restrictor position s in the semantic representation of a binary modal operator is filled by the empty context ε (which needs to be syntactically realized by a silent element). Although the derivations for examples like (18) and (19) will become more complicated, we can get the desirable semantic representations for all the constructions we examined so far.

7 Conclusion

In this paper, we extended the framework of DTS with a mechanism to handle modality and its interaction with anaphora. In doing so, we integrated the findings of possible world semantics with a proof-theoretic formal semantics based on dependent type theory. This enabled us to give the semantic representations of modals and conditionals using the expressive type structures provided by dependent type theory, and thereby to broaden the empirical coverage of DTS.

There are other important constructions that are relevant to MS but are not discussed in this paper, including negation (Frank and Kamp [7]; Geurts [8]), the so-called *Veltman's asymmetry* (Veltman [19]; Asher and McCready [1]), and generics (Carlson and Spejewski [4]). These issues are left for future work.

Acknowledgments. This paper is a revised and expanded version of [18]. We thank the two reviewers of LENLS11 for helpful comments and suggestions on an earlier version of this paper. I also thank the audiences at LENLS11, in particular, Chris Barker and Matthew Stone, for helpful comments and discussion. Special thanks to Nicholas Asher, who gave constructive comments and advice, and to Antoine Venant, Fabio Del Prete, and Márta Abrusán for their feedback and discussions. This research was supported by JST, CREST.

References

1. Asher, N., McCready, E.: Were, would, might and a compositional account of counterfactuals. J. Semant. **24**(2), 93–129 (2007)
2. Bekki, D.: Representing anaphora with dependent types. In: Asher, N., Soloviev, S. (eds.) LACL 2014. LNCS, vol. 8535, pp. 14–29. Springer, Heidelberg (2014)
3. Bekki, D., McCready, E.: CI via DTS. In: Proceedings of the 11th International Workshop on Logic and Engineering of Natural Language Semantics (LENLS11), Kanagawa, Japan, pp. 110–123 (2014)
4. Carlson, G.N., Spejewski, B.: Generic passages. Nat. Lang. Semant. **5**(2), 101–165 (1997)

5. Chatzikyriakidis, S., Luo, Z.: Natural Language Reasoning Using Proof-assistant Technology : Rich Typing and Beyond. In: Proceedings of the EACL 2014 Workshop on Type Theory and Natural Language Semantics (TTNLS), Gothenburg, Sweden, pp. 37–45 (2014)
6. Clark, H.H.: Bridging. In: Schank, R.C., Nash-Webber, B.L. (eds.) Theoretical Issues In Natural Language Processing, pp. 169–174. Association for Computing Machinery, New York (1975)
7. Frank, A., Kamp, H.: On Context Dependence in Modal Constructions. In: Proceedings of SALT (1997)
8. Geurts, B.: Presuppositions and Pronouns. Elsevier, Oxford (1999)
9. Kratzer, A.: Modals and Conditionals: New and Revised Perspectives. Oxford University Press, Oxford (2012)
10. Martin-Löf, P.: Intuitionistic Type Theory. Bibliopolis, Naples (1984)
11. Muskens, R.: An analytic tableau system for natural logic. In: Aloni, M., Bastiaanse, H., de Jager, T., Schulz, K. (eds.) Logic, Language and Meaning. LNCS, vol. 6042, pp. 104–113. Springer, Heidelberg (2010)
12. Ranta, A.: Type-theoretical Grammar. Oxford University Press, Oxford (1994)
13. Roberts, C.: Modal subordination and pronominal anaphora in discourse. Linguist. Philos. **12**, 683–721 (1989)
14. Roberts, C.: Anaphora in intensional contexts. In: Lappin, S. (ed.) The Handbook of Contemporary Semantic Theory, pp. 215–246. Blackwell, Oxford (1996)
15. van Rooij, R.: A modal analysis of presupposition and modal subordination. J. Semant. **22**(3), 281–305 (2005)
16. Steedman, M.: The Syntactic Process. MIT Press/Bradford Books, Cambridge (2000)
17. Sundholm, G.: Proof Theory and Meaning. In: Gabbay, D., Guenthner, F. (eds.) Handbook of Philosophical Logic. Synthese Library, vol. 166, pp. 471–506. Springer, Netherlands (1986)
18. Tanaka, R., Mineshima, K., Bekki, D.: Resolving modal anaphora in Dependent Type Semantics. In: Proceedings of the 11th International Workshop on Logic and Engineering of Natural Language Semantics (LENLS11), Kanagawa, Japan, pp. 43–56 (2014)
19. Veltman, F.: Defaults in update semantics. J. Philos. Logic **25**, 221–261 (1996)

Canonical Constituents and Non-canonical Coordination

Simple Categorial Grammar Account

Oleg Kiselyov[(✉)]

Tohoku University, Sendai, Japan
oleg@okmij.org

Abstract. A variation of the standard non-associative Lambek calculus with the slightly non-standard yet very traditional semantic interpretation turns out to straightforwardly and uniformly express the instances of non-canonical coordination while maintaining phrase structure constituents. Non-canonical coordination looks just as canonical on our analyses. Gapping, typically problematic in Categorial Grammar–based approaches, is analyzed like the ordinary object coordination. Furthermore, the calculus uniformly treats quantification in any position, quantification ambiguity and islands. It lets us give what seems to be the simplest account for both narrow- and wide-scope quantification into coordinated phrases and of narrow- and wide-scope modal auxiliaries in gapping.

The calculus lets us express standard covert movements and anaphoric-like references (analogues of overt movements) in types – as well as describe how the context can block these movements.

1 Introduction

Non-canonical coordination and in particular gapping (2) challenge the semantic theory [2,3].

(1) John gave a book to Mary and a record to Sue.
(2) I gave Leslie a book and she a CD.
(3) John gave a present to Robin on Thursday and to Leslie on Friday.
(4) Mrs. J can't live in Boston and Mr. J in LA.

Further challenges are accounting for both narrow- and wide-scope reading of "a present" in (3) and for the two readings in (4), with the wide-scope "can't" and the wide-scope coordination.

Combinatory Categorial Grammar (CCG) answers the challenge of non-constituent coordination [6], but at the price of giving up on phrase structure constituents. The CCG analysis of gapping requires further conceptually problematic (and eventually, not fully adequate) postulates. Kubota and Levine [3] present a treasure trove of empirical material illustrating the complexities of coordination (from which we drew our examples). They develop a variant of type-logical categorial grammar with both directed and undirected implications

© Springer-Verlag Berlin Heidelberg 2015
T. Murata et al. (Eds.): JSAI-isAI 2014 Workshops, LNAI 9067, pp. 99–113, 2015.
DOI: 10.1007/978-3-662-48119-6_8

and higher-order phonology. Here we demonstrate that the challenges of coordination and gapping can be met within the standard non-associative Lambek calculus **NL**, one of the basic categorial grammars. Our calculus represents arbitrary discontinuous constituents, including multiple discontinuities and the discontinuous displaced material. Hence our calculus seems both simpler and more expressive than other type-logical grammar approaches [5].

The main idea is a slightly non-standard semantic interpretation of **NL** derivations (the phonology remains standard). It opened the way to represent what looks like typical overt and covert movements. Being strictly within **NL**, we add no modes or structural rules to the calculus per se. No lexical items are ever moved. The antecedent of a sequent, describing the *relevant* phrase structure, can still be manipulated by the standard rules, given constants of particular types. These constants have the empty phonology but the clear semantic (computational) interpretation. One may say that the structural rules are all lexicalized and computational. Although our approach may be reminiscent of QR and dynamic semantics, we stay squarely within logic: the semantic interpretation is compositionally computed by combining closed formulas using the standard operations of higher-order logic. Unbound traces, free variables, any movement or rearrangement of lexical items are simply not possible in our approach.

After presenting the calculus in Sect. 2 we illustrate its various features in Sect. 3.1 on very simple examples of coordination. That section describes two techniques that underlie the analyses of more complex non-constituent coordination and gapping in Sect. 3.3. The same techniques also account for quantification: in Sect. 4 we analyze not only the simple QNP in subject and object positions but also the quantification ambiguity and scoping islands. Finally, Sect. 5 treats the seemingly anomalous wide scope of QNP and modals in sentences with non-canonical coordination, which proved most difficult to account for in the past. We answer the challenges posed at the beginning of this section. All presented examples have been mechanically verified: see the accompanying code http://okmij.org/ftp/gengo/HOCCG.hs for the verification record and more examples.

2   Non-associative Lambek Calculus and Theory

After reminding of the non-associative Lambek calculus **NL**, we derive convenient rules and introduce types and constants to be used throughout the paper. The semantic interpretation is in Sect. 3.2.

Figure 1 is our presentation of **NL**. Types A, B, C are built with the binary connectives / and \; antecedent structures Γ, Δ are built with $(-, -)$ and the empty structure •. The antecedent structure is an ordered tree and hence $(A, •)$ is different from just A. Labeled sequents have the form $\Gamma \vdash t : A$, to be read as the term t having the type A assuming Γ. The figure presents the standard introduction and elimination rules for the two binary connectives (slashes). To ease the notational burden, we write $t_1 t_2$ for both left and right applications.

As basic (atomic) types we choose NP and S plus a few others used for quantification and to mark context boundaries. They will be introduced as needed.

$$\frac{\Gamma \vdash t_1 : B/A \quad \Delta \vdash t_2 : A}{(\Gamma, \Delta) \vdash t_1 t_2 : B}\ /E \qquad \frac{\Gamma \vdash t_1 : A \quad \Delta \vdash t_2 : A\backslash B}{(\Gamma, \Delta) \vdash t_1 t_2 : B}\ \backslash E$$

$$\frac{(\Gamma, A) \vdash t : B}{\Gamma \vdash \lambda t : B/A}\ /I \qquad \frac{(A, \Gamma) \vdash t : B}{\Gamma \vdash \lambda t : A\backslash B}\ \backslash I$$

$$\frac{}{A \vdash x : A}\ Var$$

Fig. 1. The non-associative Lambek calculus **NL**

$$\frac{\Gamma \vdash t_1 : A \quad (A, \Delta) \vdash t_2 : B}{(\Gamma, \Delta) \vdash t_1 \cdot t_2 : B}\ HypL \qquad \frac{(\Delta, A) \vdash t_2 : B \quad \Gamma \vdash t_1 : A}{(\Delta, \Gamma) \vdash t_1 \cdot t_2 : B}\ HypR$$

$$\frac{\Delta\backslash A \vdash t_1 : \Gamma\backslash A \quad C[\Delta] \vdash t_2 : A}{C[\Gamma] \vdash t_1 \uparrow t_2 : A}\ Hyp$$

Fig. 2. Convenient derived rules (In the rule Hyp, Γ must be a full structure type)

We also use a parallel set of types, which we write as $\overline{NP}$, $\overline{NP\backslash S}$, etc. They are not special in any way: one may regard $\overline{A}$ as an abbreviation for $A\backslash\bot$ where $\bot$ is a dedicated atomic type.

The introduction rules $\backslash I$ and $/I$ almost always[1] appear in combination with the corresponding elimination rules. To emphasize the pattern and to save space in derivations we introduce in Fig. 2 admissible cut-like rules: HypL is $\backslash I$ immediately followed by $\backslash E$; HypR is similar.

The calculus per se has no structural rules. Still, the existing rules may manipulate the antecedent structure Γ using constants of appropriate types. For example, consider the sequent $(\bullet, (\bullet, A)) \vdash t : B$. Applying $\backslash I$ twice we derive $A \vdash \lambda\lambda t : \bullet\backslash(\bullet\backslash B)$. The derivation shows that we are not really distinguishing types and structures: structures are types. We will call types that are structures or include structures (that is, contain $\bullet$ and commas) full structure types. Assuming the constant ooL with the type $\bullet \vdash ooL : B/(\bullet\backslash(\bullet\backslash B))$ gives us $(\bullet, A) \vdash ooL(\lambda\lambda t) : B$. In effect, ooL transformed $(\bullet, (\bullet, A)) \vdash B$ into $(\bullet, A) \vdash t : B$. Since this and other similar structural transformations will appear very frequently, we will abbreviate the derivations through the essentially admissible rule Hyp[2] in Fig. 2.

The Hyp rule can replace a structure type Δ within the arbitrary context $C[]$ of another structure. The replacement must be a full structure type. Although

---

[1] A notable exception is quantification, see Sect. 4.

[2] We call this rule essentially admissible because it cannot transform $(\bullet, \bullet) \vdash A$ to $\bullet \vdash A$. Therefore, we will have many derivations and sequents that differ only in $(\bullet, \bullet)$ vs. $\bullet$. Since they are morally the same, it saves a lot of tedium to treat them as identical, assuming that $(\bullet, \bullet)$ can always be replaced by $\bullet$. We will use this assumption throughout.

our calculus has no structural rules whatsoever, Hyp lets us rearrange, replace, etc. parts of the antecedent structure, *provided* there is a term t of the suitable type for that operation. One may say that our structural rules are lexicalized and have a computational interpretation. For now we introduce the following schematic constants.

$$(\bullet, \bullet)\backslash A \vdash oo : \bullet\backslash A$$
$$(\bullet, (\bullet, \Gamma))\backslash A \vdash oL : (\bullet, \Gamma)\backslash A$$
$$((\Gamma, \bullet), \bullet)\backslash A \vdash oR : (\Gamma, \bullet)\backslash A$$
$$(C, \bullet)\backslash S \vdash resetCtx : \bullet\backslash S$$

The types of these terms spell out the structural transformation. We will later require all such terms, whose type is a full structure type, be phonetically silent. In $resetCtx$, C is any context marker (such as those that mark the coordinated or subordinated clause). The corresponding structural rule drops the context marker when the context becomes degenerate.

Section 3.1 illustrates the calculus on many simple examples. Throughout the paper, we write VP for $NP\backslash S$, VT for VP/NP and PP for $VP\backslash VP$.

3   Coordination

This section describes analyses of coordination in our calculus, from canonical to non-canonical. We show two approaches. The first is less general and does not always apply, but it is more familiar and simpler to explain. It builds the intuition for the second, encompassing and general approach. The approaches are best explained on very simple examples below. Section 3.3 applies them to non-canonical coordination and gapping; Sect. 5 to scoping phenomena in coordination.

3.1   Two Approaches to Coordination

We start with the truly trivial example "John tripped and fell." VP coordination is the simplest and the most natural analysis, assuming the constant $and : (VP\backslash VP)/VP$. Here is another analysis, with the coordination at type S rather than VP:

$$
\cfrac{
\cfrac{NP \vdash x : NP \qquad \bullet \vdash \text{tripped} : VP}{(NP, \bullet) \vdash (x \text{ tripped}) : S}\ \backslash E
\qquad
\cfrac{And \vdash \text{and} : (S\backslash S)/S}{}
\qquad
\cfrac{NP \vdash y : NP \qquad \bullet \vdash \text{fell} : VP}{(NP, \bullet) \vdash (y \text{ fell}) : S}\ \backslash E
}{((NP, \bullet), (And, (NP, \bullet))) \vdash (x \text{ tripped}) \text{ and } (y \text{ fell}) : S}
$$

Here, *And* is an atomic type, used to mark the coordination in the antecedent structure[3]. Clearly the derivation can be reconstructed from its conclusion. To save space, we will only be writing conclusions. To proceed further, we assume the "structural constant" (schematic)

$$(\Gamma, (And, \Gamma))\backslash A \vdash and_C : \Gamma \backslash A$$

It has the full structure type and is hence phonetically silent. (All terms in the same font as *and* are silent.) The Hyp rule then gives

$$(NP, \bullet) \vdash and_C \uparrow (x \text{ tripped}) \text{ and } (y \text{ fell}) : S$$

We now can apply the HypL rule with $\bullet \vdash$ John : NP obtaining

$$(\bullet, \bullet) \vdash \text{John} \cdot and_C \uparrow (x \text{ tripped}) \text{ and } (y \text{ fell}) : S$$

Finally Hyp is used once again, with the *oo* constant to contract $(\bullet, \bullet)$ to just $\bullet$ producing the final conclusion:

$$\bullet \vdash oo \uparrow \text{John} \cdot and_C \uparrow (x \text{ tripped}) \text{ and } (y \text{ fell}) : S$$

The phonology is standard; hypotheses, *and* and other constants in the same font are silent. The semantic interpretation is described in Sect. 3.2.

There is yet another analysis of the same example. Although patently overkill in this case, it explains the most general technique to be used for non-canonical coordination. The intuition comes from the apparent similarity of our analysis of quantification in Sect. 4 to the quantifier raising (QR). The key idea is to paraphrase "John tripped and fell" as "John tripped; he fell" with the silent 'pronoun'. The paraphrase is analyzed in the manner reminiscent of dynamic logic. We assume the axiom schema

$$\overline{A} \vdash ref : A/A$$

where $\overline{A}$ is intended to signify that *ref provides* the value of the type A. We derive

$$((\overline{NP}, \bullet), \bullet) \vdash (ref \text{John}) \text{ tripped} : S$$

and, as before, $(NP, \bullet) \vdash x$ fell : S. Coordinating the two gives:

$$(((\overline{NP}, \bullet), \bullet), (And, (NP, \bullet))) \vdash (ref \text{John}) \text{ tripped and } (x \text{ fell}) : S$$

Matching the reference $x : NP$ with its referent $\overline{NP}$ is done by the following two structural constants.

$$(((\overline{A}, B), \Gamma), (And, (A, \Delta)))\backslash S \vdash and_L : ((B, \Gamma), (And, (\bullet, \Delta)))\backslash S$$
$$((\Gamma, A), (And, (\Delta, (\overline{A}, B))))\backslash S \vdash and_R : ((\Gamma, \bullet), (And, (\Delta, B)))\backslash S$$

---

[3] Incidentally, such a mark restricts the use of the structure constants such as and_L and especially and_D below – in effect restricting gapping to coordination.

The constant and_L matches up the reference A and the referent $(\overline{A}, B)$, provided both occur at the left edge, and the referent is in the left conjunct; A is replaced with $\bullet$ and $(\overline{A}, B)$ with B. The motivation for this choice comes from the HypL rule, which and_L is meant generalize. Suppose we have the derivations

$$\bullet \vdash t_1 : A \qquad \Gamma_2 \vdash t_2 : A \backslash S \qquad \Gamma_3 \vdash t_3 : A \backslash S$$

Hypothesizing $A \vdash x : A$ and the similar y gives

$$((A, \Gamma_2), (And, (A, \Gamma_3))) \vdash (x\ t_2) \text{ and } (y\ t_3)\ : S$$

If Γ_2 and Γ_3 happen to be the same Γ, we can apply Hyp with and_C:

$$(A, \Gamma) \vdash and_C \uparrow ((x\ t_2) \text{ and } (y\ t_3))\ : S$$

which, followed-up with the HypL rule, finally yields

$$(\bullet, \Gamma) \vdash t_1 \cdot and_C \uparrow ((x\ t_2) \text{ and } (y\ t_3))\ : S$$

The structural constant and_L is designed to complete the above derivation in a different way: from

$$(((\neg A, \bullet), \Gamma), (And, (A, \Gamma))) \vdash ((ref\ t_1)\ t_2) \text{ and } (y\ t_3)\ : S$$

obtaining first

$$((\bullet, \Gamma), (And, (\bullet, \Gamma))) \vdash and_L \uparrow ((ref\ t_1)\ t_2) \text{ and } (y\ t_3)\ : S$$

and finally

$$(\bullet, \Gamma) \vdash and_C \uparrow and_L \uparrow ((ref\ t_1)\ t_2) \text{ and } (y\ t_3)\ : S$$

The derivation no longer uses HypL, which is subsumed into and_L. This structural constant, unlike and_C, can be applied repeatedly and in the circumstances when Γ_2 is not the same as Γ_3.

Coming back to our example, its conclusion is now clear:

$$\bullet \vdash and_C \uparrow and_L \uparrow (ref \text{John}) \text{ tripped and } (x \text{ fell})\ : S$$

(we shall elide the final step of reducing $(\bullet, \bullet)$ to $\bullet$ from now on). Again, the read-out is standard, keeping in mind that all italicized items are silent. The semantic interpretation is described in Sect. 3.2.

The subject coordination, as in "John and Mary left." is similar. Besides the straightforward NP coordination (if we assume and : $(NP \backslash NP)/NP$), we can coordinate at type S. Our first approach leads to

$$(\bullet, VP) \vdash and_C \uparrow (\text{John } x) \text{ and } (\text{Mary } y) : S$$

followed by the application of the HypR rule with left : VP on the right. The more general approach derives $(\bullet, VP) \vdash (\text{John } x) : S$ with the VP hole for the left conjunct and $(\bullet, (\overline{VP}, \bullet)) \vdash$ Mary (*ref* left) : S for the right conjunct. The right-edge hole is matched up with the right-edge referent by and_R, which requires the referent to be in the right conjunct. Again, the motivation is to generalize the HypR rule, which comes from the $/I$ rule of the Lambek calculus.

Object coordination "John saw Bill and Mary." lets us introduce the final structural constant of our approach, which matches a hole with a referent in the medial position rather than at the edge. Our first approach invariably leads to

$$(NP, (TV, \bullet)) \vdash and_C \uparrow (x \ (y \ \text{Bill})) \text{ and } (u \ (v \ \text{Mary}))$$

which is the dead-end: since TV is not at the edge of the antecedent structure, neither HypL nor HypR can eliminate it. The first approach, although simple, clearly has limitations – which should be obvious considering it is just the standard Lambek calculus. The latter too has trouble with eliminating hypotheses far from the edges.

The second approach produces

$$((\bullet, ((\overline{TV}, \bullet), \bullet)), (And, (\bullet, (TV, \bullet)))) \vdash$$
$$and_L \uparrow (ref \ \text{John}) ((ref \ \text{see}) \ \text{Bill}) \text{ and } (x \ (y \ \text{Mary})) : S$$

Since the referent is now deep in the structure, neither and_L nor and_R can get to it. We have to use the more general constant

$$((\Gamma_1, ((\overline{TV}, \bullet), \Gamma_2)), (And, (\Delta_1, (TV, \Delta_2)))) \backslash S \vdash$$
$$and_D : ((\Gamma_1, (VB, \Gamma_2)), (And, (\Delta_1, (VB, \Delta_2)))) \backslash S$$

(it is actually a family for different shapes of contexts). Although this constant seems to give rise to an unrestricted structural rule, it is limited by the type of the hole. Since the hole is for a term deep inside a formula, that term must denote a relation, that is, have at least two arguments. The hole is thus restricted to be of the type TV, VP/PP and similar. Kubota and Levine discuss in detail a similar restriction for their seemingly freely dischargeable hypotheses in [4, Sect. 3.2]. The context marker VB tells that the verb has been gapped.

One may be concerned that plugging the hole is too loose a feature, letting us pick any word in the left conjunct and refer to it from the right conjunct, or vice versa. We could then derive "*John tripped and." (with the hole in the right conjunct referring to "tripped"). Such a derivation is not possible however. We can get as far as

$$(((\overline{NP}, \bullet), (\overline{VP}, \bullet)), (And, (NP, VP))) \vdash$$
$$(ref \ \text{John}) \ (ref \ \text{tripped}) \text{ and } (x \ y) : S$$

Although we can plug the NP hole at the left edge with and_L, after that we are stuck. Since the VP hole is at the right edge in the antecedent structure of the conjunct, it can be eliminated only if we apply and_R – which however requires the referent to be in the right conjunct rather than the left one. On the other hand, VP does not denote a relation and and_D does not apply either. (The latter also does not apply because it targets holes in the middle of the structure rather than at the edge.)

The prominent feature of our second approach is a peculiar way of eliminating a hypothesis: a hypothetical NP phrase introduced in the derivation of one coordinated clause is eliminated by "matching it up" with the suitable referent in the other coordinated clause. This hypothetical phrase is strongly reminiscent of a trace, or discontinuity (as in Morrill et al. [5]). It is also reminiscent of anaphora, especially of the sort used in Montague's PTQ. To be sure, this 'anaphora' differs notably from overt pronouns or even null pronouns. When targeted by the HypR rule, the hypothesis acts as a pure cataphora rather than anaphora. Mainly, the rules of resolving our 'pronoun', such as and_R, are quite rigid. They are syntactic rather than pragmatic, based on matching up two derivations, one with the hole and the other with the referent. The derivations should be sufficiently similar, 'parallel', for them to match. To avoid confusion, we just call our hypothetical phrase a hole, a sort of generalized discontinuity. Unlike Morrill, this hole may also occur at the edge.

3.2   Semantic (Computational) Interpretation

The phonological interpretation of a derivation – obtaining its yield – is standard. We read out the fringe (the leaves) of the derivation tree in order, ignoring silent items. This section expounds the conservative, traditional and yet novel semantic interpretation. It is this new interpretation that lets us use **NL** for analyzing phenomena like gapping, quantifier ambiguity and scope islands that were out of its reach before.

The semantic interpretation of a grammatical derivation maps it to a logical formula that represents its meaning. The mapping is compositional: the formula that represents the meaning of a derivation is built from the formulas for sub-derivations. In our interpretation, the meaning of every derivation, complete or incomplete, is represented by an always *closed* formula in the higher-order logic: the simply-typed lambda calculus with equality and two basic types e and t. Although pairs are easy to express in lambda calculus (using Church encoding), for notational convenience we will treat pairs, the pair type (A, B) and the unit type () as primitives. Another purely notational convenience is pattern-matching to access the components of a pair; for example, to project the first component we write $\lambda(x, y).x$. (The accompanying code instead of pattern-matching uses the projection functions, which are expressible in lambda-calculus.) We write underscore for the unused argument of the abstraction.

Our interpretation maps every **NL** sequent $\Gamma \vdash A$ to a closed formula of the type $\lceil \Gamma \backslash A \rceil$ where the homomorphic map $\lceil - \rceil$ from the **NL** types to the semantic types is given in Fig. 3. (The interpretation of a sequent confirms that a structure is really treated as a type.)

A	$\mapsto$	$\ulcorner A \urcorner$
NP	$\mapsto$	e
S	$\mapsto$	t
$A \backslash B$	$\mapsto$	$\ulcorner A \urcorner \rightarrow (\ulcorner B \urcorner, t)$
B / A	$\mapsto$	$\ulcorner A \urcorner \rightarrow (\ulcorner B \urcorner, t)$
$\bullet$	$\mapsto$	$()$
(A, B)	$\mapsto$	$(\ulcorner A \urcorner, \ulcorner B \urcorner)$
$\overline{A}$	$\mapsto$	$\ulcorner A \urcorner$
And	$\mapsto$	$()$

the same for all other context markers

$\mathcal{U}$	$\mapsto$	e
$\mathcal{E}$	$\mapsto$	e

Fig. 3. Mapping **NL** types to semantic types

As expected, the **NL** type NP maps to the type of entities e, S maps to the type of propositions t, and the contextual markers such as And, VB, etc. have no semantic significance. The mapping of directional implications and sequents is, on one hand, is traditional. Intuitively, an implication or a sequent are treated as a computation which, when given the term representing its assumptions will produce the term representing the conclusion. Uncommonly, our interpretation of an implication (sequent) produces two terms. The first represents the conclusion, and the second, always of the type t, represents side-conditions. For example, the sequent representing the complete sentence $\bullet \vdash S$ is mapped to the formula of the type $() \rightarrow (t, t)$. It is the computation which, when applied to $()$ (the trivial assumption, the synonym for $\top$) will produce *two* truth values. Their conjunction represents the truth of the proposition expressed by the original sentence. This splitting off of the side conditions (which are composed separately by the inference rules) is the crucial feature of our interpretation.

Each axiom (sequent) is mapped to a logical formula of the corresponding semantic type. Each rule of the calculus combines the formulas of its premises to build the formula in the conclusion. For example, the rule $\backslash I$ takes the sequent $(A, \Gamma) \vdash B$ (whose interpretation, to be called f, has the type $(\ulcorner A \urcorner, \ulcorner \Gamma \urcorner) \rightarrow (\ulcorner B \urcorner, t))$ and derives the sequent $\Gamma \vdash A \backslash B$. Its interpretation is the formula $\lambda g.((\lambda a.f(a, g)), \top)$ of the type $\ulcorner \Gamma \urcorner \rightarrow (\ulcorner A \urcorner \rightarrow (\ulcorner B \urcorner, t), t)$. More interesting are the eliminations rules, for example, $\backslash E$. Recall, given the formula (to be called x) interpreting the sequent $\Gamma \vdash A$ and the formula f for the sequent $\Delta \vdash A \backslash B$, the rule builds the interpretation of $(\Gamma, \Delta) \vdash B$ as follows.

$$\lambda(g, d). \text{ let } (f_v, f_s) = f \; d \text{ in}$$
$$\text{let } (x_v, x_s) = x \; g \text{ in}$$
$$\text{let } (b_v, b_s) = f_v \; x_v \text{ in}$$
$$(b_v, b_s \wedge f_s \wedge x_s)$$

(where let $x = e_1$ in e_2 is the abbreviation for $(\lambda x.e_2)e_1$.) Informally, whereas introduction rules correspond semantically to a λ-abstraction, elimination rules correspond to applications. In our interpretation, the elimination rules also combine the side-conditions.

As an illustration we show the semantic interpretation of two derivations from the previous section. The first is rather familiar

$$\bullet \vdash oo \uparrow \text{John} \cdot and_C \uparrow (x \text{ tripped}) \text{ and } (y \text{ fell}) : S$$

and its interpretation is unsurprising. The sequent $\bullet \vdash \text{John} : NP$ is assigned the logical formula $\lambda\_.(\text{john}, \top)$ where $\text{john} : e$ is the domain constant. The semantic interpretation of the other axiom sequents is similar. The constant

$$(\Gamma, (And, \Gamma))\backslash A \vdash and_C : \Gamma\backslash A$$

corresponds to the formula

$$\lambda f.(\lambda d.f \ (d, (0, d)), \top) : (([\ulcorner\Gamma\urcorner, (0, \ulcorner\Gamma\urcorner)) \rightarrow (\ulcorner A\urcorner, t)) \rightarrow (\ulcorner\Gamma\urcorner \rightarrow (\ulcorner A\urcorner, t))$$

All side-conditions are $\top$; the sequent for the sentence is interpreted as $\lambda\_.(\text{tripped john} \wedge \text{fell john}, \top)$.

More interesting is the derivation with holes and referents:

$$\bullet \vdash and_C \uparrow and_L \uparrow (ref \text{John}) \text{ tripped and } (x \text{ fell}) : S$$

The referent maker $\overline{A} \vdash ref : (A/A)$ is interpreted as $\lambda\overline{x}.(\lambda x.(x, x = \overline{x}), \top)$ That is, ref John corresponds to a formula that receives an $\overline{NP}$ assumption (the value of the type e) and produces john, what John by itself would have produced, *along with* the side-condition that the received assumption must be john. The structured constant and_L matches up the hole and the referent. Recall, it converts the structure of the type $((((\overline{A}, B), \Gamma), (And, (A, \Delta)))$ into $(((B, \Gamma), (And, (\bullet, \Delta)))$ which no longer has the assumption A in it (nor the matching $\overline{A}$). This assumption is eliminated 'classically' by assuming it with the existential quantifier.

$$(\lambda f. \ (\lambda((0, g), (0, (0, d))).$$
$$\exists x.\text{let } (b_v, b_s) = f \ (((x, 0), g), (0, (x, d))) \text{ in } b_v \wedge b_s,$$
$$\top), \top)$$

The quantified variable is passed as the A assumption and as the $\overline{A}$ assumption. The latter will later be converted to the side condition that the quantified variable must be equal to the referent. The final truth condition is represented by the formula $\exists x.(\text{tripped john} \wedge \text{fell } x) \wedge x = \text{john}$: the hole in the conjunct fell x is filled by the referent john.

We have just demonstrated a dynamic-semantic–like approach that uses only the traditional means (rather than mutation, continuation and other powerful features) to accomplish the filling of a hole with a referent deep in the tree.

Our side-condition plays the role of the constraint store, or of the current substitution in unification algorithms. We prefer to view our semantics in structural rather than dynamic terms, as matching up/unifying derivations (trees) rather than mutating the shared 'discourse context'. In general, mutations, continuations and other effects may be just as artifact of the constructive, computational approach to building logical formulas representing the meaning of a sentence. In a more high-level, declarative approach advocated here, the meaning can be derived non-constructively, "classically" and more clearly.

3.3   Non-canonical Coordination and Gapping

We now apply the two methods from the previous section to analyses of non-canonical coordination. Our first example is "John liked and Mary hated Bill."

$$(and_C \uparrow \text{John (liked } x) \text{ and (Mary (hated } y))) \cdot \text{Bill}$$

The last significant rule in the derivation is HypR. The very similar derivation can be given with the hole-referent approach. Unlike CCG, we do not treat 'John liked' as a constituent. In fact, the latter is not derivable in our calculus.

We stress that "*John liked Bill and Mary hated ϕ" is not derivable: since Bill occurs at the right edge of the conjunct, it can only be targeted by and_R. However, that rule requires the referent to be in the right conjunct rather than the left conjunct.

Next is gapping, for example "Mary liked Chicago and Bill Detroit."

$$and_D \uparrow \text{Mary }((ref \text{ liked}) \text{ Chicago}) \text{ and (Bill } (x \text{ Detroit}))$$

The analysis is almost identical to the object coordination analysis in Sect. 3.1. It may seem surprising that in our calculus such a complex phenomenon as gapping is analyzed just like the simple object coordination.

As presented, the analysis can still overgenerate, for example, erroneously predicting coordination with a lower clause:

*John bought a book and Bill knows that Sue a CD.

The problem can be easily eliminated with the mechanism that used in Sect. 4 for quantification islands. Kubota and Levine [4] however argue against the island conditions in gapping. They advocate that we should accept the above sentence *at the level of derivation (combinatorial grammar)* and rule it out on pragmatical grounds. Although Kubota and Levine's argument applies as it is in our case, we should stress that our calculus does offer a mechanism to express the requirement that the coordinated clauses must be 'parallel' (see the type of and_C for example). We could specify, at the level of combinatorial grammar, exactly what it means.

4   Quantification

The techniques introduced to analyze non-canonical coordination turn out to work for quantification in any position, quantification ambiguity and scope islands.

We start with the deliberately simple example "John liked everyone", where 'everyone' is typed as $\mathcal{U} \vdash NP$ where $\mathcal{U}$ is an atomic type. From its simple NP type, 'everyone' looks like an ordinary NP, letting us easily derive

$$(\bullet, (\bullet, \mathcal{U})) \vdash \text{John (liked everyone)} : S$$

'Everyone' however has the assumption $\mathcal{U}$, which has to be eventually discharged. The only way to do it in our system is to find a structural rule (structural constant) that moves $\mathcal{U}$ to the left edge of the antecedent structure, where it can be abstracted by the rule $\backslash I$. We do in fact posit such a structural constant, which, in combination with the Hyp rule, converts $(\bullet, (\bullet, \mathcal{U}))$ to $(\mathcal{U}, (\bullet, \bullet))$. We call it $float_U$ (actually, it is a combination of constants, each responsible for smaller-step $\mathcal{U}$ 'movements'). Once $\mathcal{U}$ is floated to the left edge of the antecedent, the $\backslash I$ rule gives $\bullet \vdash \mathcal{U}\backslash S$. The final step applies $/E$ with the silent constant $\bullet \vdash forall : (S/(\mathcal{U}\backslash S))$ producing

$$\bullet \vdash forall \ (\curlywedge(float_U \uparrow (\text{John (liked everyone)}))) : S$$

The analysis wrote itself: each step is predetermined by the types and the available constants. It only works if $\mathcal{U}$ is allowed to float to the top of the assumption structure, which has been the case.

The semantic interpretation maps $\mathcal{U}$ (and the similar $\mathcal{E}$ below) to e and interprets 'everyone' as essentially the identity function. The semantics of $forall$ is $\lambda_-.(\lambda k.(\forall x.\text{let } (b_v, b_s) = k \ x \text{ in } b_s \Rightarrow b_v, \top), \top)$ The meaning of the phrase is hence given by the formula $(\lambda k.\forall x.kx) \ (\lambda x. \ \text{like } x \ \text{john})$.

Existential quantification uses the $\mathcal{E}$ hypothesis. Quantification in the subject and even medial positions are just as straightforward. The key is the ability to float $\mathcal{U}$ or $\mathcal{E}$ to the left edge of the antecedent.

Our approach treats the quantification ambiguity. Consider "Someone likes everyone". At an intermediate stage we obtain

$$(\mathcal{E}, (\bullet, \mathcal{U})) \vdash \text{someone (like everyone)} : S$$

Since $\mathcal{E}$ is already at the left edge, it can be abstracted by $\backslash I$ and discharged by the application of $exists$. The hypothesis $\mathcal{U}$ still remains; it can be floated and then discharged by $forall$. The resulting truth condition reflects the inverse reading of the sentence. If $\mathcal{U}$ is floated first, the linear reading results.

We stress that the analysis of quantification crucially relies on the ability to float the hypotheses $\mathcal{U}$ or $\mathcal{E}$ to the left edge, permuting them with the other components of the structure. We can easily block such moves by introducing context markers, for example, *Clause*:

$$Clause \vdash \text{TheFactThat} : NP/S$$

We can then posit that $\mathcal{U}$ is not commutable with *Clause*, which explains why "That every boy left upset a teacher" is not ambiguous. The hypothesis $\mathcal{E}$ may still be allowed to commute with *Clause*, hence letting existentials take scope beyond the clause.

It is instructive to compare the present analyses of the quantification ambiguity and islands with the continuation analyses of [1]. The latter rely on the so called quantifier strength (the position of quantifiers in the continuation hierarchy) to explain how one quantifier may outscope another. Here we use essentially the structural rules (programmed as special lexical items to be applied by the Hyp rule). In the continuation analyses, the clause boundary acts as a 'delimiter' that collapses the hierarchy and hence prevents the quantifiers within from taking a wider scope. Essentially the same effect is achieved here by simply not permitting reassociation and commutation with *Clause*.

For the sake of the explanation we have used the very simple QNP "everyone" and "someone" with the unrealistically trivial restrictors. Our approach handles arbitrary restrictors. It turns out the 'side-conditions' in the semantic representation are exactly the restrictors of the quantification. For the lack of space we can only refer to the accompanying source code for details.

5   Anomalous Scoping in Non-canonical Coordination

We now combine the analyses of quantification and coordination and apply them to the wide scoping of quantifiers in gapped sentences, for example: "I gave a present to Robin on Thursday and to Leslie on Friday."

We start with the straightforward derivation

$$(\bullet, \mathcal{E}) \vdash \text{gave (a present)} : VP/PP$$

and extend it to

$$((\overline{NP}, \bullet), (((\overline{VP/PP}, (\bullet, \mathcal{E})), (\bullet, \bullet)), (\bullet, \bullet))) \vdash$$
$$(\textit{ref } \text{John}) (((\textit{ref}(\text{gave (a present)}))) (\text{to Bill})) (\text{on Monday}))$$

The derivation for the second conjunct assumes the 'holes' $x_j : NP$ and $x_g : VP/PP$:

$$(NP, ((VP/PP, (\bullet, \bullet)), (\bullet, \bullet))) \vdash x_j ((x_g (\text{to Leslie})) (\text{on Friday}))$$

The NP hole x_j is at the left edge of the antecedent and is filled with the referent "John", using Hyp and the constant and_L. The hole x_g is filled with the referent (gave (a present)), using Hyp and and_D. The type VP/PP corresponds to a relation (between NP and PP) and hence can be used with and_D. Finally, $\mathcal{E}$ in the remaining antecedent structure is floated to the left edge and eliminated with *exists*. The semantic interpretation shows the wide scope of "a present".

The derivation of the narrow-scope reading also starts with

$$(\bullet, \mathcal{E}) \vdash \text{gave (a present)} : VP/PP$$

but proceeds by first floating $\mathcal{E}$ to the left and abstracting with $\backslash I$:

$$\bullet \vdash \angle\text{gave (a present)} : \mathcal{E}\backslash(VP/PP)$$

(for clarity, we omit the structural constant needed floating $\mathcal{E}$). This term is used as a referent. Assuming $\mathcal{E} \vdash x_E : \mathcal{E}$ leads to

$$(\mathcal{E}, (\overline{\mathcal{E}\backslash(VP/PP)}, \bullet)) \vdash x_E \ (ref(\angle\text{gave (a present)})) : (VP/PP)$$

and, similarly to the above, to the derivation for the left conjunct:

$$((\overline{NP}, \bullet), (((\mathcal{E}, (\overline{\mathcal{E}\backslash(VP/PP)}, \bullet)), (\bullet, \bullet)), (\bullet, \bullet))) \vdash$$
$$(ref \text{ John}) \ (((x_E \ (ref(\angle\text{gave (a present)})))) \ (\text{to Bill})) \ (\text{on Monday}))$$

The marker $\mathcal{E}$ is then floated to the left edge, abstracted by $\backslash I$ and eliminated by applying the constant *exists*:

$$((\overline{NP}, \bullet), (((\overline{\mathcal{E}\backslash(VP/PP)}, \bullet), (\bullet, \bullet)), (\bullet, \bullet))) \vdash$$
$$exists(\angle(ref \text{ John}) \ (((x_E \ (ref(\angle\text{gave (a present)})))) \ (\text{to Bill})) \ (\text{on Monday})))$$

The derivation for the right conjunct is similar, only using the holes $NP \vdash x_j : NP$ and $\mathcal{E}\backslash(VP/PP) \vdash x_g : \mathcal{E}\backslash(VP/PP)$:

$$(NP, ((\mathcal{E}\backslash(VP/PP), (\bullet, \bullet)), (\bullet, \bullet))) \vdash$$
$$exists(\angle x_j \ (((x_E \ x_g) \ (\text{to Leslie})) \ (\text{on Friday})))$$

The two conjuncts (clauses) can now be coordinated and the holes plugged using and_D and and_L constants. Each clause has its own quantifier. Exactly the same approach applies to narrow- and wide- scope modal auxiliaries such as "must" and "cannot".

6   Discussion and Conclusions

The familiar non-associative Lambek calculus with the conservative, traditional and yet novel semantic interpretation turns out capable of analyzing discontinuous constituency. Gapping, quantifier ambiguity and scope islands now fall within **NL**'s scope. The semantic interpretation is traditional in that it is compositional, assigning every (sub)derivation an always closed logical formula and using only the standard operations of the higher-order logic. The interpretation uses no dependent types, monads, effects or the continuation-passing style. The crucial feature is side-conditions to the semantic interpretation, to be combined with truth conditions at the end.

Our calculus shares many capabilities with the hybrid type-logical categorial grammar of Kubota and Levine K&L [2]. Both calculi analyze non-canonical coordination, gapping, and narrow and wide scopes of quantifiers and modal auxiliaries in coordinated structures. Our calculus uses no modes, type raising or higher-order phonology. Our coordination is always at type S and maintains phrase structure constituents.

The immediate future work is the treatment of summatives (with "total") and symmetric phrases (with the "same"). Our interpretation has the classical logic flavor, which is interesting to explore further. One may argue if structural postulates should be added directly rather than sneaked through structural constants. The resulting calculus should then undergo the formal logical investigation, including the evaluation of the complexity of decision procedures.

What gives **NL** its unexpected power is the somewhat non-standard semantic interpretation. The interpretation is classical in the sense of classical logic, involving existentials (Hilbert epsilons). If we are to insist on a computational interpretation, we have to eliminate such existentials and resort to logical variables (and hence mutation) or delimited continuations and backtracking. Could it be that the continuation semantics, dynamic semantics with mutation, monads and effects are just the artifacts of the computational, constructive approach to the syntax-semantics interface? If the interface is formulated classically, declaratively, then even the simplest **NL** will suffice?

Acknowledgments. I am very grateful to Yusuke Kubota for very helpful conversations and many suggestions.

References

1. Kiselyov, O., Shan, C.: Continuation hierarchy and quantifier scope. In: McCready, E., Yabushita, K., Yoshimoto, K. (eds.) Formal Approaches to Semantics and Pragmatics. Studies in Linguistics and Philosophy, pp. 105–134. Springer, Netherlands (2014). http://dx.doi.org/10.1007/978-94-017-8813-7_6

2. Kubota, Y., Levine, R.: Gapping as like-category coordination. In: Béchet, D., Dikovsky, A. (eds.) LACL 2012. LNCS, vol. 7351, pp. 135–150. Springer, Heidelberg (2012)

3. Kubota, Y., Levine, R.: Empirical foundations for hybrid type-logical categorial grammar. the domain of phenomena, August 2013

4. Kubota, Y., Levine, R.: Gapping as hypothetical reasoning (2014), to appear in Natural Language and Linguistic Theory, http://ling.auf.net/lingbuzz/002123

5. Morrill, G., Valentín, O., Fadda, M.: The displacement calculus. J. Logic Lang. Inf. **20**(1), 1–48 (2011). http://dx.doi.org/10.1007/s10849-010-9129-2

6. Steedman, M.J.: Gapping as constituent coordination. Linguist. Philos. **13**, 207–263 (1990)

A Good Intensifier

Elena Castroviejo[1] and Berit Gehrke[2]([✉])

[1] ILLA-CSIC, Madrid, Spain
elena.castroviejo@cchs.csic.es
[2] CNRS/Université Paris Diderot, Paris, France
berit.gehrke@linguist.univ-paris-diderot.fr

Abstract. We provide a semantic account of the Catalan ad-adjectival modifier *ben* 'well', which yields intensification by, we argue, positively evaluating a property ascription. Formally, this translates as applying the predicate **good** to the saying event available to any utterance. We treat the output of this modification as a Conventional Implicature (rather than at-issue content), which is responsible for its positive polarity behavior. Additionally, we exploit the semantic similarity between this intensifier use of WELL with other readings of WELL, including manner (*well written*) and degree (*well acquainted*), which we analyze as 'manner-in-disguise'. In our proposal, they all predicate goodness of an event.

Keywords: Degree · Manner · Intensification · Conventional implicature · Positive polarity · Vagueness

1 Introduction

This paper focuses on the use of the evaluative adverb *well* in contexts where it conveys amount/degree intensification. Specifically, we employ Catalan data, since *ben* 'well' has a wider distribution than its counterpart in languages like English and German, as we will show. Consider (1).

(1) Marxem amb el cap **ben** alt.
     we.leave with the head WELL high
     '(lit.) We leave with our head WELL high.' (We leave with dignity.)
     http://www.esport3.cat/video/4619973/futbol/
     Boadas-Marxem-amb-el-cap-ben-alt

This sentence conveys that the degree to which the head is high is considerable. Our challenge is to yield the semantics of degree intensification while maintaining its relation to other uses of *well* across languages, most prominently the manner use, as well as maintaining the lexical semantics related to goodness.

This research has been supported by project FFI2012-34170 (MINECO) and by the Ramón y Cajal program (RYC-2010-06070).

T. Murata et al. (Eds.): JSAI-isAI 2014 Workshops, LNAI 9067, pp. 114–129, 2015.
DOI: 10.1007/978-3-662-48119-6_9

We will provide a formal analysis for Catalan ad-adjectival modifier *ben* 'well' [henceforth BEN], which we argue derives intensification by positively evaluating a property ascription. It introduces the predicate **good** as part of its denotation, but in examples like (1), it applies to a saying event rather than to the event associated with the lexical verb (manner *ben*). As an extension, we suggest a unified treatment of intensifying BEN with ad-nominal modifier *bon* good.

2   The Empirical Generalizations

Among various readings or uses that the adverb *well* and its counterparts in other languages [henceforth WELL] can have, the manner reading is the most common one, where WELL can be paraphrased by 'in a good manner' (2).

(2)   He has written the article **well**. ⤳ in a good manner

Under this reading, which arises with all verbs that allow manner modification, i.e. with most eventive verbs, the adverb can be straightforwardly analyzed as predicating the property *good* over the event in question (e.g. in (2) the event of writing the article is said to be good), on a par with manner modifiers more generally (e.g. [25]).

A second reading that is less clear how to analyze is the "degree reading" (cf. [3,16]), under which the adverb can be paraphrased by 'to a good degree' (3).

(3)   They are **well** acquainted. ⤳ to a good degree

While the manner reading seems to be available across languages, we argue that what has been identified as degree WELL in the literature does not correspond to a uniform phenomenon. Rather, there is a distinction between what we will call 'manner-in-disguise' WELL, illustrated in (3), and a (degree-)'intensifying' WELL, which we will label BEN and which is absent in English and German, but present in Spanish and Catalan.

2.1   Intensifying *ben* vs. Manner-in-Disguise *well*

Whereas the examples to illustrate the degree reading of WELL in English generally involve participles, as in (3) (e.g. [3,16]), it does not seem to be possible to use WELL as a degree modifier of genuine adjectives (4-a); the same can be observed for German (4-b).

(4)   a.   *The train is **well** blue / long / beautiful.
       b.   *Der Zug ist **gut**   blau / lang / schön.
            the train is WELL blue   long   beautiful

In contrast, in languages like Catalan (and some varieties of Spanish, cf. [11,12, 14]), this is possible (5).

(5)   El tren és **ben**   blau / llarg / bonic.
       the train is WELL blue   long   beautiful
       'The train is pretty blue / long / beautiful.'

These facts suggest that English and German WELL is exclusively a VP modifier (a predicate of events, in the broadest sense, to include states), whereas in languages like Catalan it has similar uses as other degree modifiers such as *pretty*, *rather*; cf. the translation of (5).

Furthermore, [16] argue that the 'degree' reading of *well* only comes about with adjectives (= adjectival participles) associated with scales that are closed on both ends, evidenced by their compatibility with *partially* or *fully* (6).

(6) a. The truck is **well** / partially loaded.
     b. ??Marge was **well** / partially worried when she saw the flying pig.

A further condition they posit is that the standard of comparison cannot be the maximum. For example, when the argument is an incremental theme (7-a), they argue that what counts as a loaded incremental theme can only be such that the maximum standard is met (it is completely loaded); the 'degree' reading of WELL is not available, as there are no different degrees of loadedness that could be compared to one another. With other arguments, on the other hand, as in (7-b), the standard is not necessarily the maximum (e.g. a truck can also be partially loaded), and thus the 'degree' reading is available.

(7) a. The hay is **well** loaded.   ONLY MANNER
     b. The truck is **well** loaded.   DEGREE/MANNER

In contrast to WELL, BEN does not exhibit such scale structure restrictions: it can also combine with the (relative) open scale adjectives in (8-a), as well as with the (absolute) closed scale adjectives with maximum standards in (8-b).

(8) a. Open scale: ben a prop 'WELL close', ben amunt 'WELL up', ben sonat 'WELL nuts', ben simpàtic 'WELL kind', ben trist 'WELL sad', ben viu 'WELL alive', ben idiota 'WELL idiotic'
     b. Closed scale, maximum standard: ben buit 'WELL empty', ben recte 'WELL straight', ben pla 'WELL flat'

It can be shown that the scale structure restrictions on 'degree' WELL can be derived from restrictions on the kinds of events that WELL applies to. This becomes evident by the fact that the same restrictions are found in the verbal domain, a fact that is not discussed in [16,21]. For example, in German, the same verbs that do or do not give rise to a 'degree' reading with adjectival participles (which combine with the copula *sein* 'be') also do or do not so with verbal participles (which combine with the auxiliary *werden* 'become') (9).

(9) a. Der Lastwagen {ist / wurde} **gut** beladen.   DEGREE/MANNER
         the truck   is   became well AT-loaded
         'The truck {is / has been} well loaded.'
     b. Das Heu {ist / wurde} **gut**   geladen.   ONLY MANNER
         the hay   is   became WELL loaded

This difference is primarily related to event structure and only indirectly to scale structure: with (simple) incremental theme verbs such as *laden* 'load (x on y)' in (9-b), there is no scale to begin with but it is only provided by the incremental theme (as [15] argues himself; see also [30]); hence the verb itself only has an activity component that can be modified by *well*, thus giving rise to the manner reading. This is different with the prefixed verb *beladen* 'load (y with x)', which – due to its prefix – has a built-in stative component that can be modified by WELL, giving rise to what one might want to call a degree reading.

Furthermore, even verbs that do not derive adjectival participles allow for the 'degree' reading, such as the stative one in (10-a); other verbs do not, such as the necessarily agentive one in (10-b).

(10)   a.   Sie   kennen einander   **gut**.   DEGREE
            they know   each other WELL

       b.   Sie ist **gut**   in den   Baum geklettert.   ONLY MANNER
            she is   WELL in the.ACC tree   climbed
            'She has climbed into the tree well.'

Hence, whether or not we get a 'degree' reading of WELL depends entirely on the nature of the event denoted by the (underlying) verb. With verbs that only have an activity component (9-b), or whose manner/activity component cannot be absent (e.g. they cannot appear as inchoatives) (10-b), we only get the manner reading. With verbs that have a stative component (resultatives and statives) the degree reading is possible (9-a), (10-a); with statives it is even the only reading. Thus, degree WELL is an an event predicate, predicated over the stative (sub)event of non-agentive verbs or verbs that allow for a non-agentive reading.

In Sect. 3, we will propose that WELL under both readings is a VP modifier which predicates the property **good** over an event; this makes it a manner modifier in the broadest sense, under both readings, which is why we label the 'degree' reading manner-in-disguise.

Another difference between WELL and BEN is that the former can be modified by degree modification (11-a), whereas the latter cannot (11-b).

(11)   a.   They know each other (very) **well**.
       b.   En Pere és (*molt) **ben**   alt.
            the Peter is very   WELL tall

The compatibility of WELL with degree modification would be left unexplained if it were a degree modifier itself, since elements that directly operate on the degree bind off the degree argument and make it inaccessible for further degree modification (cf. [16]). However, nothing prevents degree modification of WELL, though, if it is treated as an event predicate, a kind of manner modifier, so the account we propose straightforwardly captures this fact.

The fact that BEN is incompatible with further degree modification suggests that it is one itself. However, in the following sections, we will discuss reasons why it should not be treated as a degree modifier, either.

2.2 Intensifying *ben* vs. Degree Modifiers

In this section we show that intensifying BEN is different from ordinary ad-adjectival degree modifiers (say, of type $\langle\langle d, et\rangle, \langle e, t\rangle\rangle$), such as *very* (a standard booster) and *completely* (a slack regulator). We will phrase the description in a degree-based approach to gradability of the type found in [16] but the overall point that BEN does not behave like a degree modifier could also be made within other approaches such as [31].

A degree modifier like *very*, which is a standard booster (see [16,31] and liter-ature cited therein), readjusts the standard of gradable adjectives. For example, to truthfully utter a sentence with a relative adjective like *tall*, one has to know what the standard of comparison is. In (12-a), Peter can truthfully be said to be tall if he is at least as tall as the standard (the average height set by the members of the comparison class), which can be different from context to con-text (12-ai,aii). For absolute adjectives like *full*, on the other hand, the standard of comparison is commonly (semantically) the upper bound of the closed scale ('completely full') (but see also [31] for qualifications) (12-b); pragmatically some slack might be allowed, e.g. if 20 seats are still empty.

(12)   a.   En Pere és alt.
              the Peter is tall
              (i)   for a 10-year-old boy from Barcelona: at least 1.40m
              (ii)   for an NBA basketball player: at least 2.05m
        b.   L'estadi   està ple.
              the stadium is   full
              'The stadium is full.'

With relative adjectives, a degree modifier like *very*, then, boosts the standard, i.e. it raises the standard degree on the scale associated with the adjective (13-a). It also does so with absolute adjectives after first relativizing them into having a context-dependent threshold. For instance, (13-b) is acceptable in a situation where the threshold for *full* has been readjusted to e.g. 80 % full and the sentence felicitously describes a situation in which the stadium is 85 % full.

(13)   a.   En Pere és molt alt.
              the Peter is very tall
              (i)   for a 10-year-old boy from Barcelona: at least 1.50m
              (ii)   for an NBA basketball player: at least 2.15m
        b.   L'estadi   està molt ple.
              the stadium is   very full
              'The stadium is very full.'

The effect with BEN is different from *molt*. With absolute adjectives (14), the standard degree remains unchanged and is semantically the same as with 'com-pletely full': it describes a situation in which the stadium is 100 % full.

(14)   L'estadi   està **ben**   ple.
        the stadium is   WELL   full

However, when BEN applies, no slack is allowed anymore, unlike what we find in the non-modified case in (12-b), which brings its effect closer to that of *completely* and other so-called slack regulators ([17]) or precisifiers. With *completament* 'completely', just like with BEN, no slack is allowed anymore, but every seat in the stadium has to be filled (15-a). Nonetheless, unlike BEN, slack regulators are only felicitous in case there is a pragmatic slack that can be regulated; hence they are infelicitous with open scale (relative) adjectives (15-b).

(15) a. L'estadi   està completament ple.
          the stadium is   completely   full
          'The stadium is completely full.'
      b. *En Pere és completament alt.
          the Peter is completely   tall

Thus, the meaning effect of BEN seems to be more of the type that it focuses on the most typical instances of the property in question, thus excluding borderline cases. For example, in combination with color adjectives which can apply to extended areas in the color spectrum, the application of BEN has the effect that this area is smaller and the property in question falls within the prototypical area. The same effect is reached with relative adjectives: borderline cases that are just at the average of what might count as tall will be excluded so that we indirectly get a similar effect to that of a standard booster, as in (16), in that this sentence would be odd if Peter were just average tall.

(16) En Pere és **ben** alt.
      the Peter is WELL tall

However, given that this apparent standard boosting effect is absent with absolute adjectives, we propose that it comes about only indirectly, by focusing on the core cases to which the adjective in question can apply. In Sect. 3 we will propose an account of BEN as positively evaluating a property ascription, which directly captures this effect.

We therefore conclude that the meaning effect of BEN has nothing to do with the standard per se: it neither directly boosts the standard associated with a given adjective nor does it regulate slack that pragmatically appears only with absolute adjectives. This, in essence, means that it is not a degree modifier.

As a final remark, note that nothing in this analysis explains why only gradable adjectives can be modified by BEN, cf. (17).

(17) En Joan és un arquitecte (*ben) tècnic.
      the John is an architect WELL technical
      'John is a (*WELL) technical architect.'

Tècnic is a relational adjective (cf. e.g. [23]), one of the few adjective types that can be hardly coerced into a gradable predicate (cf. e.g. [2]). This example can be used as a counterargument to our claim that intensification is obtained by indirect means, i.e. by evaluating a property ascription, and as an argument in favor of treating BEN as a degree modifier after all.

Nevertheless, we can provide a strong argument in favor of the present proposal, namely that if BEN should be a degree modifier, we would still need to explain how *goodness* is involved in deriving intensification. The alternative we propose consists in blaming the ill-formedness of (17) not on a type clash, but on a conceptual constraint. In particular, it only makes sense to self-evaluate a property ascription if vagueness arises, i.e. if its criteria of application can be different in different contexts (e.g. [5]). That is, if we cannot evoke contexts where we have different truth conditions for the predication **technical–architect(j)**, then it does not make sense to positively evaluate this property ascription. Take for instance a gradable predicate like *tall*. Depending on the context, which is determined by the choice of comparison class – i.e. all the male individuals, basketball players, kindergarten classmates, etc. – **tall(j)** can be true or false. The critical issue is that discourse participants can disagree on whether the property is well ascribed, and so the positive evaluation is acceptable. This makes *tall* a vague (and gradable) predicate which warrants that it can combine with BEN.

2.3 Intensifying *ben* Involves Subjective Evaluation

How does the apparent meaning of intensification come about then and what are the meaning effects of BEN more generally? For examples like (16), for instance, BEN has the effect that the speaker contradicts something that was said in the previous context (e.g. after someone said that Peter is short) or expresses some satisfaction or positive evaluation of Peter's tallness. This is related to the point that BEN nevertheless shows some contextual restrictions depending on the type of gradable adjective it modifies, even if it is not a degree modifier. Specifically, it does not felicitously modify an adjective out of the blue (18), unless it is a clear case of a predicate of personal taste.

(18)   A: Com és en Carles?
             how is the Charles
             'What is Charles like?'
        B: És {molt / #**ben**} intel.ligent i   {molt / #**ben**} generós.
             is very   WELL   intelligent and very   WELL   generous
             'He is {very /#pretty} intelligent and {very /#pretty} generous.'

Predicates of personal taste, on the other hand, which have a built-in evaluative character and therefore a clearly subjective meaning component (cf. e.g. [18,33]), are generally good with BEN, even out of context (19-a); the subjective nature of such predicates is evident from the fact that they can be embedded under predicates like *find* (19-b) (on which cf. [34] and literature cited therein).

(19)   a.   El pastís és **ben** bo.
             the cake is WELL good
             'The cake is WELL tasty.'
        b.   Trobo el pastís bo.
             find.I the cake good
             'I find the cake tasty.'

Other gradable adjectives that are not qua their lexical semantics predicates of personal taste become felicitous with BEN when it is under discussion and when there can be disagreement whether or not x has the property in question. The added subjective component is explicit in the use of *trobar* 'find' in (20), but it is also implicit in (21), where a contrast is established between the expectations of the speaker and her actual opinion.

(20)   A: Ahir   m'ho vaig   passar molt bé   amb en Pere. És tan
            yesterday me-it have.1SG passed very well with the Peter is   so
            divertit!
            funny
            'I had such a blast yesterday with Peter. He is so funny!'
      B: Doncs   jo el   trobo **ben** avorrit.
            actually I   him find   WELL boring
            'Actually, I find him WELL boring.'

(21)   La Mar porta un barret **ben** bonic. M'ha   sorprès   que tingui
        the Mar wears a   hat   WELL pretty me-has surprised that has.SUBJ
        tan bon   gust.
        that good taste
        'Mar is wearing a WELL pretty hat. I am surprised that she has such good taste.'

The connection to a subjective evaluation of a property ascription is further illustrated by the correlation in (in)compatibility with BEN and *trobar* in (22).

(22)   a.   *ben   just cf. *El trobo   just.
          WELL fair   him find.1SG just

       b.   *ben   solidari cf. *El trobo   solidari.
          WELL solidary   him find.1SG solidary
       c.   **ben** ridícul   cf. El   trobo   ridícul.
          WELL ridiculous   him find.1SG ridiculous

No such contextual restrictions and requirements of a subjective component are found with the degree modifier *molt* 'very', see, e.g. (23).

(23)   En Joan és molt just / solidari / ridícul.
       the John is very fair   solidary   ridiculous
       'John is very fair / solidary / ridiculous.'

Related to the observation that BEN involves a subjective evaluative component is the fact that it cannot occur under negation (24-a), and it cannot be questioned (24-b) (cf. [12,14] for Spanish).

(24)   a.   *En Pere no és **ben** simpàtic.
          the Peter not is WELL nice
       b.   *En Pere és **ben** simpàtic?
          the Peter is WELL nice

These facts have led some people ([12–14]) to argue that BEN (in Spanish and Catalan) is a positive polarity item (PPI), a proposal also made cross-linguistically for evaluative adverbs like *unfortunately* ([9,24]). However, we will argue that BEN contributes a meaning at the Conventional Implicature [CI] tier and that the PPI properties follow from it being a factive evaluative. Building on [19,20] on evaluative adverbs, we propose that BEN sides with factive adverbs in that infelicity under negation is the result of a contradiction between the meaning conveyed at the at-issue tier and the presupposition of the CI.

Let us then turn to our unified analysis of WELL and BEN as event modifiers, which at the same time captures their differences.

3   The Analysis

In this section we argue that **good** applies to events to derive both manner(-in-disguise) *well* and intensifying BEN. In the former case, the event that is targeted is the event associated with the lexical verb (the VP), while in the latter case, it is the saying event. Moreover, we provide an analysis that accounts for the PPI behavior of BEN building on the idea that CI items may bear their own presuppositions, which are not satisfied in entailment-canceling contexts.

3.1   Manner and Manner-in-Disguise *well*

We assume that the adverb WELL under both readings has the same general lexical semantics as the underlying adjective *good* (approval by some judge). Following the degree approach to gradable adjectives (e.g. [16]), we treat *good* as a relation between degrees and individuals (25-a). Combined with the standard treatment of manner modifiers (= VP modifiers) as predicates of events (e.g. [25]), we get the uniform semantics of WELL in (25-b).

(25)   a. $[\![\text{good}]\!] = \lambda d.\lambda x[\textbf{good}(x) \geq d]$
       b. $[\![\text{well}]\!] = \lambda d.\lambda e[\textbf{good}(e) \geq d]$

In the absence of additional degree morphology, d gets bound by POS, which determines the standard with respect to some comparison class, as commonly assumed in degree approaches to gradability.

[21] propose that the 'degree' reading comes about (via selective binding) when *good* modifies the event in the telic quale of the participle, whereas under the manner reading it applies to the event in its agentive quale (building on the analysis of *fast cake* vs. *fast car* in [29]). Since nothing in this paper hinges on the precise implementation of this idea, we formalize this in terms of underspecification. In particular, we follow [32], who builds on [8]'s notion of a 'big event', represented by e^*, which is a complex event consisting of smaller event objects, introduced by the **PART_OF**-relation. Abstracting away from the degree argument, which is bound off by POS at this point, *good* accesses either the big event or part of the event, as illustrated for (9-a) in (26) (ignoring Tense).

(26) $\exists e^*, x[\textbf{subject}(x, e^*) \wedge \textbf{object}(\textbf{the-cart}, e^*) \wedge \exists e[\textbf{PART\_OF}(e, e^*) \wedge$
$\textbf{load}(e) \wedge \textbf{good}(e/es^*)]]$

In the following, we will argue that intensifying BEN shares with the other two the lexical semantics of *good* and the fact that it is a predicate of events, but in this case not of the VP event but of an implicit saying event.

3.2   Intensifying *ben*: Self-Evaluation of a Property Ascription

To provide a uniform account of WELL and BEN as predicates of events, we rely on [26]'s analysis of the performative use of speech-act adverbs like *frankly* in (27-a), which is treated as a predicate of expression manners (**expression**(e)) that self-describes the utterance of the context C as a saying event (27-b).

(27)   a.   Frankly, Facebook is overrated.
        b. $\textbf{utterance}(C) = e \wedge \textbf{speaker}(C) = x \wedge \textbf{hearer}(C) = y \wedge \textbf{say}(e,$
              $\exists s(\textbf{overrated}(s, \textbf{facebook})) \wedge \textbf{now} \subseteq \tau(e) \wedge \textbf{agent}(e, x) \wedge \textbf{recip-}$
              $\textbf{ient}(e, y) \wedge \textbf{frank}(\textbf{expression}(e))$

Unlike *frankly*, BEN is not a sentential adverb, but an ad-adjectival modifier. It also does not evaluate a proposition, but a property ascription. Moreover, we do not want to claim that BEN characterizes an expression manner, but just the saying event, so we dispose of the expression manner. We propose the slightly amended denotation in (28), and the translation of (29-a) as in (29-b).

(28) $[\![\textbf{BEN}]\!] = \lambda P.\lambda z[\textbf{utterance}(C) = e \wedge \textbf{speaker}(C) = x \wedge \textbf{hearer}(C)$
        $= y \wedge \textbf{say}(e, P(z)) \wedge \textbf{now} \subseteq \tau(e) \wedge \textbf{agent}(e, x) \wedge \textbf{recipient}(e, y) \wedge$
        $\textbf{good}(e)]$

(29)   a.   En Joan és **ben** alt.
              the John is WELL tall
              'John is WELL tall.'
        b. $\textbf{utterance}(C) = e \wedge \textbf{speaker}(C) = x \wedge \textbf{hearer}(C) = y \wedge \textbf{say}(e,$
              $[\![\textbf{POS tall}]\!](\textbf{j})) \wedge \textbf{now} \subseteq \tau(e) \wedge \textbf{agent}(e, x) \wedge \textbf{recipient}(e, y) \wedge$
              $\textbf{good}(e)$

This analysis of BEN essentially encodes the notion of *emphasis* as the self-evaluation of a property ascription. Emphatic statements are not supposed to be felicitous in out of the blue contexts or question-answer pairs – i.e. where conveying emphasis is not justified – unless we can accommodate that it is a matter of taste whether or not x should be considered ADJ. Hence, we expect that BEN should be felicitous in contexts where contrastive statements are being discussed and/or with predicates of personal taste, which are clearly judge-dependent.

Other than this, note that BEN operates on the speech act event. We assume with [26] that a performative verb is semantically represented in every utterance, and we claim that its output meaning is not part of the descriptive at-issue content of the assertion, but rather conventionally implicated, in the sense of [27] (and further elaborations). This allows us to explain its resistance to embed under negation (24-a) and interrogation (24-b), and to be modified by additional degree expressions (11-b), as will be elaborated on in the following.

Previous literature on Spanish BEN (*bien*) has observed that it has a distribution analogous to that of PPIs (e.g. [12,14]). We argue that BEN's resistance to certain embeddings parallels the resistance of expressive items ([28]) such as evaluative adverbs ([19,20]) and extreme degree modifiers ([22]). Let us illustrate this point for the extreme degree modifier *downright*, as in *downright dangerous*. It too is marginal when embedded under negation and interrogation (30) (from [22]); this also holds for better known expressives, such as *fucking* (31) or other such elements (e.g. *full-on, straight-up, balls-out*).

(30)   a. ??Murderers aren't downright dangerous.
         b. ??Are murderers downright dangerous?

(31)   a. ??He isn't fucking calm.
         b. ??Is he fucking calm?

(30) and (31), and similarly our examples (24-a) and (24-b), can only be rescued under an echo-reading. This is also expected under the assumption that they are expressive items (or else, that they convey meaning through a CI). The ill-formedness of these sentences can be accounted for by saying that there is a mismatch between the two meanings conveyed (at-issue and CI), as illustrated in (32-a) and (32-b), respectively.

(32)   ??He isn't fucking calm. (= (31-a))
       a.   At-issue tier: $\neg(\mathbf{calm(he}_i))$
       b.   CI tier: Speaker expresses a negative attitude at him being calm.

Here we adopt the analysis for evaluative adverbs like German *leider* and *unglücklicherweise* ('unfortunately') by [19,20] to account for BEN's PPI distribution. In her analysis, evaluative adverbs can be of two types. While *leider* cannot occur in the semantic scope of any entailment-canceling contexts (including negation, conditionals, questions, modals, also called "non-veridical" by [10,35] a.m.o.), *unglücklicherweise* can occur in all of these contexts but negation (33).

(33)   a.   Otto ist nicht {*leider/*unglücklicherweise} krank.
              Otto is   not   unfortunately   sick
       b.   Otto ist vielleicht {*leider/unglücklicherweise} krank.
              Otto is   maybe   unfortunately   sick
       c.   Ist Otto {*leider/unglücklicherweise} krank?
              is Otto unfortunately   sick
       d.   Falls Otto {*leider/unglücklicherweise} krank ist, muss das
              if   Otto unfortunately   sick is   must the
              Seminar ausfallen.
              seminar be cancelled

The former is factive and the latter, non-factive, which correlates with a difference in their lexical semantics. Specifically, Liu proposes the following:

(34) a. $[\![\text{leider}]\!] \rightsquigarrow \lambda p.\text{unfortunate}(p)$ FACTIVE
      b. $[\![\text{unglücklicherweise}]\!] \rightsquigarrow \lambda p.\ p \rightarrow \text{unfortunate}(p)$ NON-FACTIVE

Liu furthermore shows that only factive evaluative adverbs come with their own presuppositions, which happen to match the at-issue content in affirmative sentences (35), but which yield presupposition failure in entailment-canceling contexts like (33). That is, in (35-a), *leider* conveys a CI (35-c) whose presupposition is that Otto is sick (35-d), which in contexts like (33) clashes with the asserted content that negates or questions the at-issue meaning in (35-b). This explains the infelicity of factive evaluative adverbs in contexts where the presupposition and the at-issue meaning do not coincide anymore.

(35) a. Otto ist leider   krank.
          Otto is unfortunately sick
          'Otto is unfortunately sick.'
      b. At-issue tier: Otto is sick.
      c. CI tier: It is unfortunate that Otto is sick.
      d. CI's presupposition: Otto is sick.

Non-factive adverbs receive a different explanation for their incompatibility with negation, inspired by [4]. The conditional semantics they have make the CI tier completely independent from the at-issue tier and in principle, (36) does not represent a logical contradiction.

(36) a. At-issue tier: Otto is not sick.
      b. CI tier: If Otto is sick, then it is unfortunate that he is sick.

However, it has the status of an incongruence, called a semantic clash by [20]. Note that in one tier we are stating that Otto is not sick and in the other tier we are entertaining the idea that he is sick, and what would follow from that.

Turning now to BEN, we seem to have a distribution parallel to factive evaluative adverbs (37),[1] so a plausible explanation for its PPI behavior can be spelled out as in (38). Note that the at-issue meaning in (38-a) is at odds with the presupposition in (38-b), which causes the ill-formedness of the sentence.

---

[1] The data are much more complex than we are able to show due to space limitations, though. For instance, certain adjectives whose degree can be interpreted as good for a purpose are fine in conditionals (i).

(i)   Si els pantalons són **ben** estrets, no caldrà   que et
      if the trousers are well tight   not be.necessary.fut.3sg that you
      posis   mitges.
      put.on.pres.subj.2sg stockings
      'If these trousers are well tight, it won't be necessary for you to wear stockings.'

We have also identified differences of behavior in BEN that seem to relate to the properties of the adjective it modifies, but they are not completely straightforward to us at this point. We leave this investigation for future research.

(37)   a. *Si en Pere és **ben** simpàtic, estaré   contenta.
            if the Peter is WELL nice   be.fut.1sg glad

        b. *És possible que en Pere sigui   **ben** simpàtic.
            is possible that the Peter is.pres.subj WELL nice

(38)   *En Pere no és **ben** simpàtic. (= (24-a))
        the Peter not is WELL nice

        a. At-issue tier: $\neg$(**nice(p)**)
        b. CI tier: **nice** is well ascribed to Peter.
        c. CI's presupposition: **nice(p)**

We hence propose that BEN is of type $\langle\langle e, t\rangle, \langle e, t^c\rangle\rangle$, where c indicates that the output is delivered at the CI tier, following [27].

4   Excursion: On *good*

So far, we have explored the distribution and analysis of intensifying BEN, which hinges on the lexical semantics of the predicate **good**. In this final section we reflect on the semantics of **good** and extend our proposal to intensifying BON 'good'.

On the basis of the examples in (39), [1] argues for the need of a more fine-grained typology of semantic types, similar in some respects to the system proposed in [29], which is also used by [21]. Specifically, a lunch is good if it tastes good, i.e. *good* seems to select the purpose or telic role of the modified N in (39-a), and thus it behaves like a subsective adjective.

(39)   a. a **good** lunch
        b. a **good** rock
        c. **good** children

However, rocks and children are natural kinds, for which it does not make sense to assume a given function. In such cases, Asher argues, they can be coerced into having an artifact for which the telic (polymorphic) type is well defined, and can thus be evaluated as good. The exact value of the type is left underspecified, but it can be made clear through e.g. a modifier, as in (40) (from [1]).

(40)   This is a **good** rock *for skipping/throwing/carving/chiseling*, etc.

Here we observe an additional use of *good*, which is not captured in (39) and which we illustrate in (41-a) for English, and as BON in (41-b) for Catalan.

(41)   a. a **good** while, a **good** thirty minutes
        b. una **bona** estona, un **bon** misteri, un **bon** embolic
            a   good while   a good mystery a   good mess

In (41), we do not really associate a while or a mystery or a mess with a purpose that can be positively evaluated. In fact, if we add a *for*-clause, as in (40), it is not interpreted as the purpose for which N is good (42).

(42)   ??Thirty minutes is a **good** while *for reading a squib.*

Rather, we propose to relate these uses to BEN, since they also involve intensification and cannot be embedded under negation or in questions (43).

(43)   a. *It didn't take John a **good** while to finish his homework.
        b. *En Joan s'ha   ficat en un **bon** embolic?
            the John SELF has put in a   good mess
            Intended: 'Did John get himself into a good mess?'

In spite of these differences, we argue that this intensifying use of *good* is still related to subsective *good*. Romance provides us with additional evidence that BON is closer to the subsective rather than the intersective interpretation of the adjective. Syntactically, the subsective interpretation correlates with ADJ-N order, while the intersective one correlates with N-ADJ order (but see [7] for qualifications); morphologically, *good* surfaces as *bon* when preposed and as *bo* when postposed (44).

(44)   a. un **bon** alumne
            a   good student
            ≡ good as a student
        b. un alumne **bo**
            a   student good
            ≡ a student with goodness

The contrast between (45-a) and (45-b) is evidence that, under the intensifying reading, BON has to be interpreted subsectively.

(45)   a. un **bon** misteri
            a   good mystery
        b. *un misteri **bo**
            a   mystery good

While a child, a lunch, a rock and a student can be good for a purpose, it is not sound to predicate goodness (in that sense) of a while, a mystery or a mess, which is why examples like (45-b) are unacceptable.

Analogously to BEN, we propose that BON is a modifier that applies to a property and an individual and returns a content that is delivered at the CI tier. As expected from a subsective modifier, BON is not a predicate of individuals, but a predicate modifier, a function from properties to properties. Thus, the lexical semantics of BEN and BON both share the predicate **good**, but rather than applying to an individual (evaluative *good*) or an event (manner *well*), they select good property ascriptions from a set of saying events. The reason why one use surfaces as BEN and the other as BON is purely syntactic in our account: Whereas the former is an ad-adjectival modifier, the latter is an ad-nominal modifier. One advantage of our analysis that treats BON on a par with BEN and thus not as a degree modifier, is that we do not need to posit the existence of gradable nouns (cf. the discussion in [6]) to explain the distribution of BON. We only need to assume that certain nouns are vague, and this is what licenses modification by BON.

5 Conclusions

In this paper we have provided a semantic account of the Catalan ad-adjectival modifier *ben* 'well', which yields intensification by, we argue, positively evaluating a property ascription. Formally, this translates as applying the predicate **good** to the saying event available to any utterance. Thus, the output of this modification is a manipulation of the performative rather than descriptive content of the utterance, in other words, a Conventional Implicature rather than at-issue meaning. We also suggested that this analysis can be transferred to similar uses of adnominal *bon* 'good'.

Issues for future research include a thorough analysis of the distribution of the properties of the nouns that can be modified by BON. Furthermore, it could be interesting to explore the additional inferences that may arise from the predication of **good** such as satisfaction or else irony depending on whether the adjective is positive or pejorative. Finally, it would be worth checking whether our account can be extended to elements that are used in other languages to render the meaning of BEN, such as German *ganz schön, richtig* and *schon*, or English *pretty*.

References

1. Asher, N.: Lexical Meaning in Context: A Web of Words. Cambridge University Press, Cambridge (2011)
2. Bally, C.: Linguistique générale et linguistique française. A. Francke, Bern (1944)
3. Bolinger, D.: Degree Words. Mouton, The Hague (1972)
4. Bonami, O., Godard, A.: Lexical semantics and pragmatics of evaluative verbs. In: McNally, L., Kennedy, C. (eds.) Adverbs and Adjectives: Syntax, Semantics and Discourse, pp. 274–304. Oxford University Press, Oxford (2008)
5. Burnett, H.: The grammar of tolerance: on vagueness, context-sensitivity and the origin of scale structure. Ph.D. thesis, UCLA (2012)
6. Constantinescu, C.: Gradability in the Nominal Domain. LOT Dissertation Series 288. LOT, Utrecht (2011)
7. Demonte, V.: El adjetivo: Clases y usos. La posición del adjetivo en el sintagma nominal. In: Demonte, V., Bosque, I. (eds.) Gramática descriptiva de la lengua española, pp. 129–216. Espase Calpe, Madrid (1999)
8. Eckardt, R.: Adverbs, Events, and Other Things: Issues in the Semantics of Manner Adverbs. Max Niemeyer Verlag, Tübingen (1998)
9. Ernst, T.: Speaker-oriented adverbs. Nat. Lang. Linguist. Theory **27**, 497–544 (2009)
10. Giannakidou, A.: Affective dependencies. Linguist. Philos. **22**, 367–421 (1999)
11. González-Rivera, M., Gutiérrez-Rexach, J.: On the syntax and semantics of extreme-degree modifiers in Puerto Rican Spanish. Paper presented at LSRL 42, Southern Utah University (2012)

12. González-Rodríguez, R.: Negación y cuantificación de grado. In: Villayandre, M. (ed.) Actas del XXXV Simposio Internacional de la Sociedad Española de Lingüística, pp. 853–871. Universidad de León, León (2006)
13. Hernanz, M.L.: Polaridad y modalidad en español: entorno a la gramática de BIEN. Research report GGT-99-6, Universitat Autònoma de Barcelona (1999). http://seneca.uab.es/ggt/membres/hernanz.htm
14. Hernanz, M.L.: Assertive 'bien' in Spanish and the left periphery. In: Beninca, P., Munaro, N. (eds.) Mapping the Left Periphery: The Cartography of Syntactic Structures, pp. 19–62. Oxford University Press, Oxford (2010)
15. Kennedy, C.: The composition of incremental change. In: Demonte, V., McNally, L. (eds.) Telicity, Change, State: A Cross-Categorial View of Event Structure, pp. 103–121. Oxford University Press, Oxford (2012)
16. Kennedy, C., McNally, L.: Scale structure, degree modification, and the semantics of gradable predicates. Language 81(2), 345–381 (2005)
17. Lasersohn, P.: Pragmatic halos. Linguist. Philos. 75, 522–571 (1999)
18. Lasersohn, P.: Context dependence, disagreement, and predicates of personal taste. Linguist. Philos. 28, 643–686 (2005)
19. Liu, M.: Multidimensional Semantic of Evaluative Adverbs. Brill, Leiden (2012)
20. Liu, M.: The projective meaning of evaluative adverbs. Ms. University of Osnabrück (2014)
21. McNally, L., Kennedy, C.: Degree vs. manner well: a case study in selective binding. In: Pustejovsky, J., Bouillon, P., Isahara, H., Kanzaki, K., Lee, C. (eds.) Advances in Generative Lexicon Theory, pp. 247–262. Springer, Dordrecht (2013)
22. Morzycki, M.: Adjectival extremeness: degree modification and contextually restricted scales. Nat. Lang. Linguist. Theory 30(2), 567–609 (2012)
23. McNally, L., Boleda, G.: Relational adjectives as properties of kinds. In: Bonami, O., Hofherr, P.C. (eds.) Empirical Issues in Syntax and Semantics 5, pp. 179–196. CSSP, Paris (2004)
24. Nilsen, Ø.: Domains for adverbs. Lingua 114(6), 809–847 (2004)
25. Parsons, T.: Events in the Semantics of English: A Study in Subatomic Semantics. Current Studies in Linguistics Series 19. MIT Press, Cambridge (1990)
26. Piñón, C.: Speech-act adverbs as manner adverbs. Ms. Université Lille 3 (2013)
27. Potts, C.: The Logic of Conversational Implicature. Oxford University Press, Oxford (2005)
28. Potts, C.: The expressive dimension. Theoret. Linguist. 33(2), 165–197 (2007)
29. Pustejovsky, J.: The Generative Lexicon. MIT Press, Cambridge (1995)
30. Rappaport Hovav, M., Levin, B.: Reflections on manner/result complementarity. In: Doron, E., Rappaport Hovav, M., Sichel, I. (eds.) Syntax, Lexical Semantics, and Event Structure, pp. 21–38. Oxford University Press, Oxford (2010)
31. Sassoon, G., Toledo, A.: Absolute and relative adjectives and their comparison classes. Ms. University of Amsterdam and Utrecht University (2011)
32. Schäfer, M.: Resolving scope in manner modification. In: Bonami, O., Hofherr, P.C. (eds.) Empirical Issues in Syntax and Semantics 7, pp. 351–372. CSSP, Paris (2008)
33. Stephenson, T.: Towards a theory of subjective meaning. Ph.D. thesis, MIT (2007)
34. Umbach, C.: Evaluative propositions and subjective judgments. Ms. Zentrum für Allgemeine Sprachwissenschaft, Berlin (2014)
35. Zwarts, F.: Nonveridical contexts. Linguist. Anal. 25(3–4), 286–312 (1995)

Strict Comparison and Weak Necessity: The Case of Epistemic *Yào* in Mandarin Chinese

Zhiguo Xie[✉]

The Ohio State University, Columbus OH 43210, USA
xie.251@osu.edu

Abstract. The epistemic use of *yào* in Mandarin Chinese, often translated as 'should' in English, shows certain interesting peculiarities. In this paper, I first describe the empirical properties of epistemic *yào*. The occurrence of epistemic *yào* is restricted only to certain comparative constructions, but forbidden in other degree constructions or non-degree constructions. It cannot appear above or below negation. It has a quantificational force stronger than that of existential modals, yet weaker than that of strong necessity modals. It can appear with another epistemic modal *yīnggāi*, which has a very similar modal flavor and an identical quantificational force. When co-occurring, however, the two epistemic modals have to follow a strict word order. Next, I examine whether the above empirical properties of epistemic *yào* arise as lexical idiosyncrasies, from syntax, semantics, or their interface. Wherever relevant in the discussion, I compare epistemic *yào* to the (near-)synonymous *yīnggāi*. The epistemic use of *yào* in Mandarin Chinese may constitute an interesting case of inter- and cross-linguistic variation in natural language modality.

1   Introduction

Modals in natural language often come with "peculiar" properties. To better understand the possible range of such peculiarities, it is an important and meaningful enterprise to provide both an empirical description and a theoretical analysis of interesting restrictions on the distribution and interpretation of modal elements across different languages. Certain peculiarities associated with a modal may receive a systematic explanation in syntax, semantics, and/or syntax-semantics interface, while certain other peculiarities may have to be wired in as lexical idiosyncrasies.

In this paper, I provide an empirical description as well as a theoretical analysis of the epistemic use of the modal *yào* 'should' in Mandarin Chinese. In Sect. 2, I discuss empirical characteristics regarding the use and meaning of epistemic *yào*. I pay particular attention to its distribution, quantificational force, and interaction with negation. I compare *yào* to the more commonly-used epistemic modal *yīnggāi* 'should'. In Sect. 3, I provide a formal analysis of the properties observed with epistemic *yào*. I show that certain properties of the modal arise from its syntax, semantics, and syntax-semantics interface, while certain other properties are best treated as lexical idiosyncrasies. Section 4 concludes the paper.

© Springer-Verlag Berlin Heidelberg 2015
T. Murata et al. (Eds.): JSAI-isAI 2014 Workshops, LNAI 9067, pp. 130–143, 2015.
DOI: 10.1007/978-3-662-48119-6_10

2   Empirical Properties of Epistemic *yào* in Mandarin Chinese

Like many other languages, Mandarin Chinese has a variety of modal elements. Among them, *yào*, which also can be used as a regular main verb meaning 'want, desire', is one of the most productive and versatile. For instance, it can be used as a deontic modal to express obligations, as a dynamic modal to express volitional future, or as a predictive modal (following Ren 2008). These several uses of *yào* have been studied from many different perspectives.

In addition, *yào* has an epistemic use which is, to my knowledge, typologically rare in that it carries several unique restrictions. Though this use has been mentioned by Chinese grammarians and linguists over the years, researchers have yet to provide a detailed empirical description, let alone a convincing theoretical treatise, of the properties of epistemic *yào*. My main goal in this section is to discuss empirical properties of epistemic *yào*. In my discussion, where relevant I compare epistemic *yào* to another modal *yīnggāi* 'should', which is often used to paraphrase the former modal.

2.1   Pattern of Distribution

First, the epistemic reading of *yào* is available only when it appears in certain comparative constructions. When *yào* appears in a non-comparative sentence, it cannot receive an epistemic reading. (1) is an example of the *bǐ*-comparative construction in Mandarin Chinese (e.g., Xiang 2005, Lin 2009).[1] It allows epistemic *yào* to appear in it. The speaker can use (1) to express, with high certainty, her belief that the house price in Beijing is higher than in Shanghai. The speaker also can use the modal *yīnggāi* 'should' in place of *yào* to express (roughly) the same proposition. The sentence (2), by contrast, does not involve any comparative construction, and is not compatible with epistemic *yào*. To express the intended meaning of (2) with epistemic *yào*, *yīnggāi* can be used.

(1) Běijīng   fángjià   yào/yīnggāi   bǐ   Shànghǎi   gāo.
     Beijing   house price   should   BI   Shanghai   high
     'The (average) house price in Beijing should be higher than in Shanghai.'

(2) huángjīn   jiàgé   *yào/√yīnggāi   zài   1500   yuán   shàngxià   fúdòng.
     gold   price   should   at   1500   dollar   around   fluctuate
     'The price of gold should be fluctuating around 1500 dollars (per ounce).'

Second, though previous literature has discussed the appearance of epistemic *yào* in the *bǐ*-comparative construction, few (if any) researchers have considered how epistemic *yào* fares with other comparative constructions. Like in many other languages, comparative constructions in Mandarin Chinese involve explicit

---

[1] The following abbreviations are used in this paper: CL = classifier, MOD = modifier marker, PERF = perfective marker, POS = positive morpheme, DIST = distributive marker.

or implicit comparison, depending on whether the ordering between two objects with respect to a gradable property is established by using special morphology of comparison or using the positive form of the gradable predicate (Kennedy 2007). Implicit comparative constructions are not compatible with epistemic *yào*. The *gēn x bǐ qǐlái* "compared to *x*" comparative is an implicit comparison strategy used in Mandarin Chinese (Erlewine 2007). It does not allow epistemic *yào* to appear in it. The sentence in (3), for instance, is only acceptable without the epistemically intended *yào*.

(3) gēn   tā   dìdi   bǐ-qǐlai,   xiǎomíng (*yào) suànshi   hěn   gāo.
    with   his   brother   compare-*qilai* Xiaoming should considered POS   tall
    Intended: 'Compared to his brother, Xiaoming should be considered tall.'

By contrast, many explicit comparative constructions are compatible with epistemic *yào*. The sentence in (1) already demonstrated the compatibility of epistemic *yào* with the *bǐ* comparative. Several other explicit comparative constructions in Mandarin Chinese have been discussed in the literature. The so-called transitive comparative construction, in which the standard of comparison appears right after the gradable predicate, allows epistemic *yào* to appear in it (4). Similarly for the closely-related *chū* comparative, in which the degree morpheme *chū* intervenes between the standard of comparison and the gradable predicate. Some other comparative constructions that "licenses" epistemic *yào* include the *gèng* comparative (5) and the *yu* comparative (6). Moreover, *yào* in such constructions can be changed to *yīnggāi* without any significant effect on the grammaticality judgment or intuitive meaning of the sentences.

(4) Wángjùn   yào   gāo   (chū)   Zhèngzhāng   yī   ge   tóu.
    Wangjun   should   tall   exceed   Zhengzhang   one   CL   head
    'Wangjun should be a head taller than Zhengzhang.'

(5) (?)zhè   kē   méiguī,   huā   hóng,   yèzi   yào   gèng   lǜ.
    this   CL   rose   flower   red   leaf   should   GENG   green
    'This rose, its flowers are red; its leaves should be even greener (than its flowers are red).

(6) hòuniǎo   de   shòumìng yào   cháng yu   qítā   niǎo lèi.
    migratory bird   MOD   life span   should   long   YU   other   bird kind.
    'The life span of migratory birds should be longer than that of other kinds.'

Third, though many degree constructions in Mandarin Chinese allow epistemic *yào*, not all of them do. For example, Mandarin Chinese has a degree construction which involves the possessive/existential verb *yǒu* and appears very similar to the *bǐ* comparative in the surface structure. It typically takes the form of "X + *yǒu* + Y + G," with X and Y being determiner phrases and G being a gradable predicate or a dimension noun (Xie 2014a). Epistemic *yào* cannot appear in this construction (7). Instead, *yīnggāi* can be used to express the meaning intended with *yào*.

(7) zhāngsān   de   chéngjì   *yào/√yīnggāi   yǒu   tā   gēge   hǎo.
    Zhangsan   MOD   grade   should   have   his   brother   good
    'Zhangsan's grade should be as good as his brother's'

Another degree construction in Mandarin Chinese is the so-called comparative correlative, which involves explicit comparison of the same or different individuals' degrees associated with a property (Lin 2007). The construction does not allow epistemic *yào*. The sentence in (8) is ungrammatical with *yào* appearing in it. Again, *yīnggāi* can be used before the first *yuè* to express (roughly) the same meaning as intended with epistemic *yào*.

(8) nà   ge   háizi   (*yào/√yīnggāi)   yuè   zhǎng   yuè   hǎokàn.
    that   CL   child   should   YUE   grow   YUE   good-looking
    'It should be the case that the more the child grows, the prettier she becomes.'

Fourth, the equative construction, marked with *hé/gēn/xiàng x yīyàng g* 'equally as *g* as *x*', does not allow epistemic *yào* to appear in it, either. However, it allows epistemic *yīnggāi*. This claim is illustrated by the sentence in (9), which is minimally different from (1) just in that it establishes an identity relation between the average house prices in Beijing and in Shanghai.

(9) Běijīng   de   fángjià   *yào/√yīnggāi   gēn   Shànghai   yīyàng   gāo.
    Beijing   MOD   house price   should   with   Shanghai   same   high
    'The (average) house price in Beijing should be as high as in Shanghai.'

2.2 *Yào* Co-occurring with *yīnggāi*

Fifth, I have shown above that when epistemic *yào* appears grammatically in a comparative sentence, it can be replaced with *yīnggāi*, and no significant change of grammaticality judgment or meaning is observed between the two choices. In addition, *yào* and *yīnggāi* can occur together as epistemic modals in certain explicit comparative sentences, a phenomenon that has escaped observation in previous research. The sentence in (10) illustrates the co-occurrence of the two modals, both with an epistemic reading. The subject, *jiāoqū de kōngqì*, is inanimate and non-volitional. This property of the subject rules out the deontic reading for *yīnggāi*, as well as the deontic and volitional future readings for *yào*. The sentence can be understood as describing the speaker's judgment about the *current*, not future, air quality in the suburb in relation to the city, thus ruling out the "predictive modal" reading for *yào* discussed in Ren (2008). Hence, it is safe to claim that both *yīnggāi* and *yào* in the sentence receive an epistemic reading.

(10) jīntiān   jiāoqū   kōngqì   yīnggāi   yào   bǐ   shìqū   hǎo.
     today   suburb   air   should   should   BI   city   good
     'Air in the suburb today should be better than in the city.'

For the co-occurrence of epistemic *yīnggāi* and *yào* to be grammatical, all the restrictions regarding epistemic *yào* must be observed. The co-occurrence of epistemic *yīnggāi* can never "coerce" epistemic *yào* to be acceptable in a sentence that does not allow the latter in the first place. In addition, in acceptable cases of *yào* co-occurring with *yīnggāi*, *yīnggāi* must precede *yào*; switching the order of the two epistemic modals would yield an ungrammatical sentence. This is illustrated by the acceptability contrast between (10) (see above) and (11).

(11) *jīntiān jiāoqū kōngqì yào   yīnggāi bǐ   shìqū hǎo.
     today   suburb air   should   should   BI   city   good

Co-occurrences of multiple modals are nothing rare in Mandarin Chinese. The interested reader can refer to Lin and Tang (1995) and Lin (2012), among several others, for related discussion. However, two epistemic modals of the same quantificational force are generally forbidden from occurring together. The sentence in (12), for example, involves epistemic modals *yídìng* and *bìrán* 'must' with the same universal quantificational force. It is not acceptable regardless how the two modals are ordered relative to each other. Epistemic *yào* and *yīnggāi*, as will be discussed shortly, have the same weak necessity quantificational force. In this sense, co-occurrence of epistemic *yào* and *yīnggāi* in a comparative sentence is an interesting exception that requires some independent explanation.

(12) *tā   yídìng   bìrán   xǐhuān nà   jiā   fàndiàn.
     he   must/definitely   must/definitely   like   that CL   restaurant
     Intended: 'He must like the restaurant.'

2.3   Lack of Scope Relation with Negation

Sixth, epistemic *yào* cannot enter into scope relation with negation in any way (Peng 2007). For instance, without occurrence of *bù* 'not', (13) would be grammatical. Adding *bù*, either before whether after *yào*, makes the sentence ungrammatical. In addition, epistemic *yào* cannot appear in a negative context in any other fashion. For example, it cannot participate in the A-not-A question, either, as illustrated in (14).

(13) diànzǐ   chǎnpǐn zhōngguó (*bù) yào   (*bù) bǐ   měiguó piányi.
     electronic product China   NEG   should NEG   BI   USA   cheap

(14) *hēi   zhīmá   jiàzhí yào   bù   yào   gāo   yu   bái   zhīmá.
     black sesame value   should NEG   should high YU   white sesame

In terms of interaction with negation, epistemic *yīnggāi* does not behave exactly the same as epistemic *yào*. Though epistemic *yīnggāi* cannot appear after negation or participate in the A-not-A question, it can appear before negation, whether in a comparative sentence (15) or elsewhere.

(15) diànzǐ   chǎnpǐn zhōngguó (*bù) yīnggāi (bù) bǐ   měiguó piányi.
     electronic product China   NEG   should   NEG   BI   US   cheap
     'For many electronic products, it should be the case that they are not cheaper in China than in US.'

2.4   Weak Necessity Quantificational Force

Seventh, different modals have different quantificational strengths. There is evidence to suggest that epistemic *yào* is a weak necessity modal that is comparable to the modals *should* and *ought to* in English. First, different from *kěnéng* 'possible', epistemic *yào* is not an existential modal that expresses the mere existence of relevant possibilities. For example, in the conversation in (16) between two speakers A and B, the first clause in B's responses indicates that B agrees with A's judgment about the reliability of diaries as compared to memoirs. The second clause in B's response is intended to be further elaboration of how she agrees. However, by using *kěnéng* 'possible', the second clause weakens, and as such, contradicts, the expressed agreement in the first clause. The weakening and contradiction is comparable to what is responsible for the infelicity of (17), which involves nominal quantificational phrases (cf., Copley 2006 and von Fintel and Iatridou 2006). Hence, epistemic *yào* has a stronger quantificational force than *kěnéng*.

(16) A: wǒ   juéde   rìjì   yào   bǐ   huíyìlù   kěkào.
          I   feel   diary   should   BI   memoir   reliable
          'I think that diaries should be more reliable than memoirs.'

      B: #wǒ   yě   zhème   juéde,   rìjì   kěnéng   bǐ   huíyìlù   kěkào.
          I   also so   feel   diary   possible BI   memoir   reliable
          'I think so, too. Diaries are possibly more reliable than memoirs.'

(17) A: jué   dàduōshù   rén   dōu   lái   le.
          outright   majority   people   DIST   come   PERF
          'The by far majority of people have come.'

      B: #duì,   yǒuxie   rén   lái   le.
          right   some   people   come   PERF
          'Right, some people have come.'

On the other hand, epistemic *yào* is somewhat weaker than canonical strong necessity modals like *yídìng* and *kěndìng* 'must, certainly.' This claim is evident from the fact that an epistemic modal statement expressed by *yào* can be ensued by a strong necessity epistemic statement, and reversing the order of the two statements would lead to infelicity (18). The pattern, again, is comparable to a statement involving a weaker quantifier followed by another statement involving a stronger quantifier (19). This similarity suggests that epistemic *yào* is not a strong necessity modal. Rather, it is similar to English *should* and *ought to* – as already argued by Copley (2006) and von Fintel and Iatridou 2006 – in being a weak necessity modal. Moreover, epistemic *yīnggāi* has the same quantificational force as epistemic *yào*: if *yào* in (16) and (18) is changed to *yīnggāi*, the acceptability judgment remains the same.

(18) a. tā   yào   bǐ   línju   yǒuqián,
          he   should BI neighbor rich
          shìshíshàng tā kěndìng bǐ línju   youqián.
          in fact   he certainly BI neighbor rich
          'He should be richer than his neighbors; in fact, he is certainly richer
          than his neighbors.'
      b. #tā kěndìng bǐ línju yǒuqián, shìshíshàng tā yào bǐ línju yǒuqián.

(19) a. He finished most of the tasks, in fact, he finished all of them.
      b. #He finished all of the tasks, in fact, he finished most of them.

To summarize, in this section I discussed several important properties of the epistemic use of *yào*. In my discussion, I compared epistemic *yào* to another epistemic modal *yīnggāi*. Epistemic *yào* is acceptable only in certain comparative constructions, and hence has a narrower distribution than the (near-) synonymous epistemic *yīnggāi*. The two epistemic modals can be used together, in which case *yīnggāi* must precede *yào*. Epistemic *yào* cannot appear above or under negation, while epistemic *yīnggāi* can appear above, though not under, negation. In terms of quantificational force, epistemic *yào* and *yīnggāi* both express weak necessity, comparable to English *should* and *ought to*.

3   Explaining Empirical Properties of Epistemic *yào*

In this section, I will address the question of where the above properties of epistemic *yào* each come from: whether they are lexical idiosyncrasies, or arise from syntax, semantics, or the interaction thereof.

3.1   Incompatibility with the Comparative Correlative

First, I posit that the incompatibility of epistemic *yào* with the comparative correlative construction, as illustrated by the sentence in (8), is most likely a lexical idiosyncrasy. It has been proposed by Lin (2007) that the comparative correlative construction involves a causation relation between degrees. This means that the construction involves a change of state, and is dynamic in nature. The unacceptability of (8) is due to the requirement that epistemic *yào* cannot be combined with a dynamic prejacent. Confirming this explanation is yet another observation that the degree achievement construction, which is dynamic as well (Kennedy and Levin 2008), is not compatible with epistemic *yào*. By contrast, *yīnggāi* is (at least marginally) compatible with a dynamic prejacent and can be used an epistemic modal in both comparative correlative and degree achievement constructions (20).

(20) nà   ge   háizi (*yào/?yīnggāi) měi   nián zhǎng gāo liǎng límǐ.
     that CL child should   every year grow tall two centimeter
     Intended: 'It should be the case that the child grows 2 cm taller each year.'

Some modals in other languages manifest a similar distinction regarding whether the epistemic reading is allowed with an eventive prejacent or not. For example, *must* and *cannot* in English are allowed to receive an epistemic reading only when it has a stative prejacent (21), but *may* and *might* can have an epistemic reading no matter whether it combines with a stative or eventive prejacent (22). To the best of my knowledge, the only attempt to address the distinction so far is Ramchand (2014). The basic idea of her analysis is to attribute the distinction to how (indexically vs. anaphorically) an epistemic modal anchors the denotation of the prejacent in terms of time and world. The distinction, therefore, is treated as a lexical property in her analysis. I assume that Ramchand's discussion applies to epistemic modals in Mandarin Chinese, as well. It is a lexical idiosyncrasy of epistemic *yào* that it cannot combine with dynamic comparative constructions.

(21) a. John must/cannot be in his office.   (epistemic or deontic)
  b. John must/cannot go to his office.   (deontic, ability (*cannot*))

(22) a. John may/might be in his office.   (epistemic)
  b. John may/might go to his office.   (epistemic)

3.2   Compatibility only with Certain Comparative Constructions

Epistemic *yào* is compatible only with certain explicit comparative constructions, viz. the *bǐ* comparative, the transitive comparative, the *chū* comparative, the *gèng* comparative, and the *yu* comparative. By contrast, it is not compatible with the *yào* degree construction, the equative construction marked with *hé/gēn/xiàng x yīyàng g* 'equally as *g* as *x*', or any non-degree construction.

A common characteristics among the comparative constructions in which epistemic *yào* can occur is that they all involve strict comparative morphology. For the *bǐ* comparative, different proposals have been entertained, but all of them include a strict comparative morpheme. Here, "strict comparison" means "greater/less than". Lin (2009), for instance, took a "direct" analysis of the *bǐ* comparative, and treated *bǐ* as an overt strict comparative morpheme. Xiang (2005) proposed a so-called "DegP-shell" analysis of the *bǐ* comparative. There are two degree heads in the syntactic structure, with the higher one occupied by *bǐ*, and the lower one by a covert strict comparative morpheme *exceed* that introduces an optional differential phrase. Liu (2011) posited that *bǐ* comparative contains either a strict comparative morpheme *geng* 'even-more' or its covert counterpart. It is sufficient to conclude that whatever form the currently available proposals for the syntax and semantics of the *bǐ* comparative take, they all include postulating some strict comparative morpheme, whether overtly or covertly.

The transitive comparative, along with the closely-related *chū* comparative construction, has been most extensively studied by Grano and Kennedy (2012). The transitive comparative requires the presence of a differential measure phrase. A differential measure phrase, in turn, "requires and is required by the presence of the degree morpheme" (p. 244). For the transitive comparative, the degree

morpheme contributes a strict comparative meaning. The preposition *chū* is analyzed by Grano and Kennedy (2012) to be an overt counterpart of such a strict
comparative morpheme. As for the *yu* comparative, Xie (2014b) showed that it
does not allow differential measure phrases. By capitalizing on this observation,
Xie showed *yu* in the *yu* comparative to be in complimentary distribution with
the comparative morpheme in the transitive comparative construction. Hence,
it is reasonable to claim that *yu* itself is a strict comparative morpheme. For the
gèng comparative, Liu (2010) has argued that *gèng* itself is a strict comparative
morpheme (cf., Liu 2011).

By contrast, the *yǒu* degree construction has been shown by Xie (2011, 2014b)
to be an equative construction comparable to the *as. . . as* construction in English. According to Xie's idea, its LF structure of the *yǒu* degree construction
involves a covert degree morpheme, which encodes a "greater than or equal to"
relation. It does not have a strict comparative morpheme. The equative construction marked by *hé/gēn/xiàng x yīyàng g* specifies a strict identity relation
between two entities, and does not involve a strict comparative morpheme. As for
the implicit comparative construction marked by *gēn x bǐ qǐlái* "compared with
x", it makes use of "the inherent context sensitivity of the positive (unmarked)
form" of gradable predicates (Kennedy 2007: p. 143). Its structure does not
involve a comparative morpheme at all.

Based on the above discussion, it is reasonable to posit that the presence of a
strict comparative morpheme (whether overt or covert) in the syntactic structure
of a degree construction is responsible for the acceptability of epistemic *yào* in the
construction. Those constructions without a strict comparative morpheme do not
allow epistemic *yào*. There may be more than one way to represent the restriction
in syntax. One option, within the Minimalist Program, is to say that in its epistemic use, *yào* somehow bears an uninterpretable Comp(arative) feature which
has to be checked by a matching Comp feature. Comparative constructions like
the *bǐ* and transitive comparative constructions provide such a matching feature,
while the equative constructions and implicit comparison do not.

Obviously, I have taken a syntactic approach to explaining the distribution
restriction of epistemic *yào*. The reader might ask whether a semantically-oriented
approach, say within Kratzer's (1981) possible-world semantics framework of
modality, will work. As far as I can see, the answer is negative. If we include
in the semantic definition of epistemic *yào* the "strict comparison" contexts in
which the modal can appear, a most likely component to encode the information
is in the domain of quantification, by claiming that the worlds accessible from the
speaker's epistemic state in her base world all involve strict comparison. However,
this restriction is at best vacuous, because any world can, in principle, support
strict comparison of any sort.

A second semantically-oriented option is to require, or presuppose, that the
prejacent of epistemic *yào* express a strict comparative relation. Then, the question comes down to how to take an intensional proposition, which is potentially
an indefinite set of possible worlds, and check whether the proposition expresses

a strict comparative relation. Though this option might be plausible, it is not clear to me how to represent it in a model-theoretic fashion.

3.3   Co-occurrence of *yào* and *yīnggāi*

It has been observed above that when epistemic *yīnggāi* and *yào* occur together, the former must appear before the latter. I argue that this property has to do with a very fine semantic distinction within epistemic modals as well as a structural constraint that reflects the semantic distinction. Lyons (1977) classified epistemic modals into subjective and objective sub-types. Subjective epistemic modals express the speaker's judgment based on what (she thinks) she knows. Objective epistemic modals, by contrast, express the speaker's judgment based on observable evidence often available to the speaker, the hearer, and possibly other people in the local speech community (Papafragou 2006). Despite the subjective vs. objective distinction, epistemic modals in general contribute semantic content and may have syntactic reflection thereof (Hacquard and Wellwood 2012).

Though *yào* and *yīnggāi* are both epistemic modals, the former is an objective epistemic modal, and the latter is used subjectively (Peng 2007, Peng and Liu 2012). Since they bear different sub-flavors of epistemic modality, it is not surprising that they can co-occur, in spite of the fact that they have the same quantificational force (a point to be discussed shortly). The two stacked modals express the speaker's judgment based on her private perception of relevant objective evidence available to her (and possibly to her local speech community, as well). Compared to its counterpart without *yīnggāi*, the sentence in (10) (repeated below) has an extra layer of uncertainty which arises from the speaker's indeterminacy typically associated with doxastic beliefs. By contrast, compared to its counterpart without *yào*, (10) does not express a mere guess on the part of the speaker, but conveys that the speaker actually bases her judgment on some objective evidence (e.g., the facts that there is a larger area of forest-covered hills in the suburb area, that it has just rained in the suburb but not in the city, etc.).

(10) jīntiān   jiāoqū   kōngqì   yīnggāi   yào   bǐ   shìqū   hao.
      today   suburb   air   should   should   BI   city   good
      'Air in the suburb today should be better than in the city.'

In addition, Peng (2007) and Peng and Liu (2012) posited that in Mandarin Chinese, a subjective (interpretation of an) epistemic modal should always appear before an objective (interpretation of an) epistemic modal. How to represent this structural restriction is not very material to the current paper. Presumably, the restriction arises from the syntax-semantics interface of epistemic modals. For our purpose, the most important thing to note is that Peng's (2007) and Peng and Liu's (2012) generalization is what lies behind the ordering constraint of *yīnggāi* and *yào* occurring together as epistemic modals: the former, a subjective epistemic modal, should appear before the latter, an objective epistemic modal.

3.4 Semantic Meaning of Epistemic *Yào*

I have shown above that the distribution restriction of epistemic *yào* is due to lexical and syntactic reasons. The semantic definition of the modal does not need to, and in fact cannot, encode the restriction. In Sect. 2, I also indicated that epistemic *yào* is semantically identical to epistemic *yīnggāi*, modulo the distinctions with regard to objectivity/subjectivity and scope relation with respect to negation (viz., epistemic *yào* cannot form scope relation with negation at all, whereas epistemic *yīnggāi* can scope above, but not under, negation). The objectivity/subjectivity distinction is clearly semantic in nature; it will be encoded in the modal base in the semantic definitions of the two modals. The distinction with regard to scopal relation with negation presumably has to do with the polarity properties of the two modals, and will be addressed in the next sub-section.

Copley (2006) and von Fintel and Iatridou (2006) addressed several important semantic properties, especially the weak necessity quantificational force, of English modals *should* and *ought to*. Epistemic *yīnggāi* and *yào* – ignoring the distinctions mentioned above for the moment – manifest properties that are comparable to *should* and *ought to*. In this paper, I primarily draw on Copley (2006) to define the semantics of epistemic *yào* and *yīnggāi*. The intuition is that a weak necessity epistemic modal requires: (i) the prejacent proposition of the modal be true in every world that is accessible from the speaker's knowledge/belief status in her base world and that is ranked as most highly plausible according to some ideal, and (ii) the prejacent proposition would be *allowed* (but not required) to be false if the speaker found herself in a different knowledge/belief status. The first requirement specifies that a weak necessity modal universally quantifies over a "most relevant" set of possible worlds – most relevant in the sense that the worlds are directly accessible from the speaker's base world. The second requirement keys in the possibility of the prejacent proposition being false in a world that is (potentially) only compatible with a world in which the speaker finds herself dislocated from her current being (so to speak). It is the secondary possibility – which exists only in a "stretched" domain of quantification – that contributes the perceived "weakness" in the quantificational force of weak necessity modals.

Regarding the objectivity/subjectivity distinction between epistemic *yào* and *yīnggāi*, I assume that it arises from the choice of modal base. For epistemic *yào*, the speaker's knowledge/belief is required to be based on objective evidence that is available to her, thus making the modal base objectively-oriented. By contrast, the modal base for epistemic *yīnggāi* is concerned with the speaker's subjective perception of evidence or probably even arbitrary judgment.

The semantics of epistemic *yào* (time variable ignored) is defined in (23), where MB_{obj} indicates that the modal base for epistemic *yào* is objective in nature. ALT is a function that takes an element and returns a set of alternatives to the element. The semantics of epistemic *yīnggāi* is the same as that of epistemic *yào*, except for the modal base being MB_{sub}.

(23) $[[y\grave{a}o_{\text{epistemic}}]] = \lambda w \lambda p. \forall w'(w' \in \text{HIGH-PLAUSIBILITY}(MB_{\text{obj}}(w)) \rightarrow p(w') = 1) \wedge \exists M(M \in \text{ALT}(MB_{\text{obj}}(w)) \wedge \exists w''(w'' \in \text{HIGH-PLAUSIBILITY}(M) \wedge p(w'') = 0))$

3.5   Negation and *yào*

It has been noted above that negation is not allowed to occur in an epistemic *yào* sentence, regardless of the relative position between negation and *yào*. As I will argue below, actually there are two separate yet related stories behind this restriction. One has to do with why epistemic *yào* (and epistemic *yīnggāi*, for that matter) cannot appear under negation. The other has to do with why the reverse order is not allowed, either.

Let us first address the former question. The idea that I would like to pursue is that when epistemic *yào* or *yīnggāi* appears under negation (often marked by *bù* 'not'), semantically it is equivalent to the existential epistemic modal *kěnéng* appearing above negation. It is lexical competition between *bù yào/bù yīnggāi* (epistemically intended) and *kěnéng bù*, I hypothesize, that leads to the unacceptable status of the former two phrases. The semantic definition of *bù yào* (epistemically intended) is given in (24). Among the two conjuncts linked by "∨," the second one basically states that all modal bases that are alternative to the one accessible from the speaker's base world can verify the prejacent proposition of epistemic *yào*. However, this requirement cannot hold in general, as it amounts to saying that the modal base accessible from the speaker's base world ranks the least ideal among all possible modal bases. Nothing a priori renders such an "ugly" status for the modal base accessible from the speaker's base world. Hence, the second conjunct is constantly false. The semantics of *bù yào_epistemic* is just equivalent to the first conjunct, which in turn is equivalent to the semantics of *kěnéng bù*. Due to the semantic equivalence, *bù yào_epistemic* competes with *kěnéng bù*. The former loses to the former, presumably because *yào* carries more morpho-syntactic restrictions and such restrictions do not have any semantic import or reflection.

(24) $[[b\grave{u} \ y\grave{a}o_{\text{epistemic}}]] = \lambda w \lambda p. \exists w'(w' \in \text{HIGH-PLAUSIBILITY}(MB_{\text{obj}}(w)) \wedge p(w') = 0) \vee \forall M(M \in \text{ALT}(MB_{\text{obj}}(w)) \rightarrow \forall w''(w'' \in \text{HIGH-PLAUSIBILITY}(M) \rightarrow p(w'') = 1))$.

Regarding the fact that epistemic *yào* cannot appear above negation, I propose, albeit rather tentatively, that it has to do with the polarity property of the modal. Iatridou and Zeijlstra (2013) showed that deontic and epistemic modals can be grouped as positive-polarity items (PPIs), negative polarity items (NPIs), and polarity-neutral items. The classification does not only apply to English modals, but to modals in many other languages. The three types of modals manifest rather distinguished behaviors with respect to their scope relation with respect to negation. For the purpose of this paper, it suffices to note that "all neutral and NPI modals scope under negation" (Iatridou and Zeijlstra 2013: p. 564).

Assuming that modals in Mandarin Chinese also carry polarity distinctions, epistemic *yào* cannot be an NPI, because it can occur in positive sentences.

It is very likely not a PPI, either, for it does not pass PPI-hood tests (Szabolcsi 2004). For instance, PPIs (like 'someone' and 'must') are acceptable in the scope of clause-external negation (25). However, epistemic *yào* cannot appear in such a context, as suggested by the unacceptability of the sentence in (26).

(25) a. No one says that the president found someone.

b. I do not think that he must come home tonight.

(26) wǒ   bú   rènwéi   tā   (*/??yào)   bǐ   tā   dìdi   gāo.
I   not   believe   he   should   BI   his   brother   tall
Intended: 'I do not think that he should be taller than his younger brother.'

Hence, epistemic *yào* patterns with such English (semi-)modals as *have to* and *need to* in being a polarity-neutral item. An interesting characteristic of polarity- neutral modals is that they scope under negation for semantic interpretation. Therefore, even when epistemic *yào* appears above negation on the surface, it has to end up scoping under negation semantically. It has been just established above, however, that epistemic *yào* does not allow for such a semantic scope relation.

4   Conclusions

Modals can carry all sorts of peculiarities, in terms of distribution and interpretation. In this paper, I provided both empirical description and theoretical investigation of the rarely-discussed epistemic use of *yào* in Mandarin Chinese. Epistemic *yào* can only occur in certain comparative constructions. It cannot enter into any scope relation with negation. Its quantificational force is stronger than that of existential modals, yet at the same time weaker than that of strong necessity modals. Epistemic *yào* can appear with another epistemic modal *yīnggāi*, which has the same modal flavor (broadly speaking) and quantificational force. When the two epistemic modals co-occur, however, *yīnggāi* must precede *yào*. In the theoretical analysis component, I examined where each property of *yào* comes from: lexical idiosyncrasies, syntax, semantics, or the interface between syntax and semantics. I think that the epistemic use of *yào* constitutes an interesting case in studying intra- and cross-linguistic variation in natural language modality.

References

Copley, B.: What Should Should Mean. Ms. (2006)

Erlewine, M.: A New Syntax-semantics for the Mandarin Bi Comparative. M.A. Thesis. University of Chicago (2007)

Grano, T., Kennedy, C.: Mandarin transitive comparatives and the grammar of measurement. J. East Asian Linguis. **21**, 219–266 (2012)

Hacquard, V., Wellwood, A.: Embedding epistemic modals in english: a corpus-based study. Semant. Pragmatics **5**(4), 1–29 (2012)

Iatridou, S., Zeijlstra, H.: Negation polarity and deontic modals. Linguist. Inquiry **44**, 529–568 (2013)

Kennedy, C.: Modes of comparison. In: Proceedings of 43rd Meeting of Chicago Linguistics Society, pp. 141–165. Chicago Linguistics Society, Chicago (2007)

Kennedy, C., Levin, B.: Measure of change: the adjectival core of degree achievements. In: McNally, L., Kennedy, C. (eds.) Adjectives and Adverbs: Syntax, Semantics and Discourse, pp. 156–182. Oxford University Press, Oxford (2008)

Kratzer, A.: The notional category of modality. In: Eikmeyer, H.-J., Rieser, H. (eds.) Words, Worlds, and Contexts: New Approaches in Word Semantics. Walter de Gruyter, Berlin (1981)

Lin, J.: On the semantics of comparative correlatives in mandarin chinese. J. Semant. **24**, 169–213 (2007)

Lin, J.: Chinese comparatives and their implicational parameters. Nat. Lang. Seman. **17**, 1–27 (2009)

Lin, T.: Multiple-modal constructions in mandarin chinese and their finiteness properties. J. Linguist. **48**, 1–36 (2012)

Liu, C.S.L.: The Chinese geng clausal comparative. Lingua **120**, 1579–1606 (2010)

Liu, C.S.L.: The Chinese Bi comparative. Lingua **121**, 1767–1795 (2011)

Lin, J., Tang, C.-C.J.: Modals as verbs in chinese: a GB perspective. Bull. Inst. Hist. Philology **66**, 53–105 (1995)

Lyons, J.: Semantics, vol. 2. Cambridge University Press, Cambridge (1977)

Papafragou, A.: Epistemic modality and truth conditions. Lingua **116**, 1688–1702 (2006)

Peng, L.: Xiandai Danyu Qingtai Yanjiu ['Study on Modality in Modern Chinese']. China Social Science Press, Beijing (2007)

Peng, L., Liu, Y.: Hanyu de Zhuguan Qingtai yu Keguan Qingtai ['Subjective and Objective Modality in Chinese']. Chinese as a Second Language Research **1**, 243–265 (2012)

Ramchand, G.: Stativity and present tense epistemics. In: Proceedings of 24th Meeting of Semantics and Linguistic Theory, pp. 102–121. CLC Publications, Ithaca (2014)

Ren, F.: Futurity in Mandarin Chinese. Ph.D. Dissertation, University of Texas (2008)

Szabolcsi, A.: Positive polarity - negative polarity. Nat. Lang. Linguist. Theory **22**, 409–452 (2004)

Von Fintel, K., Iatridou, S.: How to Say Ought in Foreign: The Composition of Weak Necessity Modals. In: Guéronand, J., Lecarme, J. (eds.) Time and Modality, pp. 115–141. Springer, Heidelberg (2006)

Xiang, M.: Some Topics in Comparative Constructions. Ph.D. Dissertation, Michigan State University (2005)

Xie, Z.: The Relevance of Gradability in Natural Language: Chinese and English. Ph.D. Dissertation, Cornell University (2011)

Xie, Z.: The degree use of the possessive verb you in Mandarin Chinese: a unified analysis and its theoretical implications. J. East Asian Linguist. **23**, 113–156 (2014a)

Xie, Z.: The Yu comparative construction in Mandarin Chinese. In: Piñón, C. (ed.) Empirical Issues in Syntax and Semantics, pp. 143–160. CNRS, Paris (2014b)

Computing the Semantics of Plurals and Massive Entities Using Many-Sorted Types

Bruno Mery[1,2(✉)], Richard Moot[1,2], and Christian Retoré[1,2]

[1] LaBRI-CNRS, Université de Bordeaux, Talence, France
[2] LIRMM-CNRS, Université Montpellier 2, Montpellier, France
bruno.mery@me.com

Abstract. We demonstrate how the specifics of the semantics for collective, distributive and covering readings for plurals and mass nouns can be integrated in a recent type-theoretical framework with rich lexical semantics. We also explore the significance of an higher-order type system for gradable predicates and other complex predications, as well as the relevance of a multi-sorted approach to such phenomena. All the while, we will detail the process of analysis from syntax to semantics and ensure that compositionality and computability are kept.

Keywords: Lexical Semantics · Plural Nouns · Mass Nouns · Higher-Order Logic · Syntax and Semantics Analysis · New Type Theories

The distinction between *massive* and *countable* entities is similar to a classical type/token distinction — as an example of the type/token distinction, "the bike" can refer both to a single physical bicycle (as in the sentence "the bike is in the garage") but also the the class of all bicycles (as in the sentence "the bike is a common mode of transport in Amsterdam"). However, linguists such as Brendan Gillon in [5] warn against such a generalisation (long made in the literature) and remark that, as far as the language is concerned, mass nouns are more alike to the collective readings of pluralised count nouns. Among the many similarities is, for instance, the identical behaviour of plurals and mass nouns with cumulative readings: "Both the pens on the desk and the pens in storage use black ink, so I only have black pens" and "There is red wine on display and red wine in the back, so we only have red" are logically similar (see [5] for discussion). Several different approaches have been proposed to account for the specific semantic issues of mass nouns, from Godehard Link's augmented mereological approach in [12] to David Nicolas' revision of plural logic in [25], all remarking upon this similarity.

The present article is based upon two presentations in LENLS 10 and 11, the former dealing with plurals and the latter with the semantics of massive entities in a many-sorted type system, with an emphasis on the latter. This work is supported by the CNRS with the PEPS CoLAN, and by the ANR ContInt Polymnie.The authors are indebted to all the LENLS 11 committee and organisers, to the reviewers for their comments, as well as to Nicholas Asher, Daisuke Bekki, Stergios Chatzikyriakidis, Robin Cooper, Thomas Icard, Robert Levine, Yusuke Kubota, Zhaohui Luo and Koji Mineshima (among many others) for many helpful discussions.

© Springer-Verlag Berlin Heidelberg 2015
T. Murata et al. (Eds.): JSAI-isAI 2014 Workshops, LNAI 9067, pp. 144–159, 2015.
DOI: 10.1007/978-3-662-48119-6_11

Many different formalisms, using advanced type theories for the purpose of modelling semantics, have been recently proposed, see e.g. [1,3]. Among those, we proposed a semantic framework based on a multi-sorted logic with higher-order types in order to account for notoriously difficult phenomena pertaining to lexical semantics – see [30] for a recent synthesis.

The aim of the present paper is twofold: to demonstrate that the semantics for plural readings and mass nouns can be integrated in our framework, and that having a multi-sorted logic is an advantage in tackling related issues. Our approach emphasises the *computable* aspect of such a framework. We detail the analysis from syntax to semantics (using state-of-the-art categorial grammars in the tradition of Lambek [9]), with a sound logical framework for meaning assembly and the computation of logical representations, while keeping compositionality as a basic principle, as in Montague's original semantic program [17]. As [19,33] have shown, this process can be fully and transparently implemented.

1 Compositional Lexical Semantics and New Type Theories

1.1 Related Work in Formal Lexical Semantics

Type-theoretical framework for computational semantics based on "new" type theories (that mostly stem from Martin-Löf Type Theory and a practical-oriented understanding of the Curry-Howard Correspondence) are relatively recent additions to the formal semantics scene, beginning with Aarne Ranta's seminal work, [28]. They can be used to present a solution to issues of lexical semantics such as polysemy, deriving from James Pustejovsky's Generative Lexicon – [26]. These formalisms have matured during the last 20 years to become a set of logically sound compositional frameworks that can be inserted in the Montagovian chain of analysis: see [1,3,13], and many others.

1.2 Our Formal and Computational Model, ΛTY_n

Our system, ΛTY_n, is based on System-F, using higher-order types with many sorts and coercive sub-typing for modelling different phenomena. Detailed in [2, 14], it has been constantly upgraded and aggregates many different phenomena such as deverbals [29] or the narrative of travel [10,27]. See [30] for a recent, complete, open-access synthesis.

Summarising, our system is formally based on a higher-order version of λ-calculus in the tradition of Girard's System-F (see [6]), in which types can be abstracted and applied like standard λ-terms: for example, $\Lambda\alpha\lambda P^{\alpha\rightarrow t}.(\iota\ P)$ is a functional term that requires both a type α and a λ-term P, a predicate of type $\alpha \rightarrow t$. This example is the *selection* operation, used to model definite articles such as *the*. (We use the ι operator as envisioned by Russel, where $\iota\ P$ can be thought of as "the most salient x in the current context such that $P(x)$ is true"). In addition, we use a logic with n base types in addition to propositions

Semantic Type	Implied meaning
$\alpha, \beta, \gamma, \tau$	Type variables
t	The type for propositions (or "truth values")
e	The Montagovian entities
	All subsequent sorts are subtypes of e
φ	The sort for *physical* objects
H	The sort for individual human beings
Pl	The sort for *places*, locations
Container,	Other specific sorts,
Animal,	which are given explicitly
Food, ...	as needed
$\mathbf{g}_\tau$	The sort for groups of individuals of sort τ
$\mathbf{m}_\tau$	The sort for masses of sort τ

and functional types TY_n, described in [24]. We will use the following conventions in the present paper:

The operators $\mathbf{g}_\tau$ and $\mathbf{m}_\tau$ are constructors for groups of individuals and masses of measurable quantities of other, pre-defined sorts. With details given in Sect. 3.3, the main points are as follows: the lexicon defines for what sort τ there exists a sort $\mathbf{g}_\tau$ and $\mathbf{m}_\tau$. Group types denote countable individuals ($\mathbf{g}_H$ is the type for groups of humans such as *committee* and *team*) and mass types denote measurable quantities ($\mathbf{m}_\varphi$ is the type for *water* and *sand*, while *stone* is ambiguous between types φ and $\mathbf{m}_\varphi$). $\mathbf{g}$ and $\mathbf{m}$ are, in effect, higher-order terms constructing types from other types, i.e., dependant types.

The use of multiple sorts (including many others not used here) allows us to correctly model complex phenomena such as co-predications, "dot"-types, qualia, while preserving a compositional model based on well-understood Montagovian principles. We use an analysis based on Categorial Grammars that provides the syntactic structure of the sentence and parse the semantics according to that structure, in a very classical way that happens to be used in the implementation of our parser for syntax and semantics, Grail (see [19]). In the rest of this paper, we will present the syntactic categories in the usual fashion: a/b yields an a given a b on its right while $b\backslash a$ yields an a given a b on its left, we use n for nouns, np for noun phrases, pp for prepositional phrases and s for sentences.

Our system uses the syntactic structure provided by such grammars in the tradition of Lambek, substitute the lexically-provided semantic terms and assemble their meanings to form the meaning of the sentence using β-reduction. The difference with the usual approaches is that using multiple sorts makes it possible to detect many lexical phenomena in the event of a typing mismatch. Higher-order types, together with *lexical transformations* – optional λ-terms that provide fine-tuned type-shifting operations to specific lexical entries, gives us the necessary control and flexibility to integrate lexical semantics in this process without sacrificing computationality and compositionality.

2   The Linguistic Phenomena

2.1   Plural Readings

Plurals are one of the most ubiquitous of semantic phenomena. Though not explicitly treated by Montague himself, a large body of research has been developed since then. This paper is not intended to be exhaustive in that respect, we will limit ourselves to some of the basic facts.

Some predicates (typical examples include verb phrases like "meet", "gather" and "elect a president", but also prepositions like "between") require their argument to be a group. These predicates are called *collective*. In these cases, the subject can either be a conjunction of noun phrases, a plural noun phrase or a singular group-denoting noun phrase (typical nouns of this sort are "committee", "orchestra" and "police"; in English these nouns require plural subject-verb agreement).

Contrasting with the collective predicates are the *distributive* predicates. When they apply to a set of individuals, we can infer that the predicate applies to each of these. Examples include verbs like "walk", "sleep" and "be hungry".

Many other predicates, like "record a song", "write a report" or "lift a piano" are ambiguous in that they accept both a group reading (which can be made explicit by adding the adverb "together") and a distributive reading (which can be made explicit by adding the adverb "each"). Thus "recorded a song together" talks about a single recording whereas "recorded a song each" talks about several recordings. The predicate "record a song" in and of itself only states that each of the subjects participated in (at least) one recording. We call such readings *covering* readings.

2.2   Some Issues with Mass Nouns

Mass nouns, much like collections and group nouns, have distributive, collective and covering readings: compare (examples adapted from [12]):

(1) The foliage was uniformly red.
(2) The foliage was of all kinds of bright colours.
(3) The foliage was bleak and creepy.

(1) is distributive (every individual leaf is red) while (2) is collective (individual leaves are of mostly different colors). *Foliage* is a mass noun that might be considered as denoting a group of entities, but human language does not associate the same status to leaves on a tree and members of a committee: the individuation condition is clearly different.

A more clear-cut example is the following (paraphrased from textbook entries on the water cycle):

(4) The water gathers in the lake.
(5) The water feeds several rivers.

(4) has a collective reading, (5) a distributive one, but the individuation conditions are even less clear that with *foliage*. In [12], Link proposes to distinguish as *atoms* any terms that denote individual objects; atoms comprising "the

water"in the above examples are thus *any portion* of the water that is denoted here. In (4), the mereological sum of every portion of the water is considered (thus aggregating any source of water that might have been previously mentioned), and in (5), the salient "atoms" are the volumes of water being contributed to every river.

Our operators for plurals are easily adapted for the purpose of deriving the correct semantics in that fashion, as we will now demonstrate.

3   Our Account of Plurals and Mass Nouns with Multiple Sorts

3.1   The Classical Account of Plurals

Most of the work on plurals in current theories of formal semantics is either inspired by or tempts to distinguish itself from the ideas and formalisation of Link [12], with many variations and reformulations on the basic theme. In Link's treatment, groups exist at the same level as individuals and we can form new groups out of individuals and groups by means of a lattice-theoretic join operation, written $\oplus$ — mathematically, we have a complete atomic join-semilattice without a bottom element. Groups and individuals are related by the *individual part* relation $\leq_i$, defined as $a \leq_i b \equiv_{def} a \oplus b = b$; in other words $a \leq_i b$ is true iff a contains a subset of the individual members of b.

Unlike later treatments of plurals, Link does not distinguish atomic individuals from group individuals but instead defines atomic individuals by means of a unary predicate *atom*, which is true for an entity x whenever its only individual part is itself (ie. the set of atoms are those a such that for all b, $b \leq_i a$ implies $b = a$).

For arbitrary unary predicates P, we have a corresponding distributive predicate P^* which is true if P holds for all atomic subparts of its argument. For example, $child^*(x)$ is true if x is a non-atomic group of children.[1]

We can therefore distinguish between

(6) $\exists x.child(x) \wedge sleep(x)$ (a child slept),
(7) $\exists x.child^*(x) \wedge sleep^*(x)$ (children slept), and
(8) $\exists x.child^*(x) \wedge build\text{-}raft(x)$ (children built a raft, as a group).

where (8) is shorthand for $\exists x.\forall y.((y \leq_i x \wedge atom(y)) \Rightarrow child(y)) \wedge build\text{-}raft(x)$, with $atom(y)$ in turn an abbreviation for $\forall z.z \leq_i y \Rightarrow z = y$.

3.2   Plural Readings in System-F

(A full development on plurals can be found in our previous publication, [15].)

---

[1] Technically, Link introduces two versions of this predicate: one which presupposes its argument is non-atomic and another which does not. Here, we will only consider the version which takes non-atomic individuals as an argument.

We can define operators that represent operations on sets, that can be assimilated with predicates, of type $\alpha \to t$ with α an appropriate subtype of e. (It is our opinion, due to the difficulty to provide entity semantics for negative, disjunctive and, especially, intersective types, that types cannot be used to model sets.)

For any group modelled as a (generic) predicate $\Lambda \alpha \lambda x . P(x)$, we can define its cardinality by the means of an operator $|\_|^{(\alpha \to t) \to \mathbb{N}}$, where the types of natural numbers can be defined as Church integers (as is in native System-F), or more simply as a primitive type such as is used in Gödel System-T, Martin-Löf Type Theory or, indeed, any reasonable computer implementation. This operator is only defined in context, and restriction of selection and a finite domain ensure that the satisfiability of the predicate is decidable and cardinality is finite. Similar operators have also been implemented in functional programming in [33].

Inclusion is easily defined as a function that maps pairs of predicates to a proposition: $\subseteq : \Lambda \alpha \lambda P^{\alpha \to t} \lambda Q^{\alpha \to t} \lambda x^{\alpha} P(x) \Rightarrow Q(x)$ — α usually is e. The logical operators such as $\wedge$ and $\vee$ between predicates correspond respectively to union and intersection of sets of terms, as expected (while operations on types like conjunction and disjunction do not).

We will also consider one entity type for groups called g_{α} (in our primary account, we only distinguished a single sort g_e for groups of any individuals), and a function $member$ that relates a group with its members, and which doing so turns a group in to a predicate: $member : g_{\alpha} \to \alpha \to t$ — or equivalently, says whether an entity satisfies a predicate. As a noticeable consequence, two groups can have the same members without being identical: groups are not defined by their members (as in set extensionality) they simply $have$ members.

We can have a constant $\oplus$ for group union. We prefer to avoid the complement because the set with respect to which the complement is computed is unclear, let alone whether or not the complement is a group; moreover, following Link, we will not have group intersection.

Because of $member$ we should have the following equivalences:

$$(member(g_{1\alpha}^{g} \cup g_{2\alpha}^{g}))^{\alpha \to t} = (((member(g_{1\alpha}^{g}))^{\alpha \to t} \vee (member(g_{2\alpha}^{g}))^{\alpha \to t})^{\alpha \to t}$$

They do hold in the target logic (many sorted logic). Indeed, we do have an obvious model with sets and individuals in which such equivalences hold.

We demonstrated that *lexical transformations* are well-suited to express the polysemy between collective, distributive and covering readings.

3.3   Including Multiple Sorts

Our system is multi-sorted, i.e. distinguishes various sorts of entities that are differentiated as various sub-types of e. The following is a short summary of constructions introduced in [15], adapted to a multi-sorted framework:

Thus example (4) can be analysed as

$$|\lambda x^{\varphi}.part\_of(x, \text{the\_water})| > 1 \wedge \text{gather}(\lambda x^{\varphi}.part\_of(x, \text{the\_water}))$$

This reads as: the water, as a mass, is made of several (more than one) distinguishable parts, and these parts gathered (in a lake).

Lexical item	Example	Syntactic Category	λ-Term
Individual Nouns	student	n	$\lambda x^H.student(x)$
(Abbreviated. Selection operators will yield the H (human) sort.)			
*The **group** types*			$\mathbf{g}_\tau$
For some singular sort τ.			
Group nouns	committee	n	$\lambda x^{\mathbf{g}_H}.committee(x)$
***Member** function*	[Committee] member	n/n	$\lambda y^{\mathbf{g}_\tau} x^\tau.member\_of(x,y)$
Group nouns are associated with predicates that we can build upon to create phrases such as *everyone within the committee* as well. The generic transformation specialises for any sort i.			
*The **mass** types*			$\mathbf{m}_\tau$
For some singular sort τ.			
Mass nouns	Water	n	$\lambda x^{\mathbf{m}_\varphi}.water(x)$
***PartOf** function*	Some volume of [water]	n/n	$\lambda y^{\mathbf{m}_i} \lambda x^i part\_of(x,y)$
Mass nouns are also associated with partitive transformations that can be specialised to any sort i.			
Collective verbs	Met	$np\backslash s$	$\lambda Q^{H\to t} . (|Q| > 1)$ $\wedge\ meet(Q)$
	Gathered	$np\backslash s$	$\varLambda\alpha\lambda Q^{\alpha\to t} . (|Q| > 1)$ $\wedge\ gather(Q)$
# transformation			$\varLambda\alpha\lambda R^{(\alpha\to t)\to t}$ $\lambda S^{(\alpha\to t)\to t}$ $\forall P^{\alpha\to t}.S(P) \Rightarrow R(P)$
Coerced forms	Met$^{\#}$	$np\backslash s$	$\lambda R^{(H\to t)\to t}\forall S^{H\to t}.$ $R(S) \Rightarrow |S| > 1 \wedge$ $meet(S)$

– Continued on next page   –

Operators for mass and plural readings – continued from previous page

Lexical item	Example	Syntactic Category	λ-Term
	Gathered[#]	$np\backslash s$	$\Lambda\alpha\lambda R^{(\alpha\to t)\to t}\forall S^{\alpha\to t}.$ $R(S) \Rightarrow \lvert S\rvert > 1 \wedge$ $gather(S)$
Collective verbs such as *meet* or *gather* apply to sets, modelled as predicates, of non-singular cardinality, including groups or massive terms. The # transformation, available in the lexical entries for those verbs, provides the necessary machinery to turn collections of individuals (singulars or plurals) into a set (modelled as a predicate). The transformed (coerced) terms for those verbs can apply to group or mass terms by the means of the *Member* or *PartOf* transformations (designated as lexical transformations for group and mass lexical items).			

Of course, the individuation conditions are different between group nouns and mass nouns, and it is unreasonable to hope to account exhaustively for all possible physical sub-volumes of water gathering into a lake; the *part_of* relation selects portions that have been introduced in the discourse or context. In the absence of such, we argue that $\lvert \lambda x^\varphi.part\_of(x, \mathrm{y}^{\mathbf{m}_\alpha})\rvert > 1$ always holds for any mass sort – i.e., that anything lexically considered as a massive entity has multiple parts. Distributive and covering readings are obtained in the same fashion.

3.4 The Pertinence of Multi-Sorted Semantics

Having a multi-sorted semantics with sub-typing gives us an edge in taking into account difficult lexical phenomena. First, we differentiate count and mass sorts (τ and $\mathbf{m}_\tau$), assuming that the lexicon includes the pertinent transformations from one to the other. We therefore can block infelicitous phrases such as *the water shattered* (requiring a physical, countable object) or *the bottle gathered* (requiring a mass or plural noun), and disambiguate between count and mass readings.

Secondly, this allows us to define polysemous predicates such as covered_with$^{\mathbf{m}_e\to t}$ and specialise their meaning so that *covered with rock* transforms the typing of the argument to $\mathbf{m}_\varphi$, while *covered with shame* produces the type $\mathbf{m}_{Abstract}$. The latter type is common in language, though not ontologically valid (compare *drinking from the fountain of glory*, *a man of much presence*, etc.). We can then correctly parse co-predicative sentences such as *[...] covered themselves with dust and glory* (from Mark Twain – this kind of zeugmatic expressions require specific care, as discussed and detailed by [4]).

3.5 Detailed Example: Partitive Quantification and Comparisons

Simple quantifications and comparative examples with massive entities include phrases such as *some water, some water is on the table, More water is on the table than wine is in the glass.* Our full analysis is shown in the table below.

The table gives a lexicon with syntactic and semantic types and semantic terms for a number of words in our lexicon. Some pertinent phrases have been given types and terms as well; these have been computed automatically from the lexical entries to give an idea of how complex expressions are derived from their parts.

The syntactic categories have been deliberately kept simple. The given type for the quantifier "some" is valid only for subject positions (but several solutions exist which generalise this type) and "more ... than" is treated as a unit and given a schematic type $n \Rightarrow (np\backslash s) \Rightarrow n \Rightarrow (np\backslash s) \Rightarrow s$, indicating it takes as its arguments first a noun n then a verb phrase $np\backslash s$ then a second noun and a second verb phrase to produce a sentence; the reader can consult [7,22] for the technical details of how we can produce the word order "More n_1 vp_1 than n_2 vp_2" and more complicated comparative constructions. For a more detailed account of the syntactic aspects of modern categorial grammars, see [18,20,21].

For the semantic types of polymorphic terms (namely *is* and *the*, $\Pi\alpha$ denotes a universally quantified type variable; this is implicitly included in the λ-term by the use of the abstraction $\Lambda\alpha$.

From the syntax (not detailed here) to the final formula, the whole process is computational and compositional.

Input text	Syntactic Category	Semantic Type	λ-Term
Some	$(s/(np\backslash s))/n$	$(\mathbf{m}_i \rightarrow t) \rightarrow n \rightarrow (\mathbf{m}_i \rightarrow t) \rightarrow t$	$\Lambda m_i \lambda P^{\mathbf{m}_i \rightarrow t} \lambda Q^{\mathbf{m}_i \rightarrow t}.\exists \lambda x^{m_i} \lambda a^{s_i}.$ $(\wedge(\wedge (P\,x)\,(Q\,x))(\text{amount } x\, a))$
In addition to the usual existential quantification and conjunction of the argument and a predication over it, we add a predicate *amount* associated to an existentially quantified measure of the sort of the argument.			
Some water	$s/(np\backslash s)$	$(\mathbf{m}_\varphi \rightarrow t) \rightarrow t$	$\lambda Q^{\mathbf{m}_\varphi \rightarrow t}.\exists \lambda x^{\mathbf{m}_\varphi} \lambda a^{s_\varphi}.$ $(\wedge(\wedge (\text{Water } x)\, (Q\,x))$ $(\text{amount } x\, a))$
Water is of the *physical* sort (φ), associated with a massive sort $\mathbf{m}_\varphi$ and an appropriate measure s_φ (physical mass).			
Table	n	$(\varphi \rightarrow t)$	$\lambda x^\varphi.(\text{table } x)$
Table also provide us with an optional transformation $f_{Telic}{}^{\varphi \rightarrow Pl}$, as the use of tables as places to present or store things is lexical.			

– Continued on next page –

Complete example – continued from previous page

Input text	Syntactic Category	Semantic Type	λ-Term
The	np/n	$\Pi\alpha.(\alpha \to t) \to \alpha$	$\Lambda\alpha\lambda P^{\alpha \to t}.(\iota\ P)$
The salient selection operator (see our earlier work on determiners for details). *The* specialises to any type α.			
On	pp/np	$Pl \to (\varphi \to t)$	$\lambda x^{Pl}.(\lambda y^\varphi.(\text{located\_on } x)\ y)$
On constructs a predicate from locations.			
Is	$\Pi(np\backslash s)/pp$	$\Pi\alpha.(\alpha \to t) \to (\alpha \to t)$	$\Lambda\alpha\lambda P^{\alpha \to t}\lambda x^\alpha.\ ((\text{is } P)\ x))$
The polymorphous copula (specialised for any type α) constructs a syntactically valid predicate.			
Is on the table	$np\backslash s$	$\varphi \to t$	$\lambda x^\varphi.(((\text{is (located\_on } (f_{Telic}{}^{\varphi \to Pl}$ $(\iota\text{ table})^\varphi)^{Pl}))^{\varphi \to t}\ x)$
(As is usual in our system.)			
Some water is on the table	s	t	$\exists\lambda x^m{}_\varphi\lambda a^{s_\varphi}.(\wedge(\wedge \text{ (Water } x)\ ((\text{is}$ $(\text{located\_on } (f_{Telic}\ (\iota\text{ table})))$ $(Pack^{m_\varphi \to \varphi}\ x)))$ $(\text{amount } x\ a))$
This is the final, intuitive reading. Note the use of the *universal packager*, instantiated for physical massive entities, that was used in order to represent the (implied) container of the water. Another possible reading (not detailed here), not using this packager nor the telic of the table as a storage, would simply be that some water has spilled on the table.			
More... than...	$n \Rightarrow$ $(np\backslash s) \Rightarrow$ $n \Rightarrow$ $(np\backslash s) \Rightarrow s$	$((m_i \to t) \to$ $(m_i \to t) \to$ $(m_i \to t) \to$ $(m_i \to t)) \to t$	$\Lambda m_i \lambda P^{m_i \to t}\lambda Q^{m_i \to t}\lambda R^{m_i \to t}$ $\lambda S^{m_i \to t}.\exists\lambda x^{m_i}\lambda y^{m_i}\lambda a^{s_i}\lambda b^{s_i}.$ $(\wedge(\wedge (\wedge(P\ x)\ (Q\ x))$ $(\wedge(R\ y)\ (S\ y)))$ $(\wedge(\wedge(\text{amount } x\ a)\ (\text{amount } y\ b))$ $(> a\ b)))$
The comparative asserts the existence of two massive entities, with pertinent discriminating predicates, as well as the inequalities between their measures according to their sort. (We consider common-sorted terms, so that direct comparison is possible – metalinguistic comparison is the other option.)			

– Continued on next page –

Complete example – continued from previous page

Input text	Syntactic Category	Semantic Type	λ-Term
Glass	n	$(\varphi \to t)$	$\lambda x^{\varphi}.(\text{glass } x)$
As is well-known in lexical semantics, *glass* has the telic of a *container*, that is modelled by $f_{Telic}{}^{\varphi \to Container}$ and $f_{Content}{}^{Container \to m_{\varphi}}$.			
In	pp/np	$Pl \to (\mathbf{m}_{\varphi} \to t)$	$\lambda x^{Container}.$ $(\lambda y^{\mathbf{m}}{}_{\varphi}.(\text{contained\_in } x) \; y)$
In this case, *in* constructs a predicate that applies to physical masses that can be inside a container.			
Is in the glass	$np\backslash s$	$\mathbf{m}_{\varphi} \to t$	$\lambda x^{\mathbf{m}}{}_{\varphi}.(((\text{is (contained\_in}$ $(f_{Telic}{}^{\varphi \to Container} (\iota \text{ glass})^{\varphi}$ $)^{Container}))\mathbf{m}_{\varphi} \to t \; x)$
More water is on the table than wine is in the glass	s	t	$\exists \lambda x^{\mathbf{m}_{\varphi}} \lambda y^{\mathbf{m}_{\varphi}} \lambda a^{s_i} \lambda b^{s_i}.(\wedge (\text{Water } x)$ $((\text{is (located\_on } (f_{Telic} (\iota \text{ table}))))$ $(Pack \; x))$ $\wedge(\text{Wine } y) (((\text{is (contained\_in}$ $(f_{Telic} (\iota \text{ glass})))) \; y) (\text{amount } x \; a)$ $(\text{amount } y \; b)(> a \; b))$
Note that the universal packager has to be used for the water in order to reify its container; not so for the wine, the glass being an explicit container.			

4   Scales, Measures, and Units with Many Sorts

4.1   Defining Scales in System-F

Gradable adjectives such as *tall*, as studied extensively in e.g. [8], can be represented in our system. In order to model those, as well the semantics of comparatives between quantities that can be explicit or implied by such adjectives, we need to be able to define scales or degrees in our system. Integers can be defined, and are already used for cardinality, by the means of Church numerals. It is also straightforward to define floating point numbers in scientific notation[2,3]. We thus define the type s for scales, that can be specialised for any pertinent sort. Sorts are typically associated with *measures*, with useful measures including lengths, surfaces, volumes; masses or weights; and frequencies, durations. Pairs of comparable measures can also be used to specify *ranges*, used for the semantics of gradable adjectives.

---

[2] Such a number is a simple data structure comprising l, the list of digits (in base 10) of the mantissa, s, a constant indicating its sign, e, an integer representing the exponent and r, a constant indicating the sign of the exponent. Comparison between such floating point numbers is easy, and the common operations are definable.

[3] This short point illustrates that scales (and operations) can be defined natively in pure System-F; of course, we can take floating-point numbers of sufficient precision to exist as a primitive type, as is the case in any reasonable computer implementation.

Gradable adjectives – qualificatives such as *tall, heavy* or *light* that are usually collocated with quantity adverbs such as *very* – are usually modelled in terms of *ranges*, given as a couple of scales indicating the "typical range" of values for the adjective. Using information provided by the sort, we can provide much more detailed semantics, associating predicates such as tall$^{\varphi \to t}$ with optional coercions such as $\lambda x^H.((\mathbf{R}^{(\varphi \to s_\varphi) \to \varphi} (f^{H \to \varphi} x) r_H)$. In this example for phrases such as *a tall man*, f is a coercion that considers the physical aspect of a term of human sort (*man*) and $\mathbf{R}$ associates with this term a range r_H of scales of height (typed as s_φ)corresponding to a tall man. Adverbial modifiers such as *very* will then affect that range.

Many gradable adjectives are directly linked to mass nouns and generalised quantification in their semantics: *a bit/very wet* corresponds to *some/a lot of water*, etc. The analysis of their semantics is similar, but inferring one from the other is far from trivial.

4.2   Sorts, Units and Classifiers

As having different sorts is useful in order to distinguish semantically different classes of lexical items, they also provide a means to distinguish between *units*, ways of counting and accumulating quantities of items. Terms of individual or group sorts can be counted (using natural integers), terms of massive sorts can be measured (using scales and ranges), but the actual type of the term specifies the means of doing such an operation: comparing volumes for liquids, mass for generic physical massive terms.

Having a different unit for every sort, arranged in a hierarchy provided by sub-typing (all liquids are physical objects and thus can be compared by mass as well as by volume; every countable entity can be counted as "some object" but we might distinguish the number of people, of pets and of cars in a given situation. . .) is linguistically interesting, as it fits very well with a feature prominent in many languages: the classifier system.

Classifiers are syntactic items common to many spoken and written Asian, Amerindian and West African languages, as well as most variations of Sign Languages. In other linguistic groups, some traces remain present, such as *head* in the phrase "ten heads of cattle". Classifiers are used to count individuals and measure masses.

Our intuition, as expressed in [16], is that classifiers can provide the basis for a system of sorts. They are clearly linguistically motivated, and correspond to an ontological hierarchy. Even if classifiers differ between languages, being influenced by the cultural and social evolution of each language or dialect involved (see [11,32,34] for details and discussions), the main features of the system are shared by most variations.

Of most interest to us is the hierarchisation of classifiers. For instance, in Japanese, some classifiers are generic, and commonly used by people not fluent in the language (denoting things, people, order, and broad "units" of any quantity measurable with a container – *Hai*); some are commonly used for specific categories of items (appliances, small objects, flat objects. . .); and some are very specific to an usage or trade (such as *koma* for panels). The fact that many classifiers (such as the ones counting small objects) are shared between Japanese and French Sign Language, for instance, illustrates the cognitive pertinence of using classifiers to define a system of linguistically different sorts.

Having a different way of counting items or measuring quantities is also needed to solve quantificational puzzles such as posed by Nicholas Asher in [1]. Building a complete system of sorts and units based on common aspects of the various classifier systems requires more resources than we have currently at our disposal, however.

4.3 Comparison Classes

The ranges used for gradable adjectives do not tell the whole story, as the study of comparison classes show. There are many views on what comparison classes expressed in phrases such as *tall for a seven-years-old girl* imply for the semantics of gradable adjectives: while Christopher Kennedy says that the semantics are not changed but that comparison classes modify the set of individuals that can be referred to at the discursive or pragmatic level [8], Stéphanie Solt argues that they give a specific range [31]. We believe that such phrases convey intrinsic (and not just use-specific) information, that might be modelled as predicates modifying the range pertinent to the sort of the argument.

5 Complex Cases

5.1 Mass-Count Alternations

For many mass nouns, there is a usage as a singular entity – compare *ten wild salmons* and *some raw salmon*. We, in agreement with most of the literature, have used the *grinding* transformation for such sentences. This operation (usually named the *universal grinder*) can take place on terms that have already undergone other grinding transformations, as the following example illustrates:

(9) The salmon we caught was lightning fast. It is delicious, and we preserved some of it. Wild salmon caught from this river could quickly become a source of income.

This shows that, given a suitable discourse structure, types *Animal*, *Food*, $\mathbf{m}_{Food}$ and $\mathbf{m}_{Animal}$ can coexist in referents with related senses.

5.2 Comparisons Between Different Units

The difficulties of comparing between different predicates, such as in the comparisons in (10) and (12), has been often remarked upon. While there have been some discussions of their characteristics and possible semantics such as [23], their integration within a Montagovian framework remains forthcoming. Consider:

(10) The table is longer than it is wide.

(11) The thread is bigger than the table.

(12) He his more dumb than ill-intended.

Our solution would compare different ranges together. The differentiation between sorts appears especially relevant on such examples. Comparing between two lengths in (10) is straightforward. Comparing between a length and a surface in (11) is much more difficult; if the sentence is felicitous, it seems to force the use of comparable units and reads as "The thread is *longer* than it is wide". On the other hand, comparing between different abstract and subjective values in (12) implies a strong intensional component. (12) has been called a *metalinguistic* comparison, as it constitutes a comparison in appearance only, instead establishing a relation between two different social judgements; it is hard to argue that the "degree of dumbness" is higher than the "degree of ill-intention" (or, indeed, to automatically assign any kind of numerical value to such measures); rather, it indicates that one adjective is more salient that the other in context.

6   Conclusion

We demonstrated a straightforward implementation of the semantics of mass nouns in our higher-order computational framework, which is based on Link's classical linguistic analysis, but does not commit to a specific ontological view, and can be adapted to other linguistic formulations of the phenomena, such as those from David Nicolas in e.g. [25].

We continue expanding the coverage of our framework, in order to prove that complex issues of lexical semantics need not be resolved in isolation.

Indeed, having a multi-sorted approach makes the analysis of other complex phenomena easier.

At the present time, our implementation of the syntax-semantics analyser Grail ([19]) includes an syntactical analysis using a variation of Categorial Grammars, and an analysis of the semantics of sentences that is done in $\lambda - DRT$ rather than λ-calculus, with wide coverage grammars (in French) statistically acquired from corpora. We have formulated the semantics of plurals and mass nouns, but in order to complete their implementation, some steps are yet missing: the construction of a wide-coverage system of sorts that can form the base types of a full-fledged semantic lexicon, and subsequently the acquisition of types and transformations for a sufficient number of lexical items. We are keen to provide such a lexicon, inspired by the classifier system, and are excited to see many researcher moving towards that goal.

References

1. Asher, N.: Lexical Meaning in Context: a Web of Words. Cambridge University Press, Cambridge (2011)
2. Bassac, C., Mery, B., Retoré, C.: Towards a type-theoretical account of lexical semantics. J. Lang. Logic Inf. **19**(2), 229–245 (2010)

3. Bekki, D., Asher, N.: Logical polysemy and subtyping. In: Motomura, Y., Butler, A., Bekki, D. (eds.) JSAI-isAI 2012. LNCS, vol. 7856, pp. 17–24. Springer, Heidelberg (2013)

4. Clément, L., Gerdes, K.: Analyzing zeugmas in XLFG. In: LFG 2006, Konstanz, Germany (2006)

5. Gillon, B.S.: Towards a common semantics for english count and mass nouns. Linguist. Philos. **15**(6), 597–639 (1992)

6. Girard, J.Y.: Interprétation fonctionnelle et élimination des coupures de l'arithmétique d'ordre supérieur. Université Paris VII, Thèse de Doctorat d'État (1972)

7. Hendriks, P.: Ellipsis and multimodal categorial type logic. In: Morrill, G., Oehrle, R.T., (eds.) Proceedings of Formal Grammar 1995, pp. 107–122, Barcelona, Spain (1995)

8. Kennedy, C.: Vagueness and grammar: the semantics of relative and absolute gradable adjectives. Linguist. Philos. **30**(1), 1–45 (2007)

9. Lambek, J.: The mathematics of sentence structure. Am. Math. Mon. **65**, 154–170 (1958)

10. Lefeuvre, A., Moot, R., Retoré, C.: Traitement automatique d'un corpus de récits de voyages pyrénéens : analyse syntaxique, sémantique et pragmatique dans le cadre de la théorie des types. In: Congrès mondial de linguistique française (2012)

11. Li, X.P..: On the semantics of classifiers in Chinese. Ph.D. thesis, Bar-Ilan University (2011)

12. Link, G.: The logical analysis of plurals and mass terms: a lattice-theoretic approach. In: Portner, P., Partee, B.H. (eds.) Formal Semantics - the Essential Readings, pp. 127–147. Blackwell, Oxford (1983)

13. Luo, Z., Soloviev, S., Xue, T.: Coercive subtyping: theory and implementation. Inf. Comput. **223**, 18–42 (2013)

14. Mery, B.: Modélisation de la Sémantique Lexicale dans le cadre de la Théorie des Types. Ph.D. thesis, Université de Bordeaux, July 2011

15. Mery, B., Moot, R., Retoré, C.: Plurals: individuals and sets in a richly typed semantics. In: Yatabe, S. (ed.) LENSL 2010 - 10th Workshop on Logic and Engineering of Natural Semantics of Language, Japanese Symposium for Artifitial Intelligence, International Society for AI - 2013, pp. 143–156, Hiyoshi, Kanagawa, Japan, jSAI-ISAI, Keio University, October 2013

16. Mery, B., Retoré, C.: Semantic types, lexical sorts and classifiers. In: NLPCS 2010- 10th International Workshop on Natural Language Processing and Computer Science - 2013, Marseille, France, October 2013

17. Montague, R.: The proper treatment of quantification in ordinary English. Selected Papers of Richard Montague. In: Thomason, R. (ed.) Formal Philosophy. Yale University Press, New Haven (1974)

18. Moortgat, M.: Categorial type logics. In: van Benthem, J., ter Meulen, A. (eds.) Handbook of Logic and Language, Chap.2, pp. 95–179. North-Holland Elsevier, Amsterdam (2011)

19. Moot, R.: Wide-coverage French syntax and semantics using Grail. In: Proceedings of Traitement Automatique des Langues Naturelles (TALN), Montreal (2010)

20. Moot, R., Retoré, C.: A logic for categorial grammars: Lambek's syntactic calculus. In: Moot, R., Retoré, C. (eds.) The Logic of Categorial Grammars. LNCS, vol. 6850, pp. 23–63. Springer, Heidelberg (2012)

21. Morrill, G.: Categorial Grammar: Logical Syntax, Semantics, and Processing. Oxford University Press, Oxford (2011)

22. Morrill, G., Valentín, O., Fadda, M.: The displacement calculus. J. Logic Lang. Inform. **20**(1), 1–48 (2011)
23. Morzycki, M.: Metalinguistic comparison in an alternative semantics for imprecision. Nat. Lang. Seman. **19**(1), 39–86 (2011)
24. Muskens, R.: Combining montague semantics and discourse representation. Linguist. Philos. **19**, 143–186 (1996)
25. Nicolas, D.: Mass nouns and plural logic. Linguist. Philos. **31**(2), 211–244 (2008)
26. Pustejovsky, J.: The Generative Lexicon. MIT Press, Cambridge (1995)
27. Prévot, L., Moot, R., Retoré, C.: Un calcul de termes typés pour la pragmatique lexicale - chemins et voyageurs fictifs dans un corpus de récits de voyages. In: Traitement Automatique du Langage Naturel - TALN 2011, pp. 161–166, Montpellier, France (2011)
28. Ranta, A.: Type-theoretical Grammar. Clarendon Press, Oxford (1994)
29. Real-Coelho, L.-M., Retoré, C.: A generative montagovian lexicon for polysemous deverbal nouns. In: 4th World Congress and School on Universal Logic - Workshop on Logic and Linguistics, Rio de Janeiro (2013)
30. Retoré, C.: The Montagovian Generative Lexicon Lambda Ty_n: a Type Theoretical Framework for Natural Language Semantics. In: Matthes, R., Schubert, A. (eds.) 19th International Conference on Types for Proofs and Programs (TYPES 2013), volume 26 of Leibniz International Proceedings in In-formatics (LIPIcs), pp. 202–229, Dagstuhl, Germany. Schloss Dagstuhl Leibniz-Zentrum fuer Informatik (2014)
31. Solt, S.: Notes on the comparison class. In: Nouwen, R., van Rooij, R., Sauerland, U., Schmitz, H.-C. (eds.) Vagueness in Communication. LNCS, vol. 6517, pp. 189–206. Springer, Heidelberg (2011)
32. T'sou, B.K.: Language contact and linguistic innovation. In: Lackner, M., Amelung, I., Kurtz, J. (eds.) New Terms for New Ideas. Western Knowledge and Lexical Change in Late Imperial China, pp. 35–56. Koninklijke Brill, The Netherlands (2001)
33. van Eijck, J., Unger, C.: Computational Semantics with Functional Programming. Cambridge University Press, Cambridge (2010)
34. Zwitserlood, I.: Classifiers. In: Pfau, R., Steinbach, M., Woll, B. (eds.) Sign Languages: an International Handbook, pp. 158–186. Mouton de Gruyter, Berlin (2012)

On CG Management of Japanese Weak Necessity Modal *Hazu*

Shinya Okano[1,2(✉)] and Yoshiki Mori[1]

[1] The Universty of Tokyo, Tokyo, Japan
okano@phiz.c.u-tokyo.ac.jp, mori@boz.c.u-tokyo.ac.jp
[2] Research Fellow of Japan Society for the Promotion of Science, Tokyo, Japan

Abstract. This paper deals with a different behavior of weak necessity modals (in Japanese, *hazu*) in comparison with plain necessity or possibility modals in their epistemic use. While the latter cannot be immediately followed by the negation of its prejacent, the former allows this. After reviewing some previous approaches to this fact in Kratzerian framework, we try to implement [14]'s insight in terms of an update semantics by [15], modified by [13]. We propose that *hazu* makes a (possibly) counterfactual update by revising CG with normalcy condition.

1 Introduction

In this paper, we discuss a problem posed by [1] regarding the different behavior of so-called weak necessity modals in comparison with plain necessity or possibility modals in their epistemic use . More specifically, one cannot utter a proposition modalized by *must* and then deny its unmodalized counterpart, while one can deny the proposition which is modalized by *should* just before:

(1) a. #The beer must be cold by now, but it isn't.
     b. The beer should be cold by now, but it isn't.   ([1, p. 5])

This contrast is carried over to Japanese modals *nichigainai* and *hazu*, as the following pair shows[1]:

(2) a. #Biiru=wa   imagoro   hie-tei-ru-nichigainai-ga,   hie-tei-nai.
        Beer=Top   by now   get cold-Result-Pres-NICHIGAINAI-Conj
        get cold-Result-Neg

We would like to thank Yusuke Kubota, Takumi Tagawa, Misato Ido, Katsumasa Ito, Hitomi Hirayama, Chunhong Park, Saori Takayanagi, Akitaka Yamada and three anonymous reviewers and participants of LENLS 11 for their insightful comments and discussions. This work was supported by JSPS Grant-in-Aid for JSPS Fellows Grant Number 25-9444.

[1] Abbreviations used in this paper are as follows: Conj = conjunction, Cop = copula, Loc = locative, Neg = negation, Nom = nominative, Pres = present, Result = resultative, Top = topic.

T. Murata et al. (Eds.): JSAI-isAI 2014 Workshops, LNAI 9067, pp. 160–171, 2015.
DOI: 10.1007/978-3-662-48119-6_12

"#The beer must be cold by now, but it isn't."

b. Biiru=wa   imagoro   hie-tei-ru-hazu-da-ga,   hie-tei-nai.
   Beer=Top   by now   get cold-Result-Pres-HAZU-Cop-Conj
   get cold-Result-Neg
   "The beer should be cold by now, but it isn't."

The Japanese modals, *nichigainai* and *hazu*, are expressions which attach to verbs and contribute to the whole sentence (mainly) an epistemic modal flavor. They are used when the speaker is near-certain about the truth of the rest of the sentence they attach to, though she doesn't know whether it is in fact true. So, they roughly correspond to English necessity modals *must* and *should*.

In the next section, we review previous approaches relevant to accounting for the contrast between the weak and strong necessity modals observed in (1) and (2). We argue that [14]'s subjunctive-presuppositional approach, which is based on [2], turns out to be the most plausible explanation among them. In Sect. 3, we try to implement his pragmatic conditions regarding weak necessity modals in terms of update semantics ([15]), adopting a modification by [13]. Section 4 is the conclusion of this paper.

2   Previous Approaches

The basic framework which is shared by all the previous approaches introduced here is that of Kratzer [6,7]. Her semantics for modal expressions like *must* or *should* includes two kinds of conversational backgrounds which are supplied by the context. One is called modal base, and provides either a set of relevant facts or a set of pieces of knowledge, represented as propositions, each of which is regarded as a set of possible worlds where it holds.[2] The other conversational background is called ordering source, and provides a set of ideals, according to which worlds in modal bases are ranked. Using these devices, the lexical entries of *must* and *may* can be roughly written as in (3):

(3) $[\![\mathrm{must}]\!]^{f,g}(p)(w) = 1$ iff $\forall w'[w' \in Best_{g(w)}(\cap f(w)) \to w' \in p]$
       $[\![\mathrm{may}]\!]^{f,g}(p)(w) = 1$ iff $\exists w'[w' \in Best_{g(w)}(\cap f(w)) \land w' \in p]$

where $[\![\,]\!]$ is an interpretation function, f is a modal base, g is an ordering source, w is a possible world, $Best_{g(w)}(\cap f(w))$ is a set of the worlds in $\cap f(w)$ which come closest to the ideal given by $g(w)$, and p is a prejacent[3]

What is problematic in this account is, that it doesn't predict (1a) to be bad. That is, a semantics of *must* like (3) would say, "(1a) is true if and only if in all

---

[2] We will often refer to the large intersection of a modal base, that is, a set of worlds in which all the propositions in the modal base hold, as a modal base, when there is no chance of misunderstanding.

[3] In this paper, we will make the Limit Assumption ([9]) for simplicity's sake because nothing hinges on the assumption.

the worlds compatible with the speaker's knowledge where things go most stereo-
typically, the beer is cold by now, but in the actual world, it isn't." The point
is, the *must*-sentence doesn't claim anything about the actual world, but only
about the best worlds accessible from the actual world. There is no guarantee
that the actual world is among the best worlds.

Now we'd like to introduce three proposals to cope with this problem and
argue that [14] is the most plausible to capture the modals' behavior.

2.1   Copley(2006)

First, we review Copley's proposal, which, as far as we know, is the first attempt
to incorporate into the modal semantics an additional mechanism to deal with
Modal p, but not p-sentences. More specifically, she posits a special presupposition for *must* and *should* respectively. The following (4) shows her semantics
for them:

(4)   [1, p. 10] For all C [context of utterance], p [proposition]:
  a. $[\![must]\!](C)(p)$ asserts that $highest\text{-}plausibility_C(\mathcal{E}_C) \subseteq p$, and
                        presupposes that $\forall \mathcal{E}$ more informative than $\mathcal{E}_C$:
                        $\mathcal{E} \subseteq highest\text{-}plausibility_C(\mathcal{E}_C)$
  b. $[\![should]\!](C)(p)$ asserts that $highest\text{-}plausibility_C(\mathcal{E}_C) \subseteq p$, and
                        presupposes that $\forall \mathcal{E}$ more informative than $\mathcal{E}_C$:
                        $\mathcal{E} \cap highest\text{-}plausibility_C(\mathcal{E}_C) \neq \emptyset$

The $\mathcal{E}_C$ in (4) means the epistemic state of the speaker, which corresponds
to the epistemic modal base in the standard Kratzerian terms. And *highest-
plausibility_C* is a function which retrieves the set of the most plausible worlds in
the modal base. In Copley's semantics, the assertive content of *must* and *should*
is exactly the same: they are treated just as ordinary strong necessity modals.
The difference between them lies in their presuppositions. What she intends to
express with them is the following intuition:

(5) [*M*]*ust p* presupposes that the actual world is going to be one of the most
plausible worlds, while *should p* [] presupposes merely that it (still) possible that
the actual world is (going to turn out to be) one of the favored worlds. [1, p. 9]

Unfortunately, this intuition isn't well captured in the presuppositions above,
though her idea itself is attractive. When we read the phrase "more informative
than" in (4) as "properly included in", the presupposition of *must* says that all
the proper subsets of the modal base are included in the set of best worlds. Since
must asserts that the set of best worlds is included in the set of the prejacent
worlds, this amounts to saying that all the proper subsets of the modal base are
included in the set of the prejacent worlds. Among the proper subsets is a set
consisting only of the actual world.

What is the problem if *must p* entails p? We will argue this in the next
subsection, where we examine von Fintel & Gillies [3]'s claim, because their

claim is exactly that *must p* entails *p*. Here we have to admit that with regard to the infelicity of *must p, but not p*, it is a welcome result that *must p* contradicts *not p* semantically.

More problematic about Copley's implementation is the presupposition of *should*. It turns out to be equivalent to the presupposition of *must*. We have to omit the strict proof for lack of space, but it goes roughly as follows. According to *should*'s presupposition in (4b), all the subsets of the modal base have a non-empty intersection with the set of the best worlds. Among the subsets are singleton sets consisting of only one world. So these singleton sets must also have a non-empty intersection with the set of the best worlds. This means all the singleton sets must be included in the set of the best worlds. And this amounts to saying that all the worlds in the modal base are included in the set of the best worlds, which in turn is included in the set of prejacent worlds, due to the assertive content of *should*. This situation is exactly the same as that of *must*.

To sum up, while Copley's idea itself is attractive to attribute the different behavior of *must* and *should* to the difference of the speaker's attitude toward some status of the actual world, its implementation isn't entirely exact enough to capture her intuition, especially for *should*.

2.2 Von Fintel & Gillies (2010)

von Fintel and Gillies [3], as we mentioned above, claim that *must p* entails *p*. According to them, the reason why *must p* sounds weaker than plain *p* is that *must* is subject to an evidential restriction. More specifically, the prejacent of *must* mustn't be entailed by the set of propositions representing direct information. They call this kind of set "kernel", and the large intersection of a kernel is a modal base in their terminology. The formal definitions of kernels and modal bases are shown in (5):

(5)   Kernels and Bases [3, p. 371]
         K is a kernel for B_K, B_K is determined by the kernel K, only if:
         i. K is a set of propositions (if $P \in K$ then $P \subseteq W$)
         ii. $B_K = \cap K$

And their lexical entry of *must* is quoted in (6):

(6)   (Strong *must* + Evidentiality). Fix a *c*-relevant kernel K:
         i. $[\![\text{must } \phi]\!]^{c,w}$ is defined only if K does not directly settle $[\![\phi]\!]^c$
         ii. $[\![\text{must } \phi]\!]^{c,w} = 1$ if $B_K \subseteq [\![\phi]\!]^c$ [3, p. 372]

We don't go into detail about what it means for a kernel to directly settle the prejacent.[4] Here we would rather just mention that the direct perception of prejacent events is taken to be in a kernel and directly settle the prejacent. This predicts, for example, that if the speaker sees a raining event, she cannot

---

[4] In [3], two ways of implementation are discussed. Our criticism to [3]'s approach below is independent of which way one chooses.

felicitously utter "It must be raining." And exactly this directness restriction is the source of the apparent weakness of *must*, they argue.

With regard to the definition in (6-ii), what is notable is that no ordering source is involved. In combination with the assumption that the actual world is in the modal base, this makes it possible for *must p* to entail *p*. Therefore, von Fintel & Gillies' story about *must* correctly captures the infelicity of *must p, but not p*.

Then, what about *should*? They don't discuss it in the article. But at least one of the authors, namely von Fintel, admits ordering sources in the lexical entry of some modals. In fact, von Fintel & Iatridou [4] propose positing multiple ordering sources in the semantics of a weak necessity modal *ought*. If this move is made for *should*, then *should p* is predicted not to entail *p*, as we saw at the beginning of this section.

Is this the end of the story? We don't think so. We have two reasons to doubt this line of analysis. The first problem is that the choice of propositions to be included in the kernel seems somewhat arbitrary. Let's see some example.

(7) [Seeing the wet rain gear]
    a. It must be raining outside.
    b. $K = \{$ "the rain gear is wet", "if the rain gear is wet, it is raining outside"$\}$

To make the sentence (7a) felicitous, we have to make sure that the prejacent "it is raining" is deduced from the propositions in the kernel. To do so, the kernel has to include the generic knowledge "if the rain gear is wet, it is raining outside". But this is somewhat counterintuitive to the notion of kernels as consisting of direct information.[5]

Second, there is an empirical problem arising when we turn our eyes on Japanese counterpart of *must*, *nichigainai*. Though this modal expression behaves like *must* in the epistemic use, it cannot be used when the truth of the prejacent is deduced from the kernel. von Fintel & Gillies argue that this is not the case with English *must*. For example, in (8a) below, where the speaker knows with full certainty that her lost ball is either in Box A or B or C, and that it is neither in A nor B, then she can utter "it must be in C". In this case, the prejacent "the ball is in C" is deduced from the kernel. At first glance, this is also the case with Japanese *nichigainai*. But if we change the situation slightly, it becomes worse. That is, if we put the sentence (8b) in a quiz show context or something, where there is no room for the ball to fail to be in Box C (8b) is worse. On second thought about the first, lost-ball searching case, we argue that this case involves still some uncertainty as to the whereabouts of the ball, at least in Japanese, and thus *nichigainai* is acceptable. We don't know whether this change of acceptability is also true of English *must*, but since uncertainty means the existence of ordering sources in Kratzerian terms, we are again faced

---

[5] This might not be a valid counterargument when one takes [10]'s strong view that all epistemic modals contribute their own evidential semantics (and all evidentials contribute modal semantics) and follows her in regarding the direct perceptual evidence and the general knowledge as forming a natural class based on "trustworthiness".

with the problem of how to distinguish between a strong necessity modal and a weak necessity modal at least in Japanese.[6]

(8)   Deduction from the kernel
       (The ball is in A or in B or in C. It is not in A... It is not in B.)
       a.   So, it must be in C.   [3, p. 362]
       b.   (#)Dakara, C=ni   hait-tei-ru-nichigainai.
             Conj   C=Loc   go into-Result-Pres-NICHIGAINAI
       (felicitous in a lost-ball searching or a magic show context,
       but not in a quiz show context)

2.3   Silk (2012)

Silk [14] tackles squarely the problem of distinguishing between weak and strong necessity modals in English. Though his main point is the treatment of teleological and deontic readings of those modals, his analysis covers their epistemic interpretation as well. As in the standard Kratzerian framework, Silk posits ordering sources both for *must* and *should*. But the form of ordering sources is somewhat different from the standard in that they are all biconditionals. This is schematically shown in (9):

(9)   Applicability Conditions (ACs) in the ordering source
       $g(w) = \{p \leftrightarrow C, p' \leftrightarrow C'...\}$ [C, C' ... are meant to be mutually exclusive ACs]
                                              [adapted from [14, p. 47]]

   In this system, propositions to be included in ordering sources in the usual sense stand in biconditional relation to their applicability conditions (ACs). This trick makes it possible that goals, preferences or stereotypes are taken into consideration only when a certain condition for them is satisfied. Assuming this, Silk proposes semantics for *must* and *should* as in (10):

(10) $[\![\text{must}]\!]^{f,g}(p)(w) = 1$ iff $\forall w'[w' \in Best_{g(w)}(\cap f(w)) \rightarrow w' \in p]$
        $[\![\text{should}]\!]^{f,g}(p)(w) = 1$ iff $\forall w'[w' \in Best_{g(w)}(\cap(f(w) \cup C_{g(w)})) \rightarrow w' \in p]$

where $C_{g(w)}$ is a (possibly improper) subset of the set of ACs for each of the premises in $g(w)$ [adapted from [14, pp. 50–51]]

Briefly put, *should* makes a conditional claim while *must* makes a categorical claim. This is reflected in the domain of quantification of each modal: in the case of *must*, it is as usual the stereotypically-best worlds compatible with the speaker's epistemic state. To make this categorical claim true, the appropriate AC(s) must be already in the modal base. In contrast, *should* adds some AC(s)

---

[6] For other arguments against the strong analysis of *must*, see [5,8] and references cited therein.

to the modal base, which results in the speaker's non-commitment to the actual truth of that condition.[7]

In Japanese, there is a piece of evidence that *hazu* involves some reference to ACs. First, *hazu* cannot be used in its "evidential" use, i.e., cases where the speaker has some perceptual evidence from which she infers that *hazu*'s prejacent is the case, when there is no *if* clause or *because* clause:

(11) [The speaker noticed that clouds are low and the winds become strong and he says whilst watching the sky.]

#(Kumo=ga hiku-ku, kaze=ga tsuyo-i node,)   ame=ga hur-u-hazu-da.
cloud=Nom low-Conj wind=Nom strong-Pres Conj rain=Nom fall-Pres-HAZU
"#(Because clouds are low and winds are strong,) it should(HAZU) rain."

(adapted from 12)

Why are these subordinate clauses obligatory in evidential uses? We currently have no answer to this question, but the proposition in *if* and *because* clauses in these cases can be regarded as ACs involved in ordering sources.

Back to the Silk's semantics, do they explain the difference in the felicity of *Modal p, but not p*, too? The answer is unfortunately no. They predict that *Modal p, but not p* is semantically consistent, because they both make a claim that some part of the modal base is among the prejacent worlds and the actual world need not be in them.

What predicts the difference lies in the presuppositions of the modals. For *must*, Silk assumes a default presupposition that the domain of quantification is in the common ground. And for *should*, the domain of quantification might not be a subset of the common ground (he regards *should* as "subjunctive weak" and applies [2]'s analysis of subjunctive conditionals to it).[8]

(12) Let CG be a common ground ($\subseteq W$), f be a modal base ($\subseteq W \times \wp(\wp(W))$), g be an ordering source ($\subseteq W \times \wp(\wp(W))$),
$Best_{g(w)}(\cap f(w))$ be the set of $g(w)$–best worlds in $\cap f(w)$ ($\subseteq W$),
$C_{g(w)}$ be a subset of the set of ACs for each of the premises in $g(w)$:
  a. *must*'s (default) presupposition: $Best_{g(w)}(\cap f(w)) \subseteq CG$
  b. *should*'s presupposition: Possibly $\neg(Best_{g(w)}(\cap(f(w) \cup C_{g(w)})) \subseteq CG)$

(adapted from [2, pp. 6–7])

Though Silk himself is formally not so clear as to how these presuppositions make *Modal p, but not p* sentences (in)felicitous, we interpret him as in (13):

(13) Pragmatic inconsistency (our interpretation of [14, p. 58])
a. *must/nichigainai p but not p* is infelicitous because $Best_{g(w)}(\cap f(w)) \subseteq CG$ because this amounts to saying "the common ground is compatible with p, but incompatible with p" in one breath.

---

[7] See [14] for arguments for this difference in presupposition.

[8] $Best_{g(w)}(\cap f(w)) = \{w \mid w \in \cap f(w) \land \forall w' \in \cap f(w)[w' \leq_{g(w)} w \rightarrow w \leq_{g(w)} w']\}$
where $w \leq_{g(w)} w'$ iff $\forall p \in g(w)[w' \in p \rightarrow w \in p]$.

b. *should/hazu p but not p* is felicitous when $Best_{g(w)}(\cap(f(w)\cup C_{g(w)}))\cap CG = \emptyset$ because the first half of the sentence doesn't have to assert anything about the common ground.

In the next section, we try to formalize this "pragmatic inconsistency" in terms of update semantics.

3 Update Semantics for *Hazu*

In this section, we try to formalize Silk's insights dynamically, using Veltman's update semantics with some necessary modifications. The motivations for using a dynamic framework are twofold. First, update semantics offers an explantation for the infelicity of *must p, but not p* and *may p, but not p* in terms of the notion of coherence. These utterances are infelicitous because there is no non-absurd state that supports them. Second, there is already a proposal to treat counterfactuals in update semantics, whose properties *should*-sentences, described as "subjunctive weak" by Silk, seem to share. From now on, we'll concentrate on *hazu* in Japanese, though we believe the same argument applies to English *should* as well.

We saw in the last section that according to Silk, ACs are relevant to the semantics of weak necessity modals and that their domain of quantification can be outside of the common ground (CG) with some ACs added to the modal base. It is worthy of note here that the shift to outside the common ground cannot be achieved unless we retract some fact in the common ground. What is that?

That cannot be evidence for the prejacent. As [11] correctly points out, we cannot make a *hazu*-claim without evidence. Instead, we propose that it is the negation of some very vague normalcy condition ("abnormalcy") that is retracted.[9] For this vague nature of normalcy, we have linguistic evidence. In the sentence below the abnormality of the situation is explicitly mentioned, but the speaker cannot specify what that is. We interpret the first *hazu*-sentence as retracting the abnormalcy.

(14)   Biiru=wa   imagoro   hie-tei-ru-hazu-da-ga,   hie-tei-nai.   Nanika=ga okashii=ga, nani=ga okashii=ka=wa wakar-anai.
"The beer should(HAZU) be cold by now, but it isn't. Something's wrong, but I don't know what that is."

Based on the observation and Silk's subjunctive analysis above, we understand *hazu*'s semantic properties as follows: (i) *hazu* (possibly) revises CG with some normalcy condition used to infer the prejacent by retracting the negation

---

[9] As the formalization below shows, the "retraction" doesn't occur when there is no abnormalcy in CG to be retracted. In that case, *hazu* comes close to an ordinary necessity modal without a "subjunctive" (counterfactual) flavor.

of this normalcy. (ii) *hazu p* amounts to saying "If some normalcy condition had held, *p* would have held".

To formalize this, we use Veltman's ([15]) update semantics for counterfactuals, but with modifications made by Schulz (2007) [13]. More specifically, to model CG, we adopt her notion of belief states which involves a set of propositions, which makes possible a more articulated analysis than a set of worlds (see below).

First, we define our language for counterfactuals. It is an ordinary propositional one except that it includes counterfactual sentences represented as $\phi > \psi$.

(15) Language
Let $\mathcal{P}$ be a set of propositional letters. The language $\mathcal{L}^0$ is the closure of $\mathcal{P}$ under $\neg, \wedge, \vee, \rightarrow$. The language $\mathcal{L}^>$ is the union of $\mathcal{L}^0$ with the set of expressions $\phi > \psi$ for $\phi, \psi \in \mathcal{L}^0$. ($\phi > \psi$ expresses counterfactuals)   (adapted from [13, p. 131])

Then, worlds and models are defined. In this framework, a possible world is an interpretation function from propositional letters to truth values.

(16) Worlds and models
A possible world for $\mathcal{L}^>$ is an interpretation function $w: \mathcal{P} \rightarrow \{0,1\}$. A model for $\mathcal{L}^>$ is a tuple $\langle W, K \rangle$, where W is a set of possible worlds and K is a belief state.
For $\phi \in \mathcal{L}^0$, $M, w \models \phi$ iff ϕ is true with respect to M and w.
$\llbracket \phi \rrbracket^M = \{w \in W | M, w \models \phi\}$ (adapted from [13, p. 131])

Next, we define satisfiablity of sentences in a set of worlds, then belief states.

(17) Satisfiability
A set of sentences A is satisfiable in a set of worlds $W' \subseteq W$ of a model $M = \langle W, K \rangle$, if $\exists w \in W' \forall \phi \in A(w, M \models \phi)$.   (adapted from [13, p. 131])

(18) Belief state
A belief state K is a tuple $\langle B, U \rangle$, where B is a set of non-counterfactual sentences and U is a subset of W such that B is satisfiable in U.
$\llbracket \langle B, U \rangle \rrbracket^M = \{w \in U | \forall \phi \in B(M, w \models \phi)\}$ (adapted from [13, p. 131])

Using the tuple $\langle B, U \rangle$, we distinguish between particular facts and general laws. B is called the basis of K and consists of sentences for which all the agents of K has independent external evidence and which are mutually accepted.[10] U is called the universe of K and consists of worlds where all general laws are true. The final clause defines the set of worlds compatible with this belief state.

---

[10] The condition of mutual acceptance is our original proposal, interpreting Schulz's belief state as representing common ground.

We call this informally "belief worlds." The distinction between belief states and belief worlds turns out to be crucial.

In (19) below, we define belief revision induced by counterfactuals and *hazu*-sentences.

(19) Let $M = \langle W, K \rangle$ be a model for $\mathcal{L}^0$ and $K = \langle B, U \rangle$.

- Order induced by a belief state

$\forall u_1, u_2 \in U : u_1 \leq^{\langle B, U \rangle} u_2$ iff $\{\phi \in B | M, u_1 \models \phi\} \supseteq \{\phi \in B | M, u_2 \models \phi\}$

- The minimality operator

Let D be any domain of objects and $\leq$ an order on D.

$Min(\leq, D) = \{d \in D | \neg \exists d' \in D : d' \leq d\}$

- Belief revision

Let $\phi \in \mathcal{L}^0$

$Rev_M(\langle B, U \rangle, \phi) = Min(\leq^{\langle B, U \rangle}, \llbracket \phi \rrbracket^M \cap U)$ (adapted from [13, p. 135])

We still have to define an update semantics and counterfactual updates, based on [15]'s framework.

(20) Update semantics

Let $M = \langle W, K \rangle$ be a model for $\mathcal{L}^0$ and $K = \langle B, U \rangle, \phi, \psi \in \mathcal{L}^0$.

- Propositional update

$\langle B, U \rangle[\phi] = \langle B \cup \{\phi\}, U \rangle$ if $\llbracket \langle B \cup \phi \rangle, U \rrbracket^M \neq \emptyset$;

$\qquad\qquad = \langle \emptyset, \emptyset \rangle$ otherwise.

- Law update (L-update)

$\langle B, U \rangle[L\phi] = \langle B \cup \{\phi\}, U \cap \llbracket \phi \rrbracket^M \rangle$ if $\llbracket \langle B \cup \phi \rangle, U \rrbracket^M \neq \emptyset$;

$\qquad\qquad = \langle \emptyset, \emptyset \rangle$ otherwise.

(Let $\mathcal{L}^{>+}$ be the union of $\mathcal{L}^>$ with the set of expressions $L\phi, \Box\phi, \Diamond\phi$ for $\phi \in \mathcal{L}^0$.)

(21) Counterfactual update

$\langle B, U \rangle[\phi > \psi] = \langle B, U \rangle$ if $\forall w \in Rev_M(\langle B, U \rangle, \phi)M, w \models \psi$;

$\qquad\qquad = \langle \emptyset, \emptyset \rangle$ otherwise.

Like modals, counterfactual sentences impose a test on the input belief state. Roughly speaking, they test whether the consequent always follows after the revision with the antecedent.

Finally, we come to the definition of *hazu*'s update. If we let e be some evidence for the prejazent, n normalcy condition, p the prejazent, then $[hazu\ p]$ update means updating first with the law "if the evidence e and normalcy n holds, then p holds" and then with a counterfactual "if normalcy had held, p would have held."

(22) *Hazu*'s update

Let e be evidence for the prejazent, n normalcy condition.

$\langle B, U \rangle[hazu\ p] = \langle B, U \rangle[L((e \wedge n) \rightarrow p)][n > p]$

Why do we use belief states along with belief worlds? The original motivation of Schulz is to explain the felicity/infelicity of some kind of counterfactuals.[11] Crucial here is that even when abnormality (represented as $\neg n$) is in B and the law $e \wedge \neg n \rightarrow \neg p$ (i.e., if evidence and abnormality hold then the prejacent doesn't) is assumed, we don't have to include $\neg p$ in B. This makes it possible that $\neg p$ is supported (see the definition below) by this belief state, but doesn't count as a fact under consideration, so that the counterfactual update with *hazu p* is successful. If we used a set of possible worlds which is a deductive closure of B in order to model CG, this wouldn't be possible. To be precise, we define the notion of support which is ordinary in update semantics (e.g., [15, 16]):

(23) Support: $\langle B, U \rangle$ supports ϕ, $\langle B, U \rangle \models \phi$, iff $[\![\langle B, U \rangle [\phi]]\!]^M = [\![\langle B, U \rangle]\!]^M$

If we assume as a law "if evidence e and abnormality $\neg\, n$ hold, then $\neg\, p$ holds", and as the basis $\{e, \neg n\}$, then this belief state supports *hazu p, but not p* ($= [n > p \wedge \neg p]$). Once we adopt the notion of coherence in the standard update semantics ("ϕ is coherent iff there is some non-absurd state by which ϕ is supported" [16, p. 192]), this means *hazu p, but not p* is coherent.

In contrast, there is no non-absurd belief state that supports *must p, but not p* or *may p, but not p*, given their ordinary update semantics:

(24) Incoherence of *must/may p, but not p*
- Given an ordering $\leq^g$,
$$\langle B, U \rangle [\Box \phi] = \langle B, U \rangle \text{ if } Min_M(\leq^g, [\![\langle B, U \rangle]\!]^M) \subseteq [\![\phi]\!]^M$$
$$= \langle \emptyset, \emptyset \rangle \quad \text{otherwise}$$
$$\langle B, U \rangle [\Diamond \phi] = \langle B, U \rangle \text{ if } Min_M(\leq^g, [\![\langle B, U \rangle]\!]^M) \cap [\![\phi]\!]^M \neq \emptyset$$
$$= \langle \emptyset, \emptyset \rangle \quad \text{otherwise}$$

The only states that support $\neg p$ are the ones whose denotations are included in $[\![\neg p]\!]^M$, which cannot pass the test imposed by $[\Box \phi]$ or $[\Diamond \phi]$.

4   Conclusion

We argued in this paper (i) that *hazu* makes a "subjunctive" update by retracting the negation of some normalcy condition and (ii) that the distinction between a belief state and the set of worlds compatible with the state makes room for *hazu p, but not p* to be coherent, while keeping the incoherence of *must/may p, but not p*.

References

1. Copley, B.: What should "should" mean?. Language Under Uncertainty Workshop, Kyoto University, January 2005 ⟨halshs − 00093569⟩ (2006)

---

[11] See [15] and [13, p. 135ff.] for discussions.

2. von Fintel, K.: The presupposition of subjunctive conditionals. In: Sauerland, U., Percus, O. (eds.) The interpretive tract (MIT Working Papers in Linguistics 25), pp. 29–44. MITWPL, Cambridge, MA (1998)
3. von Fintel, K., Gillies, A.: Must.stay.strong!. Nat. Lang. Seman. **18**(4), 351–383 (2010)
4. von Fintel, K., Iatridou, S.: How to say ought in foreign: the composition of weak necessity modals. In: Guéron, J., Lecarme, J. (eds.) Time and Modality, pp. 115–141. Springer, Netherlands (2008)
5. Giannakidou, A., Mari, A.: Future and unviersal epistemic modals:reasoning with nonveridicality and partial knowledge. In: Blaszack, J. et al. (eds.) Tense, Mood, and Modality: New Perspectives on OldQuestions. University of Chicago Press. (forthcoming) Retrieved from 27 May 2014. http://home.uchicago.edu/~giannaki/pubs/Giannakidou_Mari_Future.MUST_Chicago_volume.pdf
6. Kratzer, A.: The notional category of modality. In: Eikmeyer, H., Rieser, H. (eds.) Words, Worlds and Contexts: New Approaches in Word Semantics, pp. 38–74. Walter de Gruyter, Berlin, New York (1981)
7. Kratzer, A.: Modals and Conditionals. Oxford University Press, Oxford (2012)
8. Lassiter. D.: The weakness of must: In defense of a Mantra. In: Proceedings of SALT vol. 24, pp. 597–618 (2014)
9. Lewis, D.: Counterfactuals. Blackwell, Oxford (1973)
10. Matthewson, L.: Evidence Type, Evidence Location, Evidence Strength. In: Lee, C., Park, J. (eds.) Evidentials and Modals (Current Research in the Semantics/pragmatics Interface). Emerald (forthcoming)
11. McCready, E., Asher, N.: Modal subordination in Japanese: Dynamics and evidentiality. In: Eilam, A., Scheffler, T., Tauberer, J. (eds.) Penn working papers in linguistics, pp. 237–249. Penn Linguistics Club, University of Pennsylvania, Pennsylvania (2006)
12. Mori, Y., Park, C.: On Some Aspects of the Deictic/Evidential Component in Korean -(u)l kesita and Japanese hazuda. In: Kuno, S., et al. (eds.) Harvard Studies in Korean Linguistics XV, pp. 119–133. Hankuk Publishing Co, Seoul (2014)
13. Schulz, K.: Minimal Models in Semantics and Pragmatics. Ph.D. Thesis, University of Amsterdam, Amsterdam (2007)
14. Silk, A.: Modality, weights, and inconsistent premise sets. In: Proceedings of SALT, vol. 22, pp. 43–64 (2012)
15. Veltman, F.: Making counterfactual assumptions. J. Seman. **22**, 159–180 (2005)
16. Groenendijk, J., Stokhof, M., Veltman, F.: Coreference and modality. In: Lappin, S. (ed.) Handbook of Contemporary Semantic Theory, pp. 179–213. Blackwell, Oxford (1996)

Using Signatures in Type Theory to Represent Situations

Stergios Chatzikyriakidis$^{(\boxtimes)}$ and Zhaohui Luo

Department of Computer Science, Royal Holloway, University of London, Egham, UK
stergios.chatzikyriakidis@cs.rhul.ac.uk, zhaohui.luo@hotmail.co.uk

Abstract. Signatures have been introduced to represent situations in formal semantics based on modern type theories. In this paper, we study the notion of signature in more details, presenting it formally and discussing its use in representations of situations. In particular, the new forms of signature entries, the subtyping entries and the manifest entries, are formally presented and studied. Besides being signature entries, these two forms of entries may be introduced to form contextual entries as well and this may have interesting implications in applications of the notion of context to, for example, belief contexts.

1 Introduction

Signatures are introduced to represent situations (or incomplete possible worlds) by the second author in [15], where it has been argued that, with the new forms of subtyping entries and manifest entries, signatures are very useful in representing situations in a formal semantics based on modern type theories (MTTs). In this paper, we shall study the notion of signature in a more formal and detailed way.

The notion of signature has been used in describing algebraic structures. Its use in type theory can be found in the Edinburgh Logical Framework [8]. There, signatures are used to describe constants (and their types) in a logical system. This is in contrast to contexts in type theory that describe variables (and their types) which can be abstracted by means of quantification or λ-abstraction. We shall study the notion of signature in MTTs by extending the logical framework LF (Chapter 9 of [10]) with signatures to obtain the system LF$_\Sigma$, which can be used similarly as LF in specifying type theories such as Martin-Löf's type theory [19] and UTT [10].

Signatures as proposed in [15] may contain two new forms of entries: subtyping entries and manifest entries. A subtyping entry $A <_\kappa B$ declares that A is a subtype of B via. coercion κ. This localises a coercive subtyping relationship as studied in the coercive subtyping framework [11, 17] that was developed for type theory based proof assistants. Subtyping has been proved useful in formal semantics and, specifically for MTT-semantics, it is crucial partly because

This work is partially supported by the research grants from Leverhulme, Royal Academy of Engineering and the CAS/SAFEA International Partnership Program for Creative Research Teams.

T. Murata et al. (Eds.): JSAI-isAI 2014 Workshops, LNAI 9067, pp. 172–183, 2015.
DOI: 10.1007/978-3-662-48119-6_13

CNs are interpreted as types rather than predicates (as in Montague semantics). It is very useful to introduce subtyping entries in signatures when they are used to represent situations. Also, we shall explain that the introduction of coherent subtyping entries to signatures preserves the nice properties of the original type theory.

The other new form of signature entries is that of manifest entries. These have the form $c \sim a : A$, which introduces the constant c and assumes that it behaves exactly like the object a of type A.[1] Formally, a manifest entry is just the abbreviation of an ordinary membership entry together with a subtyping entry. The latter enforces the abbreviation: $c \sim a : A$ abbreviates $c : \mathbf{1}_A(a)$, where $\mathbf{1}_A(a)$ is the inductive unit type, together with the subtyping entry $\mathbf{1}_A(a) <_{\xi_{A,a}}$ A with the coercion $\xi_{A,a}$ that maps the object of the unit type to a. Such an extension with manifest entries is sound: meta-theoretically, the extension preserves all of the nice properties of the original type theory. We shall make this clear in more detail in the paper.

Both subtyping and manifest entries can be considered as contextual entries for declaring variables. This makes contributions to the application of contexts. One such example can be found in Ranta's treatment of belief contexts [23]. We show how to extend this notion of belief context with these new entries. We shall also point out that, if we introduce these to form contextual entries, we should allow the corresponding move to the right of the turnstile: by quantification and λ-abstraction for manifest entries and by local coercions for subtyping entries. In particular, for subtyping entries, this requires the introduction of the new form of terms, **coercion A $<_{\mathbf{c}}$ B in M**, to express local coercions[2] and this may make the meta-theoretical study more sophisticated.

The notion of signature is formally introduced in Sect. 2, where we present the system LF$_\Sigma$ and give an example to illustrate its use in representations of situations. In Sect. 3, the subtyping and manifest entries in signatures are studied: they are shown to be useful in expressing situations and, with further meta-theoretic studies, the extensions with them can be shown to preserve the nice meta-theoretic properties. The potential of adding such new forms of entries as contextual entries is considered in Sect. 4, where we use belief contexts as an example to illustrate that this can be useful.

2   Signatures for Representing Situations

Situations, or incomplete possible worlds, are proposed by the second author to be representable by *signatures* in MTT-semantics, i.e., when modern type theories are used to give formal semantics [15].

---

[1] Contextual manifest entries were first proposed by the second author in [12], where they are studied in a different context, focussing on its intensional nature, as compared with traditional extensional definition entries in proof assistants.

[2] Local coercions are useful in formal semantics based on MTTs. See, for example, [1,14] for discussions.

The use of possible worlds in set theory has been a central mechanism within Montagovian approaches of formal semantics, especially, in dealing with intensional phenomena including, for example, belief intensionality among other things. However, the use of set-theoretical possible worlds has given rise to the well-known hyperintensional problem, with various paradoxes associated with it (e.g., the Paris Hilton paradox and the woodchuck-groundhog paradox) [22]. When intensional type theories are employed for formal semantics, types rather predicates over sets are used to interpret CNs and significantly different mechanisms are available in representing and dealing with such phenomena. Using signatures to represent situations is such a proposal.

We shall describe the notion of signature formally, compare it with that of context, and give a simple example of its use in representing situations. In this section. we shall only describe signatures with the traditional membership entries. Contexts with such traditional entries have been used by Ranta [23] and others [3,6] to represent situations, where they do not consider the issue of difference between variables and constants. We consider signatures rather than contexts here. Note that signatures may contain other forms of entries which are studied in the next section Sect. 3.

2.1   Signatures in Type Theory: A Formal Presentation

Type theories can be specified in a logical framework such as Martin-Löf's logical framework [19] or its typed version LF [10]. We shall extend LF with signatures to obtain LF_Σ.

Informally, a signature is a sequence of entries of several forms, one of which is the form of membership entries $c : K$, which is the traditional form of entries as occurred in contexts (we shall add two other forms of entries in the next section). If a signature has only membership entries, it is of the form $c_1 : K_1, \ldots, c_n : K_n$.

LF is a dependent type theory whose types are called *kinds* in order to be distinguished from types in the object type theory. It has the kind *Type* of all types of the object type theory and dependent Π-kinds of the form $(x : K)K'$ (we omit their details here – see [10]). In LF, there are five forms of judgements:

- $\vdash \Gamma$ (or written as 'Γ *valid*'), which asserts that Γ is a valid context.
- $\Gamma \vdash K$ *kind*, which asserts that K is a kind in Γ.
- $\Gamma \vdash k : K$, which asserts that k is an object of kind K in Γ.
- $\Gamma \vdash K_1 = K_2$, which asserts that K_1 and K_2 are equal kinds in Γ.
- $\Gamma \vdash k_1 = k_2 : K$, which asserts that k_1 and k_2 are equal objects of kind K in Γ.

To extend LF with signatures, we amend each form of judgement with a signature Σ and add another form of judgements saying that a signature is valid. In other words, LF_Σ has the following six forms of judgements:

- Σ *valid*, which asserts that Σ is a valid signature.
- $\vdash_\Sigma \Gamma$, which asserts that Γ is a valid context under Σ.

Signature Validity and Assumptions

$$\frac{}{\langle\rangle \; valid} \qquad \frac{\langle\rangle \vdash_\Sigma K \; kind \quad c \notin dom(\Sigma)}{\Sigma, \; c : K \; valid} \qquad \frac{\vdash_{\Sigma, c:K, \Sigma'} \Gamma}{\Gamma \vdash_{\Sigma, c:K, \Sigma'} c : K}$$

Context Validity and Assumptions

$$\frac{\Sigma \; valid}{\vdash_\Sigma \langle\rangle} \qquad \frac{\Gamma \vdash_\Sigma K \; kind \quad x \notin dom(\Gamma)}{\vdash_\Sigma \Gamma, \; x : K} \qquad \frac{\vdash_\Sigma \Gamma, x : K, \Gamma'}{\Gamma, x : K, \Gamma' \vdash_\Sigma x : K}$$

Fig. 1. Rules for signatures/contexts in LF$_\Sigma$.

- $\Gamma \vdash_\Sigma K \; kind$, which asserts that K is a kind in Γ under Σ.
- $\Gamma \vdash_\Sigma k : K$, which asserts that k is an object of kind K in Γ under Σ.
- $\Gamma \vdash_\Sigma K_1 = K_2$, which asserts that K_1 and K_2 are equal kinds in Γ under Σ.
- $\Gamma \vdash_\Sigma k_1 = k_2 : K$, which asserts that k_1 and k_2 are equal objects of kind K in Γ under Σ.

All of the inference rules of LF (those in Figs. 9.1 and 9.2 of Chapter 9 of [10]) become inference rules of LF$_\Sigma$ after replacing $\vdash$ by $\vdash_\Sigma$ (and changing the judgement form 'Γ *valid*' to '$\vdash_\Sigma \Gamma$'). For instance, the following rule for λ-abstraction[3] in LF

$$\frac{\Gamma, x : K \vdash b : K'}{\Gamma \vdash [x : K]b : (x : K)K'}$$

becomes, in LF$_\Sigma$,

$$\frac{\Gamma, x : K \vdash_\Sigma b : K'}{\Gamma \vdash_\Sigma [x : K]b : (x : K)K'}$$

In addition, in LF$_\Sigma$, we have the rules in Fig. 1 for signatures (and contexts), concerning their validity and their roles of making basic assumptions, where $\langle\rangle$ is the empty sequence and $dom(p_1 : K_1, \; \dots \; p_n : K_n) = \{p_1, \dots, p_n\}$.

Note that the assumptions in a signature or in a context can be derived – this is characterised by the third rule and the last rule in Fig. 1, respectively.

Remark 1. The membership entry $c : K$ in a signature declares that c is a *constant* of kind K. This is different from a contextual entry $x : K$ that declares x to be a *variable*. Note that a variable can be abstracted by, for example, quantification or λ-abstraction as exemplified by a rule like the one below, where *Prop* is the universe of logical propositions:

$$\frac{\Gamma, \; x : K \vdash_\Sigma P : Prop}{\Gamma \vdash_\Sigma \forall x : K.P : Prop}$$

However, constants in signatures can never be abstracted in this way – that is why they are called constants. Therefore, signatures can adequately be used to represent situations. Also, because the constants in signatures cannot be

---

[3] In LF, we use the notation $[x : K]b$ for $\lambda x : K.b$ and $(x : K)K'$ for $\Pi x : K.K'$.

abstracted, it is easier meta-theoretically to add new forms of entries to signatures than to contexts (see later).

2.2   Use of Signatures to Represent Situations: A Simple Example

Signatures can adequately be used to represent situations, or incomplete possible worlds, in the MTT-semantics. This possibility can easily be understood whenever one realises that types represent collections of objects just like sets, although types are syntactic (or, better, proof-theoretic) entities different from sets in set theory. Intuitively, the similarity between types and sets is one of the crucial reasons that MTT-semantics can be viewed as model-theoretic, while the differences between types and sets and, especially that the former are proof-theoretically defined, are why MTT-semantics can be also viewed as proof-theoretic (see [15] for more details).

That signatures can be used to represent situations is the other facet that the MTT-semantics is model-theoretic. Here, we use an example given in [15] to illustrate how signatures can be used to represent situations.

Example 1. The example, taken from Chapter 10 of [24], is about an (imagined) situation in the Cavern Club at Liverpool in 1962 where the Beatles were rehearsing for a performance. This situation can be represented as follows.

1. The domain of the situation consists of several peoples including the Beatles (John, Paul, George and Ringo), their manager (Brian) and a fan (Bob). This can be represented be means of the following signature Σ_1:

$$\Sigma_1 \equiv D : Type,$$
$$John : D, \ Paul : D, \ George : D, \ Ringo : D, \ Brian : D, \ Bob : D$$

2. The assignment function assigns, for example, predicate symbols such as B and G to the propositional functions expressing 'was a Beatle' and 'played guitar', respectively. We can introduce the following in our signature to represent such an assignment function:

$$\Sigma_2 \equiv B : D \rightarrow Prop, \ b_J : B(John), \ \ldots, \ b_B : \neg B(Brian), \ b'_B : \neg B(Bob),$$
$$G : D \rightarrow Prop, \ g_J : G(John), \ \ldots, \ g_G : \neg G(Ringo), \ \ldots$$

The signature that represents the situation will be of the form $\Sigma \equiv \Sigma_1, \Sigma_2, \ldots, \Sigma_n$. We shall then have, for instance,

$$\vdash_\Sigma G(John) \ true \ \ and \ \ \vdash_\Sigma \neg B(Bob) \ true.$$

where $G(John)$ and $B(Bob)$ are the semantic interpretations of **John played Guitar** and **Bob was a Beatle**, respectively.

3 Subtyping and Manifest Entries in Signatures

In the last section, we introduced signatures with only traditional membership entries. In this section, we consider two other forms of entries – the subtyping entries and manifest entries: introducing them into signatures, discussing meta-theoretic implications and illustrating their uses in representing situations.

In earlier work, these forms of entries were considered contextual entries: contextual manifest entries were first studied in [12] and contextual subtyping entries (in so-called coercion contexts) in [14]. Here in this section, we consider them as entries in signatures, as proposed in [15]. For this reason, they are not only useful in representing situations, but are also simpler meta-theoretically, since they are introducing constants rather than variables and, as a consequence of the subtyping entries, one does not need to introduce corresponding terms for the purpose of making abstraction operations possible (see Sect. 4 for further discussion in this last respect).

3.1 Subtyping Entries and Their Uses

Coercive subtyping has been studied for subtyping and abbreviations in MTTs and the associated proof assistants [11,17].[4] Introducing subtyping entries (to either signatures or contexts) is to localise the coercive subtyping mechanism, which has been studied globally in earlier research.

Syntactically, the system LF_Σ is extended with the judgement forms $\Gamma \vdash_\Sigma A <_\kappa B : Type$ (we shall often just write $A <_\kappa B$ even when A and B are types) and $\Gamma \vdash_\Sigma K <_\kappa K'$. A subtyping entry to signatures can be introduced by means of the first rule in Fig. 2, where $(A)B$ is the kind of functional operations from A to B. The second rule in Fig. 2 expresses that the subtyping assumptions in a signature are derivable. Then the rules for coercive subtyping [17], albeit extended for judgements with signatures, are all applicable. For instance, if signature Σ contains $A <_\kappa B$ and $B <_{\kappa'} C$, we can derive $A <_{\kappa' \circ \kappa} C$ under Σ.

It is worth pointing out that *validity* of a signature is not enough anymore when we consider subtyping entries in signatures. For signature Σ to be legal, we need the subtyping assumptions in Σ to be *coherent* in the sense that, informally, all coercions between any two types are equal, i.e., in some appropriate subsystem,[5] if $\Gamma \vdash_\Sigma A <_\kappa B$ and $\Gamma \vdash_\Sigma A <_{\kappa'} B$, then $\Gamma \vdash_\Sigma \kappa = \kappa' : (A)B$.

---

[4] The word 'coercion' has been used for related but maybe different things including coercions in programming languages and coercions in linguistics. See Asher and Luo [1] for a use of coercive subtyping in modelling linguistic coercions and Retoré et al. [2] for another proposal of using coercions to deal with some linguistic coercions in lexical semantics.

[5] It is important that the condition is not stated for the whole system of coercive subtyping, for otherwise it would become trivial. Here we do not detail the description of the subsystem because we would then have to make explicit some technical details we feel unnecessary for this paper. An interested reader may look at [17] for details how coherence is defined in a global case.

$$\frac{\vdash_\Sigma A : Type \quad \vdash_\Sigma B : Type \quad \vdash_\Sigma c : (A)B}{\Sigma, A <_c B \; valid} \qquad \frac{\vdash_{\Sigma, A <_c B, \Sigma'} \Gamma}{\Gamma \vdash_{\Sigma, A <_c B, \Sigma'} A <_c B}$$

Fig. 2. Rules for subtyping entries in signatures.

It is then possible to show that the conservativity result in [17] can be carried over to the current setting[6] and, in particular, if the original type theory is strongly normalising, so is the type theory extended with the subtyping entries. As a consequence, the extension with subtyping entries preserves logical consistency – a basic requirement for a type theory to be employed for formal semantics.

Introducing subtyping entries makes using type theory for formal semantics much more convenient. First of all, it is now possible for one to *localise* subtyping assumptions. In some specific situations, some special subtyping relations may reasonably be assumed, which may not be reasonable in general. For instance, only in a cafe or restaurant would it be reasonable to say

(1) The ham-sandwich left without paying the bill.

In representing a situation in a cafe, we might reasonably assume the following subtyping entry:

$$Ham\text{-}sandwich < Human,$$

which will then allow the sentence (1) to be semantically interpreted as intended. Such reference transfers are studied by Nunberg [20] among others.

3.2   Manifest Entries and Their Uses

A *manifest entry* is of the form

$$c \sim a : A \qquad (2)$$

Informally, it assumes that c behaves exactly like a of type A. Alternatively, one can think that in any place that we could use an object of type A, we could use c which actually plays the role of a. Signatures can be extended with manifest entries:

$$(*) \qquad \frac{\vdash_\Sigma A : Type \quad \vdash_\Sigma a : A \quad c \notin dom(\Sigma)}{\Sigma, c \sim a : A \; valid}$$

where $Type$ is the kind of all types (in the object type theory). In fact, such manifest entries can be introduced by means of special membership entries with the help of the coercive subtyping mechanism. We now proceed with its formal description.

---

[6] At the moment, this is only a conjecture: although the authors do not see any real problems in doing so, tedious and careful work is needed to carry such a proof out (work in progress).

Manifest entries can be regarded as abbreviations of special membership entries [12] with the help of the coercive subtyping mechanism [11,17]. Formally, to add the above manifest entry (2) to a signature is to add the following two entries:

$$c : \mathbf{1}_A(a), \quad \mathbf{1}_A(a) <_{\xi_{A,a}} A \tag{3}$$

where $\mathbf{1}_A(a)$ is the inductive unit type parameterised by $A : Type$ and $a : A$, whose only object is $*_A(a)$, and $\xi_{A,a}(x) = a$ for every $x : \mathbf{1}_A(a)$. It is now easy to see that, if an expression has a hole that requires a term of type A, we can use c to fill that hole; then the whole expression is equal to that with the hole filled by a. For example, if the expression is $f(\_)$, then $f(c)$ is equal to $f(a)$.

Note that the subtyping entries involving ξ form coherent signatures; in particular, if for two manifest entries $c \sim a : A$ and $d \sim b : B$ we have $\mathbf{1}_A(a) = \mathbf{1}_B(b)$ and $A = B$, then $\xi_{A,a} = \xi_{B,b}$, as coherence requires. Put in another way, if the subtyping entries in a signature are coherent, the signature is coherent since its manifest entries do not cause incoherence. Therefore, the extension with manifest entries in signatures preserves the nice properties of the original type theory such as strong normalisation and logical consistency.

Manifest entries can considerably reduce the complexity of representation, as the following example shows.

Example 2. With manifest entries, the situation in Example 1 can be represented as the following signature:

$$D \sim a_D : Type, \; B \sim a_B : D \to Prop, \; G \sim a_G : D \to Prop, \; \dots \dots \tag{4}$$

where

- $a_D = \{John, \; Paul, \; George, \; Ringo, \; Brian, \; Bob\}$ is a finite type,
- $a_B : D \to Prop$, the predicate 'was a Beatle', is an inductively defined function such that $a_B(John) = a_B(Paul) = a_B(George) = a_B(Ringo) = True$ and $a_B(Brian) = a_B(Bob) = False$, and
- $a_G : D \to Prop$, the predicate 'played guitar', is an inductively defined function such that $a_G(John) = a_G(Paul) = a_G(George) = True$ and $a_G(Ringo) = a_G(Brian) = a_G(Bob) = False$.

In other words, Σ_1 in Example 1 is now expressed by the first entry of (4) and Σ_2 in Example 1 by the second and third entries of (4).

Manifest entries in signatures can be used to represent infinite situations such as those with infinite domains. With traditional membership entries (as in the traditional notion of context), we can only describe finite domains as we have done in Example 1. What if the domain D is infinite? This can be done by using a manifest entry – as in Example 2, we can assume that

$$D \sim Inf : Type,$$

where Inf is some inductively defined type with infinitely many objects. Similarly, one can assume an infinite predicate over the domain, represented as:

$$P \sim \text{P-defn} : D \rightarrow Prop,$$

where P-defn is also inductively defined.

4   Subtyping and Manifest Entries in Contexts

The subtyping or manifest entries may be introduced in contexts as well. If this were done, it would further widen the uses of contexts in their applications. However, before introducing them and illustrating their uses be means of belief contexts, we should make clear that introducing contextual subtyping entries (and manifest entries, which have associated subtyping entries via ξ) complicates meta-theoretic studies. Until now, although the proposal of introducing contextual subtyping entries was already made in 2009 [13], the corresponding meta-theoretic studies have not been carried out in detail (for an initial study of this, see [16]) and further studies are needed.

Because a contextual entry should be able to be abstracted or moved to the right of turnstile (see Remark 1), it is necessary to introduce a new form of terms so that subtyping assumptions in a context can be represented as *local coercions* in terms. An term with a local coercion is of the form

$$\textbf{coercion A} <_\kappa \textbf{B in M},$$

which indicates that the scope in which subtyping $A <_\kappa B$ takes effects is term M – it does not take effect outside M. Local coercions are introduced the rules like the following:

$$\frac{\Gamma, A <_\kappa B \vdash_\Sigma k : K}{\Gamma \vdash_\Sigma (\textbf{coercion A} <_\kappa \textbf{B in k}) : (\textbf{coercion A} <_\kappa \textbf{B in K})}$$

where the parentheses are there for readability, but not necessary.

Ranta [23] has proposed an account of belief intensionality in which he uses contexts to model agents' beliefs as a sequence of membership entries.[7] The idea is simple and it is based on the assumption that contexts can be seen as the equivalent type theoretic notion of a (partial) world as found in the traditional Montagovian semantics. Ranta introduces an agent's belief context: for agent p, p's belief context may be:

$$\Gamma_p = x_1 : A_1, \ldots, x_n : A_n.$$

---

[7] Similar ideas have been put forth in [5] and [4] to deal with intensional adjectives and adverbs.

A belief operator is then introduced: for a proposition A, $B_p(A)$ is true just in case that A is true in p's belief context Γ_p, which is equivalent to saying that $\Pi x_1 : A_1 ... \Pi x_n : A_n.A$ is true.[8]

In a case like (5):

(5) John believes that all woodchucks are woodchucks $\Rightarrow$ John believes all wood-chucks are groundhogs.

the sentences are evaluated against the agent's belief context. If, from John's belief context, one cannot derive the belief that 'all woodchucks are ground-hogs',[9] the unwanted entailment (5) does not go through. Similar considerations apply to the Hesperus/Phosphorus examples shown below:

(6) The Ancients believed that Hesperus is Hesperus.

(7) The Ancients believed that Hesperus is Phosphorus.

In a coarse grained system like Montague Semantics [7,18], (7) follows from (6) [21,22]. For the account as sketched here, this is not the case. If a logical equality between Hesperus and Phosphorus cannot be derived from the belief context of the Ancients, say B_A, then (7) does not follow from (6).

Similar considerations apply to the Paris-Hilton paradox which says that if Paris Hilton knows she is Paris Hilton, then she also knows either (a) that every nontrivial zero of the zeta-function has real part $1/2$ (if this is indeed the case) or (b) knows that this is not the case (if it is not). Let us call this disjunction R. In effect, the Paris Hilton paradox says that if Paris Hilton knows that she is Paris Hilton, she also knows whether the Riemann hypothesis is true [21]. This is because in the set-theoretical semantics, necessary true propositions have the same meaning since they are functions from possible worlds to truth values. Both sentences, i.e. Paris Hilton knows that Paris Hilton is Paris Hilton and Paris Hilton knows that R, are analytically true, i.e. true in every world and as such have the same meaning, which means that one can be substituted for the other! In the above approach, this is not true and the analysis does not suffer from this problem.[10]

Remark 2. Another interesting note is that the above approach to belief contexts allows us to represent embedded beliefs. For example, the following sentence (8) can be expressed (9):

---

[8] Here, we do not discuss the issue whether such a proposal is adequate to represent intensional beliefs. For instance, one might argue against such proposals simply by arguing that ordinary logical inference does not capture the intended inference concerning beliefs. We are simply take this as an example to show that the usefulness of subtyping/manifest entries in contexts.

[9] For example, using the heterogenous equality Eq, this belief can be expressed as $\forall x : G \forall y : W.Eq(G, W, x, y)$. We do not get into the formal details here.

[10] A similar problem due to Kripke is the Pierre problem, according to which Pierre thinks that Londres is beautiful but London is not [9]. It is obvious how this can be handled given what we have said.

(8) John believes that George believes that John is handsome.

(9) $B_J(B_G(A))$, where A is the proposition expressing 'John is handsome'.

It is interesting to note that, from the above, we cannot conclude that $B_J(A)$ ('John believes that John is handsome').

If we introduce subtyping entries and manifest entries into contexts, we would then be able to make the above mechanism for beliefs more powerful. Here are some examples:

- In one's belief context, there can be subtyping entries like $Man < Human$ (or even unreasonably $Human < Man$).
- Infinite beliefs can be expressed by manifest entries. In particular, we can use inductive definitions to capture infinitely many entries by means of finitely many entries.

Formally, when contexts are extended with subtyping (and manifest) entries, the belief operator $B_p(P)$ can be defined as follows.

Definition 1. *First, define B_Γ for arbitrary context Γ as follows.*

- *If $\Gamma = \langle\rangle$, then $B_\Gamma(P) = P$.*
- *If $\Gamma = x : A, \Gamma_0$, then $B_\Gamma(P) = \Pi x : A. \ B_{\Gamma_0}(P)$.*
- *If $\Gamma = A <_\kappa B, \Gamma_0$, then $B_\Gamma(P) = $ **coercion $A <_\kappa B$ in $B_{\Gamma_0}(P)$.***

Then, let p be an agent and P a Γ_p-proposition. Define the belief operator as

$$B_p(P) = B_{\Gamma_p}(P).$$

Remark 3. In the above definition, we have not considered manifest entries because a manifest entry can be represented by an ordinary membership entry together with a subtyping entry and, therefore, the above definition covers manifest entries as well.

References

1. Asher, N., Luo, Z.: Formalisation of coercions in lexical semantics. Sinn und Bedeutung 17, Paris (2012)
2. Bassac, C., Mery, B., Retoré, C.: Towards a type-theoretical account of lexical semantics. J. Log. Lang. Inf. **19**(2), 229–245 (2010)
3. Boldini, P.: Formalizing contexts in intuitionistic type theory. Fundamenta Informaticae **4**(2), 1–23 (2000)
4. Chatzikyriakidis, S.: Adverbs in a modern type theory. In: Asher, N., Soloviev, S. (eds.) LACL 2014. LNCS, vol. 8535, pp. 44–56. Springer, Heidelberg (2014)
5. Chatzikyriakidis, S., Luo, Z.: Adjectives in a modern type-theoretical setting. In: Morrill, G., Nederhof, M.-J. (eds.) FG 2012 and 2013. LNCS, vol. 8036, pp. 159–174. Springer, Heidelberg (2013)
6. Dapoigny, R., Barlatier, P.: Modelling contexts with dependent types. Fundamenta Informaticae **104**, 293–327 (2010)

7. Dowty, D.R.: Introduction to Montague Semantics. Studies in linguistics and philosophy, vol. 11. Springer, Netherlands (1981)
8. Harper, R., Honsell, F., Plotkin, G.: A framework for defining logics. J. Assoc. Comput. Mach. **40**(1), 143–184 (1993)
9. Kripke, S.A.: A puzzle about belief. In: Margalit, A. (ed.) Meaning and Use. Studies in Linguistics and Philosophy, pp. 239–283. Springer, Netherlands (1979)
10. Luo, Z.: Computation and Reasoning: A Type Theory for Computer Science. Oxford University Press, New York (1994)
11. Luo, Z.: Coercive subtyping. J. Log. Comput. **9**(1), 105–130 (1999)
12. Luo, Z.: Manifest fields and module mechanisms in intensional type theory. In: Berardi, S., Damiani, F., de'Liguoro, U. (eds.) TYPES 2008. LNCS, vol. 5497, pp. 237–255. Springer, Heidelberg (2009)
13. Luo, Z.: Type-theoretical semantics with coercive subtyping. Semantics and Linguistic Theory 20 (SALT20), Vancouver (2010)
14. Luo, Z.: Formal semantics in modern type theories with coercive subtyping. Linguist. Philos. **35**(6), 491–513 (2012)
15. Luo, Z.: Formal Semantics in modern type theories: is it model-theoretic, proof-theoretic, or both? In: Asher, N., Soloviev, S. (eds.) LACL 2014. LNCS, vol. 8535, pp. 177–188. Springer, Heidelberg (2014)
16. Luo, Z., Part, F.: Subtyping in type theory: coercion contexts and local coercions (extended abstract). In: TYPES 2013, Toulouse (2013)
17. Luo, Z., Soloviev, S., Xue, T.: Coercive subtyping: theory and implementation. Inf. Comput. **223**, 18–42 (2012)
18. Montague, R.: The proper treatment of quantification in ordinary English. In: Hintikka, J., Moravcsik, J., Suppes, P. (eds.) Approaches to Natural Languages. Synthese Library, vol. 49, pp. 221–242. Springer, Netherlands (1973)
19. Nordström, B., Petersson, K., Smith, J.: Programming in Martin-Löf's Type Theory: An Introduction. Oxford University Press, New York (1990)
20. Nunberg, G.: Transfers of meaning. J. Seman. **12**(2), 109–132 (1995)
21. Pollard, C.: Hyperintensional questions. In: Hodges, W., de Queiroz, R. (eds.) WoLLIC 2008. LNCS (LNAI), vol. 5110, pp. 272–285. Springer, Heidelberg (2008)
22. Pollard, C.: Hyperintensions. J. Log. Comput. **18**(2), 257–282 (2008)
23. Ranta, A.: Type-Theoretical Grammar. Oxford University Press, Oxford (1994)
24. Saeed, J.: Semantics. Wiley-Blackwell, Malden (1997)

Scope as Syntactic Abstraction

Chris Barker$^{(\boxtimes)}$

New York University, New York, NY 10003, USA
chris.barker@nyu.edu
http://files.nyu.edu/cb125/public/

Abstract. What is the logic of scope? By "scope", I mean scope-taking in natural languages such as English, as illustrated by the sentence *Ann saw everyone*. In this example, the quantifier denoted by *everyone* takes scope over the rest of the sentence, that is, it takes the denotation of the rest of the sentence as its semantic argument: **everyone**($\lambda x.$**saw**(x)(**ann**)). The answer I will give here will be to provide a substructural logic whose two modes are related by a single structural postulate. This postulate can be interpreted as constituting a kind of lambda-abstraction over structures, where the abstracted structures are interpreted as delimited continuations. I discuss soundness and completeness results, as well as cut elimination. I also compare the logic to a number of alternative approaches, including the standard technique of Quantifier Raising, and mention applications to scope ambiguity and parasitic scope.

Keywords: Scope · Continuations · Substructural logic · Quantifier raising · Parasitic scope · Natural language quantification

1 What is the Logic of Scope?

Just as we might ask "What is the logic of negation?", we might ask "What is the logic of scope?". And just as the first question has many answers, so too will the second. The answer I will give here will take the form of a substructural logic containing a single structural postulate. I will suggest this logic characterizes a kind of scope-taking that has applications in the analysis of natural language.

1.1 Scope in Natural Language

Many natural languages have scope-taking expressions, including English:

(1) Ann saw everyone.

In (1), the denotation of the quantifier *everyone* takes the rest of the sentence in which it occurs as its semantic argument. That is, the denotation of the sentence as a whole is given by **everyone**($\lambda x.$**saw**(x)(**ann**)).

There are three important properties of scope-taking in natural language that I will discuss here: unbounded scope displacement, embedded scope-taking, and scope ambiguity (see [4] for a more complete discussion).

© Springer-Verlag Berlin Heidelberg 2015
T. Murata et al. (Eds.): JSAI-isAI 2014 Workshops, LNAI 9067, pp. 184–199, 2015.
DOI: 10.1007/978-3-662-48119-6_14

(2) Ann saw the mother of everyone's lawyer.

In (2), despite being embedded inside of two layers of possessive constructions, the quantifier still takes scope over the entire sentence. In general, there is no upper limit to the structural distance over which an expression can take scope.

(3) a. Bill thinks [Ann saw everyone].
    b. **thinks**$(\forall x.$**saw**$(x)($**ann**$))($**bill**$)$

However, in (3a), the quantifier takes scope only over the [bracketed] embedded clause *Ann saw everyone*, which is a proper subpart of the complete sentence. The fact that scopal elements can take embedded scope is what makes *undelimited* continuations unsuited to modeling scope (see Chap. 18 of [5] for discussion); *delimited* continuations are a better fit.

(4) a. Someone loves everyone.
    b. $\exists x \forall y.$**loves**$(y)(x)$
    c. $\forall y \exists x.$**loves**$(y)(x)$

Scope ambiguity can arise when there is more than one quantifier in the sentence. There can in general be as many as $n!$ distinct denotations, where n is the number of quantifiers.

1.2 Quantifier Raising

By far the dominant way to think about scope-taking is Quantifier Raising (QR), as discussed in detail in [8]. Quantifier Raising accounts for unbounded scope displacement, embedded scope-taking, and scope ambiguity.

From a logical point of view, Quantifier Raising can be seen as a structural relation. That is, Quantifier Raising reconfigures a logical structure by moving the quantifier to adjoin to its scope domain, placing a variable in the original position of the quantifier, and abstracting over the variable at the level of the scope domain.

$$[\text{Ann [called everyone]]} \overset{\text{QR}}{\Rightarrow} [\text{everyone}(\lambda x[\text{Ann [called } x]])]$$

Here, the scope domain of *everyone* is the entire clause.

Because the QR operation can target embedded S nodes, embedded scope falls out naturally. Just as naturally, QR easily accounts for scope ambiguity by allowing QR to target quantifiers in any order.

Linear scoping : $[\text{someone [called everyone]}]$
$\Rightarrow [\text{everyone}(\lambda x[\text{someone [called } x]])]$
$\Rightarrow [\text{someone}(\lambda y[\text{everyone}(\lambda x[y \text{ [called } x]])])]$

Inverse scoping : $[\text{someone [called everyone]}]$
$\Rightarrow [\text{someone}(\lambda y[y \text{ [called everyone]}])]$
$\Rightarrow [\text{everyone}(\lambda x[\text{someone}(\lambda y[y \text{ [called } x]])])]$

Raising the direct object first and then the subject gives linear scope, and raising the subject first and then the direct object gives inverse scope.

So far, so good. What remains to be done is to characterize Quantifier Raising from a logical point of view. This is what the remainder of this paper sets out to do (see especially the discussion in Sect. 6.4).

1.3   The q Type Constructor

[14] extends Lambek grammar with a type constructor q ('q' for 'quantification') which takes three categories as parameters and has the following logical behavior:

$$\frac{\Gamma[A] \vdash B \qquad \Sigma[C] \vdash D}{\Sigma[\Gamma[q(A,B,C)]] \vdash D} \; q \tag{5}$$

An expression in category $q(A, B, C)$ functions locally (i.e., with respect to the context $\Gamma[\ ]$) as an A, takes scope over a structure in category B, and allows the structure over which it takes scope to function in the larger context (i.e., with respect to $\Sigma[\ ]$) as an expression of category C. This is exactly what a scope-taking expression needs to do, and any adequate account of scope in natural language should account for the ground covered by q.

However, from a logical point of view, q is problematic. For instance, although it is easy to write a left rule (a rule of use) for q, as in (5), a general right rule (a rule of proof) remains elusive (see [16]). As I will explain below in Sect. 6.1, the resolution of this puzzle here will be to factor the q inference into the interaction of the structural postulate with two independent logical inferences, each of which has its own left and right rules.

1.4   What this Logic for Scope Will not Account for

The account here seeks only to characterize an idealized, unconstrained version of quantifier scope. In any natural language, scope-taking will be constrained by syntactic and lexical factors. See [6] or [10] for formal grammars (also based on delimited continuations) that propose principled constraints on scope-taking.

2   NL$_\lambda$

The substructural grammar for characterizing scope discussed here is based on the non-associative Lambek grammar NL (see, e.g., [13,17]). Since NL rejects all structural rules, including exchange, there will be two versions of implication: \, in which the argument is on the left, and /, in which the argument is on the right.

NL characterizes the logic of function/argument combination when the functor is linearly adjacent to the argument. However, for scope-taking, linear adjacency is not sufficient. After all, a scope-taker is not adjacent to its argument—it is contained *within* its argument. What we need is a syntactic notion of 'surrounding' and 'being surrounded by'. Therefore the grammar here will provide two modes: not only a merge mode (already introduced), for ordinary function/argument combination, with implications \ and /; but also a continuation mode, which will govern scope-taking, with implications \\ and //. (The interpretation of the continuation mode will be explained shortly).

The logical rules for these connectives are identical to the rules given in [17]:129. They constitute the logical core of a two-mode type-logical grammar:

$$\frac{}{A \vdash A} \text{ Axiom} \qquad (6)$$

$$\frac{\Gamma \vdash A \quad \Sigma[B] \vdash C}{\Sigma[\Gamma \cdot A \backslash B] \vdash C} \backslash L \qquad \frac{A \cdot \Gamma \vdash B}{\Gamma \vdash A \backslash B} \backslash R \qquad \frac{\Gamma \vdash A \quad \Sigma[B] \vdash C}{\Sigma[B/A \cdot \Gamma] \vdash C} /L \qquad \frac{\Gamma \cdot A \vdash B}{\Gamma \vdash B/A} /R$$

$$\frac{\Gamma \vdash A \quad \Sigma[B] \vdash C}{\Sigma[\Gamma \circ A \backslash\!\backslash B] \vdash C} \backslash\!\backslash L \qquad \frac{A \circ \Gamma \vdash B}{\Gamma \vdash A \backslash\!\backslash B} \backslash\!\backslash R \qquad \frac{\Gamma \vdash A \quad \Sigma[B] \vdash C}{\Sigma[B /\!/ A \circ \Gamma] \vdash C} /\!/ L \qquad \frac{\Gamma \circ A \vdash B}{\Gamma \vdash B /\!/ A} /\!/ R$$

The sequents in the logical rules above have the form $\Gamma \vdash A$, where A is a category and Γ is a structure. All categories are structures, and if Γ and Δ are structures, then $\Gamma \cdot \Delta$ (merge mode) and $\Gamma \circ \Delta$ (continuation mode) are also structures.

In order to allow expressions to combine with material that surrounds it (or that it surrounds), we need to add a structural rule. In order to state this structural rule, we will need to enlarge the set of structures to include **gapped structures**: if $\Sigma[\Delta]$ is a structure containing a distinguished substructure Δ, then $\lambda\alpha\, \Sigma[\alpha]$ is also a structure, where α is a variable taken from the set $x, y, z, \ldots$. For instance, $\lambda x\, x$, $\lambda y\, y$, $\lambda x\, (x \cdot \text{left})$, $\lambda x\, (\text{John} \cdot (\text{saw} \cdot x))$, and $\lambda x\, \lambda y\, (y \cdot (\text{saw} \cdot x))$ are gapped structures.

Although gapped structures have important predecessors, including [7, 19], they are not standard in discussions of substructural logics. One of the main goals of this paper is to explain how to understand gapped structures. A crucial part of achieving this goal will be to introduce a second substructural logic in the next section, NL_{CL}, which will be equivalent to (a restricted version of) NL_λ. NL_{CL} is a standard substructural logic, and does not involve any gapped structures.

With gapped structures in hand, we can state the following structural inference rule:

$$\frac{\Gamma[\Sigma[\Delta]] \vdash A}{\Gamma[\Delta \circ \lambda\alpha\, \Sigma[\alpha]] \vdash A} \lambda \qquad (7)$$

In words: if a structure Σ contains within it a structure Δ, then Δ can take scope over the rest of Σ, where 'the rest of Σ' is represented as the gapped structure $\lambda\alpha\, \Sigma[\alpha]$.

Schematically, we have:

$$(8)$$

The postulate says that if Δ (the small grey triangle) is some structure embedded within a larger structure Σ (the complete larger triangle), we can view these components in a completely equivalent way by articulating them into a foreground and a background, that is, into a plug and a context—an expression and its continuation. Then Δ will be the foregrounded expression, and the clear notched triangle will be its context, the continuation $\lambda\alpha\Sigma[\alpha]$.

An expression in a category with the form $A\backslash\!\backslash B$ is a *continuation*: something that would be a complete expression of category B, except that it is missing an expression of category A somewhere inside of it. An expression in a category with the form $C/\!/(A\backslash\!\backslash B)$ will be something that combines with a continuation of category $A\backslash\!\backslash B$ surrounding it to form a result expression of category C.

This logic allows for unbounded scope displacement, since there are no constraints on the complexity of the scope host Σ. It also allows for embedded scope-taking, since Γ may be non-empty. As for scope ambiguity, we have the following two derivations:

$$
\cfrac{
\cfrac{
\cfrac{
\cfrac{
\cfrac{
\cfrac{
\cfrac{
\cfrac{
\cfrac{\text{DP}\cdot(\text{loves}\cdot\text{DP}) \vdash \text{S}}{\text{DP} \circ \lambda x(\text{DP}\cdot(\text{loves}\cdot x)) \vdash \text{S}}\ \lambda
}{\lambda x(\text{DP}\cdot(\text{loves}\cdot x)) \vdash \text{DP}\backslash\!\backslash\text{S}}\ \backslash\!\backslash R \qquad \text{S} \vdash \text{S}
}{\text{S}/\!/(\text{DP}\backslash\!\backslash\text{S}) \circ \lambda x(\text{DP}\cdot(\text{loves}\cdot x)) \vdash \text{S}}\ /\!/L
}{\text{everyone} \circ \lambda x(\text{DP}\cdot(\text{loves}\cdot x)) \vdash \text{S}}\ \text{LEX}
}{\text{DP}\cdot(\text{loves}\cdot\text{everyone}) \vdash \text{S}}\ \lambda
}{\text{DP} \circ \lambda x(x\cdot(\text{loves}\cdot\text{everyone})) \vdash \text{S}}\ \lambda
}{\lambda x(x\cdot(\text{loves}\cdot\text{everyone})) \vdash \text{DP}\backslash\!\backslash\text{S}}\ \backslash\!\backslash R \qquad \text{S} \vdash \text{S}
}{\text{S}/\!/(\text{DP}\backslash\!\backslash\text{S}) \circ \lambda x(x\cdot(\text{loves}\cdot\text{everyone})) \vdash \text{S}}\ /\!/L
}{\text{someone} \circ \lambda x(x\cdot(\text{loves}\cdot\text{everyone})) \vdash \text{S}}\ \text{LEX}
$$
$$
\text{someone}\cdot(\text{loves}\cdot\text{everyone}) \vdash \text{S}\qquad \lambda
$$

The Curry-Howard labeling for this derivation (see [5]) is $\exists x\forall y.\mathbf{loves}\,y\,x$. In general, the scope-taker that is focussed (i.e., targeted by the structural postulate) lower in the proof takes wider scope.

$$
\cfrac{
\cfrac{
\cfrac{
\cfrac{
\cfrac{
\cfrac{
\cfrac{
\cfrac{
\cfrac{\text{DP}\cdot(\text{loves}\cdot\text{DP}) \vdash \text{S}}{\text{DP} \circ \lambda x(x\cdot(\text{loves}\cdot\text{DP})) \vdash \text{S}}\ \lambda
}{\lambda x(x\cdot(\text{loves}\cdot\text{DP})) \vdash \text{DP}\backslash\!\backslash\text{S}}\ \backslash\!\backslash R \qquad \text{S} \vdash \text{S}
}{\text{S}/\!/(\text{DP}\backslash\!\backslash\text{S}) \circ \lambda x(x\cdot(\text{loves}\cdot\text{DP})) \vdash \text{S}}\ /\!/L
}{\text{someone} \circ \lambda x(x\cdot(\text{loves}\cdot\text{DP})) \vdash \text{S}}\ \text{LEX}
}{\text{someone}\cdot(\text{loves}\cdot\text{DP}) \vdash \text{S}}\ \lambda
}{\text{DP} \circ \lambda x(\text{someone}\cdot(\text{loves}\cdot x)) \vdash \text{S}}\ \lambda
}{\lambda x(\text{someone}\cdot(\text{loves}\cdot x)) \vdash \text{DP}\backslash\!\backslash\text{S}}\ \backslash\!\backslash R \qquad \text{S} \vdash \text{S}
}{\text{S}/\!/(\text{DP}\backslash\!\backslash\text{S}) \circ \lambda x(\text{someone}\cdot(\text{loves}\cdot x)) \vdash \text{S}}\ /\!/L
}{\text{everyone} \circ \lambda x(\text{someone}\cdot(\text{loves}\cdot x)) \vdash \text{S}}\ \text{LEX}
$$
$$
\text{someone}\cdot(\text{loves}\cdot\text{everyone}) \vdash \text{S}\qquad \lambda
$$

In this case, the semantic labeling gives the universal wide scope: $\forall y\exists x.\mathbf{loves}\,y\,x$.

3   Soundness and Completeness via NL_{CL}

The proofs of soundness and completeness for NL_λ will proceed by defining NL_{CL}, a more standard substructural logic whose soundness and completeness follows from the general results of [20]. I will then give conditions under which NL_λ and NL_{CL} are equivalent.

NL_{CL} has the same logical rules as NL_λ. Instead of the structural postulate λ, however, NL_{CL} has the following three structural postulates:

$$\frac{p}{p \circ \mathsf{I}} \mathsf{I} \qquad\qquad \frac{p \cdot (q \circ r)}{q \circ ((\mathsf{B} \cdot p) \cdot r)} \mathsf{B} \qquad\qquad \frac{(p \circ q) \cdot r}{p \circ ((\mathsf{C} \cdot q) \cdot r)} \mathsf{C} \qquad (9)$$

These postulates are identical to the ones given in [2]. [20]:30 considers I (which he writes '0') as "a zero-place punctuation mark," where punctuation marks (p. 19) "stand to structures in the same way that connectives stand to formulae." Likewise, B and C are also zero-place punctuation marks. The double horizontal line indicates that these rules are bi-directional, i.e., inference in the top-to-bottom direction and in the bottom-to-top direction are both valid. Restall calls the top-to-bottom inference for the I postulate Push, and the other direction Pop.

In the form of an official inference rule, the I postulate (for instance) is written

$$\frac{\Sigma[p] \vdash A}{\Sigma[p \circ \mathsf{I}] \vdash A}, \qquad (10)$$

and similarly for the other rules.

An example derivation will show how these postulates work together to achieve in-situ quantification for the sentence *John saw everyone*:

$$
\begin{array}{c}
\dfrac{\dfrac{\dfrac{\dfrac{\dfrac{\dfrac{\dfrac{\dfrac{\dfrac{DP \vdash DP \quad S \vdash S}{DP \cdot DP\backslash S \vdash S}\backslash L}{DP \cdot ((DP\backslash S)/DP \cdot DP) \vdash S}/L}{\text{john} \cdot (\text{saw} \cdot DP) \vdash S}\text{LEX}}{\text{john} \cdot (\text{saw} \cdot (DP \circ \mathsf{I})) \vdash S}\mathsf{I}}{\text{john} \cdot (DP \circ ((\mathsf{B} \cdot \text{saw}) \cdot \mathsf{I}))) \vdash S}\mathsf{B}}{DP \circ ((\mathsf{B} \cdot \text{john}) \cdot ((\mathsf{B} \cdot \text{saw}) \cdot \mathsf{I})) \vdash S}\mathsf{B}}{(\mathsf{B} \cdot \text{john}) \cdot ((\mathsf{B} \cdot \text{saw}) \cdot \mathsf{I}) \vdash DP\backslash\backslash S}\backslash\backslash R \quad S \vdash S}{S /\!\!/ (DP\backslash\backslash S) \circ ((\mathsf{B} \cdot \text{john}) \cdot ((\mathsf{B} \cdot \text{saw}) \cdot \mathsf{I})) \vdash S}/\!\!/ L \\
\vdots
\end{array}
\qquad (11)
$$

DP ⊢ DP   S ⊢ S
——————————— \L
DP ⊢ DP   DP·DP\S ⊢ S
——————————————————— /L
DP·((DP\S)/DP·DP) ⊢ S
——————————————— LEX
john·(saw·DP) ⊢ S
——————————————— I
john·(saw·(DP ∘ I)) ⊢ S
——————————————————— B
john·(DP ∘ ((B·saw)·I))) ⊢ S
——————————————————— B
DP ∘ ((B·john)·((B·saw)·I)) ⊢ S
——————————————————— \\R
(B·john)·((B·saw)·I) ⊢ DP\\S   S ⊢ S
————————————————————————————— //L
S//(DP\\S) ∘ ((B·john)·((B·saw)·I)) ⊢ S
——————————————————————————— LEX
everyone ∘ ((B·john)·((B·saw)·I)) ⊢ S
———————————————————————— B
john·(everyone ∘ ((B·saw)·I)) ⊢ S
———————————————————— B
john·(saw·(everyone ∘ I)) ⊢ S
———————————————— I
john·(saw·everyone) ⊢ S

NL_{CL} is sound and complete with respect to the usual class of relational models. This follows directly from the proofs given in [20], Chap. 11. In particular, [20]:249 provides an algorithm for constructing frame conditions corresponding to the structural postulates.

Theorem (Soundness and Completeness): $X \vdash A$ is provable in NL_{CL} iff for every model $\mathfrak{M} = \langle \mathscr{F}, \models \rangle$ that satisfies the frame conditions, $\forall x \subset \mathscr{F}, x \models X \to x \models A$.

Proof: given in [20], theorems 11.20, 11.37.

Furthermore, NL_{CL} is conservative with respect to NL. That is,

Theorem (Conservativity): Let an NL sequent be a sequent built up only from the formulas and structures allowed in NL: $/, \backslash, \cdot$. An NL sequent is provable in NL_{CL} iff it is provable in NL.

See [5] for details.

4   The Connection Between NL_{λ} and NL_{CL}

This section investigates the conditions under which a derivation in NL_{λ} has an equivalent derivation in NL_{CL}.

I define the following class of structures:

$$\Gamma\lceil p\rceil ::= \quad p \quad | \quad p \circ q \quad | \quad q \cdot \Gamma\lceil p\rceil \quad | \quad \Gamma\lceil p\rceil \cdot q \quad | \quad \lambda y.\, \Gamma\lceil p\rceil \qquad (12)$$

Given a structure p, a $\lceil\ \rceil$-context will consist either of the empty context, or else the entire left element at the top level of a $\circ$ structure, or else a larger context built up from $\cdot$ and λ. We can impose these restrictions on NL_{λ} by replacing the original lambda postulate with one that mentions $\lceil\ \rceil$-contexts:

$$\Sigma\lceil\Delta\rceil \equiv \Delta \circ \lambda\alpha\, \Sigma\lceil\alpha\rceil \qquad (13)$$

To illustrate, the following (bidirectional) inferences are licensed by (13):

$$\frac{A}{A \circ \lambda xx} \qquad \frac{A \circ B}{A \circ \lambda x(x \circ B)} \qquad \frac{A \cdot B}{A \circ \lambda x(x \cdot B)} \qquad \frac{\lambda x.(x \cdot B)}{B \circ \lambda y \lambda x(x \cdot y)} \qquad (14)$$

But not these:

$$\frac{(A \cdot B) \circ C}{A \circ \lambda x((x \cdot B) \circ C)} \qquad \frac{A \circ B}{B \circ \lambda y(A \circ y)} \qquad (15)$$

The reason these last two inferences are not allowed is that abstraction across $\circ$ is forbidden unless the abstractee is the complete left element connected by $\circ$.

The inspiration for NL_{CL} comes from the well-known equivalence between the lambda calculus and Combinatory Logic. More specifically, the postulates of NL_{CL} implement a version of Shönfinkel's embedding of λ-terms into Combinatory Logic. Adapting the presentation in [1]:152, [5] define $\langle\cdot\rangle$, which maps an arbitrary gapped structure into a NL_{CL} structure:

$$\langle x \rangle \equiv x$$
$$\langle p{\cdot}q \rangle \equiv \langle p \rangle{\cdot}\langle q \rangle$$
$$\langle p \circ q \rangle \equiv \langle p \rangle \circ \langle q \rangle$$
$$\langle \lambda x.p \rangle \equiv \mathbb{A}(x, \langle p \rangle)$$

$$\mathbb{A}(x, x) \equiv \mathsf{I}$$
$$\mathbb{A}(x, p{\cdot}q) \equiv (\mathsf{B}{\cdot}p){\cdot}\mathbb{A}(x, q) \quad (x \text{ not free in } p)$$
$$\mathbb{A}(x, p{\cdot}q) \equiv (\mathsf{C}{\cdot}\mathbb{A}(x, p)){\cdot}q \quad (x \text{ not free in } q)$$
$$\mathbb{A}(x, x \circ q) \equiv (\mathsf{C}{\cdot}\mathsf{I}) \circ q \quad (x \text{ not free in } q)$$

(16)

With this mapping defined, I can state the following three theorems given in [5]) characterizing the relationship between NL_λ and NL_{CL}:

Theorem (Faithfullness of the $\langle \cdot \rangle$ mapping from λ-structures into CL-structures): For any structure p and context $\Gamma \lceil\ \rceil$,

$$\cfrac{\langle p \circ \lambda x \Gamma\lceil x \rceil \rangle}{\langle \Gamma\lceil p \rceil \rangle}\ CL \tag{17}$$

Here, CL schematizes over some series of structural inferences allowable in NL_{CL}.

Theorem (Embedding of λ-free theorems of NL_λ in NL_{CL}): For any derivation in NL_λ (with abstraction restricted to $\lceil\ \rceil$-contexts) whose final sequent does not contain any λ-structures, there is an equivalent derivation in NL_{CL}.

Here, two derivations are equivalent if they differ only in the application of structural rules. They must have the same axiom instances, the same conclusion, and the Curry-Howard labeling must be the same up to α-equivalence.

Theorem (Embedding of IBC-free theorems of NL_{CL} in NL_λ): for any derivation in NL_{CL} whose conclusion does not contain the structures I, B, or C, there is an equivalent derivation in NL_λ.

The equivalence involves replacing each instance of I, B, and C with instances of the lambda postulate as follows:

$$\cfrac{p}{p \circ \mathsf{I}}\,\mathsf{I} \qquad \sim \qquad \cfrac{p}{p \circ \lambda xx}\,\lambda$$

$$\cfrac{p{\cdot}(q \circ r)}{q \circ ((\mathsf{B}{\cdot}p){\cdot}r)}\,\mathsf{B} \qquad \sim \qquad \cfrac{p{\cdot}(q \circ r)}{q \circ \lambda x(p{\cdot}(x \circ r))}\,\lambda \tag{18}$$

$$\cfrac{(p \circ q){\cdot}r}{p \circ ((\mathsf{C}{\cdot}q){\cdot}r)}\,\mathsf{C} \qquad \sim \qquad \cfrac{(p \circ q){\cdot}r}{p \circ \lambda x((x \circ q){\cdot}r)}\,\lambda$$

Note that each of these applications of the lambda postulate obeys the restriction to $\lceil\ \rceil$-contexts.

Thus NL_λ (with the lambda-postulate restricted to $\lceil\ \rceil$-contexts) and NL_{CL} are equivalent: any sequent containing only structures built from $\cdot$ and $\circ$ will be a theorem of one just in case it is a theorem of the other. Furthermore, for each derivation in one system, there will be a matching derivation in the other that differs only in the application of structural rules, which means that the semantic values of the two derivations will be identical. Since NL_{CL} is conservative with respect to the non-associative Lambek grammar NL, NL_λ is too. As a result, NL_λ with restricted abstraction contexts can be used with full confidence that it is equivalent to an ordinary and well-behaved substructural grammar.

5   Cut elimination and decidability

5.1   Cut Elimination for NL_λ

The cut rule characterizes transitivity of the logical system:

$$\frac{\Gamma \vdash A \qquad \Sigma[A] \vdash B}{\Sigma[\Gamma] \vdash B} \text{ CUT} \tag{19}$$

The cut rule says that if Γ is a proof of A, and Σ is a proof of B that depends on proving A, then we can construct a new proof of B in which A has been replaced with Γ. The formula A has been 'cut out' of the derivation.

The proof strategy, just as it was above for completeness, will be to rely on Restall's general proof of cut elimination for Gentzen-style sequent systems. This strategy emphasizes the ordinariness and the standardness of the logics here, and how they fit into a larger landscape of substructural logics.

In order for Restall's proof to apply, we need to demonstrate that the cut rule, the structural rule, and the logical rules conform to certain conditions. This is perfectly straightforward (see [5] for full details). Therefore we have:

Theorem (Cut Elimination): given that the parameter conditions, the eliminability of matching principal constituents, and the regularity condition hold, if $\Gamma \vdash A$ and $\Delta[A] \vdash B$ are provable, then $\Delta[\Gamma] \vdash B$ is also provable.

Proof: see [20]: Sect. 6.3.

5.2   Decidability of NL_λ

Decidability is a property a logic has if it is always possible to figure out whether a sequent is a theorem (has a proof, has a derivation) in a bounded amount of time, where the bound is some concrete function of the complexity of the sequent to be proved.

The structural postulate given above in (7) is a reversible inference, that is, it is bidirectional. In the discussion that follows, it will be helpful to keep track of the two directions separately:

$$\frac{\Sigma[\Delta\lceil A\rceil] \vdash B}{\Sigma[A \circ \lambda x \Delta\lceil x\rceil] \vdash B} \text{ REDUCTION} \qquad \frac{\Sigma[A \circ \lambda x \Delta\lceil x\rceil] \vdash B}{\Sigma[\Delta\lceil A\rceil] \vdash B} \text{ EXPANSION} \tag{20}$$

Since in proof search we are starting with the conclusion and trying to find appropriate premises, the names 'reduction' and 'expansion' are relative to the bottom-to-top direction of reading proofs. The main challenge for decidability is that there is no limit to the opportunities for expansion, since $B \equiv B \circ \lambda xx \equiv (B \circ \lambda xx) \circ \lambda xx \equiv \ldots$.

I will leave a full account of the decidability of NL_λ for another occasion. Nevertheless, I will discuss a strategy that handles the vast majority of cases. The goal will be to push each Expansion inference upwards in the proof until one of two things happens: either it encounters a matching Reduction instance, in which case the two rules cancel each other out, and can be eliminated from the proof; or else the expansion is adjacent to a logical rule that introduces the focussed occurrence of $\circ$.

It turns out that the only candidate for such a logical rule is $/\!\!/L$.

$$
\cfrac{\cfrac{\lambda x \Gamma\lceil x\rceil \vdash A \qquad \Sigma[B] \vdash C}{\Sigma[B /\!\!/ A \circ \lambda x \Gamma\lceil x\rceil] \vdash C}\ /\!\!/L}{\Sigma[\Gamma\lceil B /\!\!/ A\rceil] \vdash C}\ \mathrm{EXP} \quad \equiv \quad \cfrac{\lambda x \Gamma\lceil x\rceil \vdash A \qquad \Sigma[B] \vdash C}{\Sigma[\Gamma\lceil B /\!\!/ A\rceil] \vdash C}\ /\!\!/L_\lambda \quad (21)
$$

We can replace the adjacent pair of inferences on the left with the derived inference on the right, which we can call $/\!\!/L_\lambda$. By repeated application of this reasoning, almost every instance of Expansion can either be eliminated, or replaced with an instance of $/\!\!/L_\lambda$. (There are exceptions that include certain parasitic scope configurations that will not be discussed here).

Having eliminated almost all expansion inferences, we can eliminate Reduction inferences in a similar fashion. That is, reasoning dually, Reduction inferences can be pushed *downwards* until the Reduction encounters an instance of $\backslash\!\backslash R$ that targets the $\circ$ connective introduced by Reduction. And once again, we can replace the combination of the reduction and the instance of $\backslash\!\backslash R$ with a derived rule that captures their net effect:

$$
\cfrac{\cfrac{\Gamma\lceil A\rceil \vdash B}{A \circ \lambda x \Gamma\lceil x\rceil \vdash B}\ \mathrm{RED}}{\lambda x \Gamma\lceil x\rceil \vdash A \backslash\!\backslash B}\ \backslash\!\backslash R \quad \equiv \quad \cfrac{\Gamma\lceil A\rceil \vdash B}{\lambda x \Gamma\lceil x\rceil \vdash A \backslash\!\backslash B}\ \backslash\!\backslash R_\lambda \quad (22)
$$

At this point, we have two derived logical inferences: $\backslash\!\backslash R_\lambda$, and $/\!\!/L_\lambda$. The $\backslash\!\backslash R_\lambda$ rule says that in-situ elements can take scope directly from embedded positions, without needing to first be abstracted leftwards. Dually, the $/\!\!/L_\lambda$ rule says that a context can surround a scope-taker even when the scope-taker is embedded in a still larger surrounding context. Adding the two derived logical rules to the standard logical rules leads to derivations of in-situ scope-taking, illustrated here for the sentence *Ann saw everyone*:

$$
\cfrac{\cfrac{\mathrm{ann}\cdot(\mathrm{saw}\cdot\mathrm{DP}) \vdash \mathrm{S}}{\lambda x.\mathrm{ann}\cdot(\mathrm{saw}\cdot x) \vdash \mathrm{DP}\backslash\!\backslash \mathrm{S}}\ \backslash\!\backslash R_\lambda \qquad \mathrm{S} \vdash \mathrm{S}}{\cfrac{\mathrm{ann}\cdot(\mathrm{saw}\cdot \mathrm{S} /\!\!/ (\mathrm{DP}\backslash\!\backslash \mathrm{S})) \vdash \mathrm{S}}{\mathrm{ann}\cdot(\mathrm{saw}\cdot\mathrm{everyone}) \vdash \mathrm{S}}}\ /\!\!/L_\lambda \quad (23)
$$

In effect, we have compiled both parts of the structural rule into the logical rules. If we add these two derived logical rules to the grammar, we can consider an approximation of NL_λ that consists entirely of logical rules.

Note that in this modified logic, each inference rule, including the derived inference rules, eliminates exactly one logical connective. As a result, no part of the proof can have a depth greater than the number of logical connectives in the final sequent. Since there is at most a finite number of ways to apply each rule to a given occurrence of a logical connective, decidability of the modified logic follows immediately.

5.3   Proof Search with Gaps

The treatment of scope-taking can be extended to a treatment of overt syntactic movement (see [5]). From the point of view of decidability, gaps are a challenge, since they allow us to posit new structure during the course of a proof search, in which case we lose the subformula property. An extension of the technique developed in the previous section allows derivations with gaps without giving up decidability.

$$\frac{\Gamma[B{\cdot}A] \vdash C}{\Gamma[A] \vdash B\backslash\!\!\backslash C}\, \backslash\!\!\backslash R_{lgap} \qquad \frac{\Gamma[A{\cdot}B] \vdash C}{\Gamma[A] \vdash B\backslash\!\!\backslash C}\, \backslash\!\!\backslash R_{rgap} \qquad (24)$$

Since each of these inferences has the subformula property, and moreover, eliminates a logical connective, adding them to the logic will not compromise decidability.

To illustrate these logical rules in action, here is a derivation of the wh-question *Who did Ann see* (with *did* suppressed for simplicity):

$$\frac{\dfrac{\dfrac{\text{ann}{\cdot}(\text{see}{\cdot}\text{DP}) \vdash S}{\text{ann}{\cdot}\text{see} \vdash DP\backslash\!\!\backslash S}\, \backslash\!\!\backslash R_{lgap} \qquad Q \vdash Q}{Q/(DP\backslash\!\!\backslash S){\cdot}(\text{ann}{\cdot}\text{see}) \vdash Q}\, /L}{\text{who}{\cdot}(\text{ann}{\cdot}\text{see}) \vdash Q}\, \text{LEX} \qquad (25)$$

6   Comparisons with Other Approaches

The participants at LENSL11 kindly suggested a number of other approaches to the logic of scope-taking that it would be useful to compare with the approach presented here. In this section, I will discuss Moortgat's [14] q type constructor (mentioned above); a multi-modal analysis also due to Moortgat [15]; Morrill et al.'s notion of scope-taking as discontinuity [16,18]; and, finally, standard Quantifier Raising.

6.1   Deriving the q Type Constructor

If we carry the strategy in Sect. 5 of fusing inferences into derived inferences one step further, we derive the rule of use for Moortgat's q type constructor, given above in (5):

$$\frac{\dfrac{\Gamma\lceil A\rceil \vdash B}{\lambda x \Gamma\lceil x\rceil \vdash A\backslash B}\backslash R_\lambda \qquad \Sigma[C]\vdash D}{\Sigma[\Gamma\lceil C /\!\!/ (A\backslash B)\rceil]\vdash D}/\!\!/L_\lambda \qquad \approx \qquad \frac{\Gamma\lceil A\rceil \vdash B \qquad \Sigma[C]\vdash D}{\Sigma[\Gamma[q(A,B,C)]]\vdash D}q \quad (26)$$

We now have an explanation for why it was impossible to find a general right rule for the q type constructor: it is because the q inference represents the fusion of two logically distinct inferences, each with their own left and right rules.

In support of the usefulness of factoring the q into independent components, consider 'parasitic scope', a technique proposed in [2] to account for the scope-taking behavior of adjectives such as *same* and *different*. Parasitic scope requires the inferences to be interleaved in a way that cannot be duplicated by the q inference alone:

$$\frac{\dfrac{\dfrac{(\text{the}\cdot(N/N\cdot\text{waiter}))\cdot(\text{served}\cdot DP)\vdash S}{\lambda x.(\text{the}\cdot(N/N\cdot\text{waiter}))\cdot(\text{served}\cdot x)\vdash DP\backslash\!\backslash S}\backslash\!\backslash R_\lambda}{\dfrac{\lambda y\lambda x.(\text{the}\cdot(y\cdot\text{waiter}))\cdot(\text{served}\cdot x)\vdash (N/N)\backslash\!\backslash(DP\backslash\!\backslash S)\quad\quad DP\backslash\!\backslash S\vdash DP\backslash\!\backslash S}{\dfrac{\lambda x.(\text{the}\cdot((DP\backslash\!\backslash S)/\!\!/((N/N)\backslash\!\backslash(DP\backslash\!\backslash S))\cdot\text{waiter}))\cdot(\text{served}\cdot x)\vdash DP\backslash\!\backslash S}{\lambda x.(\text{the}\cdot(\text{same}\cdot\text{waiter}))\cdot(\text{served}\cdot x)\vdash DP\backslash\!\backslash S}\text{LEX}}/\!\!/L_\lambda}\backslash\!\backslash R_\lambda\quad\quad S\vdash S}{\dfrac{\dfrac{(\text{the}\cdot(\text{same}\cdot\text{waiter}))\cdot(\text{served}\cdot S/\!\!/(DP\backslash\!\backslash S))\vdash S}{(\text{the}\cdot(\text{same}\cdot\text{waiter}))\cdot(\text{served}\cdot\text{everyone})\vdash S}\text{LEX}}{}}/\!\!/L_\lambda$$

$$(27)$$

Although the innermost pair of $/\!\!/L_\lambda$ and $\backslash\!\backslash R_\lambda$ can be fused into a single instance of the q inference, the outermost pair cannot.

6.2   Comparison with a Unary Modality Strategy

Moortgat, in [15], gives an analysis that at first glance is strikingly similar to NL_{CL}. The heart of the approach is a set of three structural postulates, lined up here in (29) underneath the corresponding NL_{CL} postulates.

$$NL_{CL}: \qquad \frac{p}{p\circ I}\ I \qquad \frac{p\cdot(q\circ r)}{q\circ((B\cdot p)\cdot r)}\ B \qquad \frac{(p\circ q)\cdot r}{p\circ((C\cdot q)\cdot r)}\ C \qquad (28)$$

$$\text{Unary modalities}: \qquad \frac{p}{p\circ I}\ P0 \qquad \frac{p\cdot(q\circ r)}{q\circ\langle r\rangle(p\cdot r)}\ P2 \qquad \frac{(p\circ q)\cdot r}{p\circ\langle l\rangle(q\cdot r)}\ P1 \qquad (29)$$

Both sets of postulates regulate the interaction of two binary modalities, $\cdot$ and $\circ$. The postulates in (29) make use in addition of two unary modalities, $\langle l\rangle$, and $\langle r\rangle$. (I've omitted a third unary modality, $\Diamond$, in order to emphasize the similarities between the approaches, and to simplify the discussion immediately below.) The presence or absence of the $\langle l\rangle$ and $\langle r\rangle$ modalities track the path between the in-situ position of a scope-taker and its scope position, very much like the structural punctuation marks B and C do in a NL_{CL} derivation.

The key difference between the two systems is that decorating a constituent with a unary modality blocks further abstraction from that constituent. As a result, the unary modalities are able to track at most one scope path at a time in the general case. In contrast, NL_{CL} allows multiple scope-takers to simultaneously share the same abstraction path without those paths getting confused. For example, note that in NL_λ, the structure $p \cdot (q \cdot s)$ is equivalent to $q \circ (s \circ \lambda y \lambda x (p \cdot (x \cdot y)))$. A derivation of the corresponding equivalence in NL_{CL} is given on the left:

$$
\frac{\dfrac{\dfrac{\dfrac{\dfrac{\dfrac{q \circ (s \circ ((B \cdot (B \cdot p)) \cdot ((B \cdot (C \cdot I)) \cdot I)))}{q \circ ((B \cdot p) \cdot (s \circ ((B \cdot (C \cdot I)) \cdot I)))}\,B}{q \circ ((B \cdot p) \cdot ((C \cdot I) \cdot (s \circ I)))}\,B}{q \circ ((B \cdot p) \cdot ((C \cdot I) \cdot s))}\,I}{p \cdot (q \circ ((C \cdot I) \cdot s))}\,B}{p \cdot ((q \circ I) \cdot s)}\,C}{p \cdot (q \cdot s)}\,I
\qquad
\frac{\dfrac{\dfrac{\dfrac{\dfrac{\dfrac{q \circ \langle r \rangle (p \cdot \langle l \rangle (s \circ \langle r \rangle (I \cdot I)))}{q \circ \langle r \rangle (p \cdot \langle l \rangle (I \cdot (s \circ I)))}\,P2}{q \circ \langle r \rangle (p \cdot \langle l \rangle (I \cdot s))}\,P0}{p \cdot (q \circ \langle l \rangle (I \cdot s))}\,P2}{p \cdot ((q \circ I) \cdot s)}\,P1}{p \cdot (q \cdot s)}\,P0
\tag{30}
$$

The derivation on the right using unary modalities can't be completed. The structure s gets trapped underneath an instance of the $\langle l \rangle$ operator, which prevents s from taking scope just underneath q.

This limitation prevents the unary modality strategy from accounting for the full range of scope analyses that have been proposed in the literature. In particular, the configuration derived by NL_λ and NL_{CL} in the example in (30) is an instance of parasitic scope. Parasitic scope has been advocated as a scope-taking strategy for handling a number of phenomena, including adjectives of comparison such as *same* and *different* [2], as illustrated in (27); respective and symmetrical predicates [12]; certain uses of the adjective *average* [9]; non-constituent coordination [11]; as well as for verb phrase ellipsis, sluicing, and anaphora in general [3], following [18]. NL_λ and NL_{CL} were originally proposed precisely in order to handle parasitic scope.

6.3 Comparison with Discontinuous Lambek Grammar

Morrill, Valentín and Fadda, in [18] and [21], present a type-logical grammar called Discontinuous Lambek Grammar. Though different from NL_λ in its historical development (see [2] versus [18]:11) and in form, the expressive power and the specific analyses it provides are closely parallel to those of NL_λ.

On a conceptual level, there is a dramatic difference. Discontinuous Lambek Grammar views the argument that a scope-taker combines with (its nuclear scope) as a discontinuous constituent. For instance, in the sentence *Mary claimed John wanted everyone to read the book*, the nuclear scope of *everyone* corresponds to the discontinuous string *John wanted ... to read the book*. Continuation-based grammars such as NL_λ and NL_{CL} view this portion of a linguistic tree as a unit: it is a constituent with one piece removed (in the position of the scope-taker). All of its parts are connected, so it is a contiguous, single constituent, as illustrated in (8).

The correspondence between Discontinuous Lambek Grammar and NL_λ is easiest to see at the level of categories. Following [16], Morrill et al. define a type connective '↑' such that $B \uparrow A$ means (roughly) "a discontinuous expression that would be of category B if one of its gaps were filled with an expression of category A". This is functionally equivalent to our $A\backslash\backslash B$ (note the reversal of the order of the subcategories). Likewise, they define a complementary connective '↓' such that $D \downarrow C$ means "an expression that would be of category C, if only it were first substituted into a discontinuous expression of category D", which is functionally equivalent to our $C/\!\!/D$. (Note again the reversal of the categories.) So their category for a generalized quantifier is $((S_1 \uparrow DP) \downarrow S_2)$, which is equivalent to our $S_2/\!\!/(DP\backslash\backslash S_1)$. As a result of this correspondence, multiple levels of discontinuity in Discontinuous Lambek Grammar can be handled as different varieties of parasitic scope in NL_λ and NL_{CL}, and vice versa.

In addition to a major difference in conceptual foundations, Morrill et al. are committed to the assumption that natural language is fully associative, that is, that the structures $p \cdot (q \cdot r)$ and $(p \cdot q) \cdot r$ are fully equivalent. Associativity is well-established as a default assumption in some varieties of categorial grammar. However, it is by no means clear that natural language is uniformly associative. Instead of building associativity into the basic definitions of the grammar, as Morrill et al. do, a more conservative strategy would be to build a non-associative grammar, and add associativity in a carefully regulated way, only where needed, as advocated in [17]. In that spirit, associativity can easily be added to NL_λ or NL_{CL} simply by adding an appropriate structural postulate, if desired.

6.4   Comparison with Quantifier Raising

Here is the structural operation of Quantifier Raising, illustrated with categories borrowed from NL_λ:

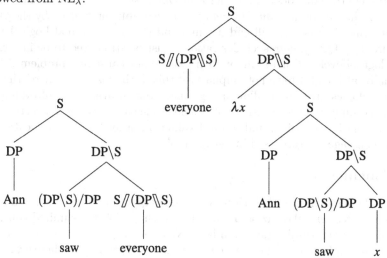

In NL_λ, this derivational step can be simulated closely using the λ postulate (reading the proof from bottom upwards):

$$\frac{\text{everyone} \circ \lambda x \, (\text{ann} \cdot (\text{saw} \cdot x)) \vdash S}{\text{ann} \cdot (\text{saw} \cdot \text{everyone}) \vdash S} \lambda \qquad (31)$$

So the logic here captures a significant portion of the insight embodied in Quantifier Raising. Is NL_λ then just the logic of Quantifier Raising? In some sense, clearly yes.

However, there are important differences between Quantifier Raising and the λ postulate of NL_λ.

For one, the lambda postulate is bidirectional. This reflects the fact that the two structures it relates are fully equivalent logically: they denote the same object in the model. In contrast, in the treatment in, e.g., [8], the pre-QR structure does not have a denotation. Thus the main motivation for executing an instance of Quantifier Raising is to produce a new meaning. In contrast, the lambda postulate here is a structural rule. Like all structural rules, the effect of the rule on the Curry-Howard labeling is null (no change to the semantic labeling). Quantifier Raising is conceived of as a rule that has a semantic effect but no syntactic effect (it constitutes 'covert' movement); the lambda postulate here has a syntactic effect, but no semantic effect. Its role in the logic is purely to characterize the syntactic operation by which a delimited continuation combines with its functor (by being surrounding by it) or its argument (by surrounding it).

For another difference, Quantifier Raising can create unbound traces.

Unbound trace: [[some [friend [of everyone]]][called]]
$\Rightarrow$ [everyone(λy[[some [friend [of y]]][called]])]
$\Rightarrow$ [[some [friend [of y]]](λx[everyone($\lambda y.x$)][called])]

If QR targets the embedded quantifier *everyone* first, and then targets the originally enclosing quantifier *some friend of* _, the variable introduced by the QR of *everyone* (in this case, y) will end up unbound (free) in the final Logical Form structure. In a QR system, such derivations must be stipulated to be ill-formed. In the logics developed here, unbound traces cannot cause any problems.

Finally, although I have not emphasized this in the discussion here, the substructural logics given here allow fine-grained control over order of evaluation, allowing accounts of order-sensitive phenomena such as crossover, reconstruction, negative polarity licensing, and more. Evaluation order and its applications in natural language are discussed in detail in [5].

7   Conclusion

What is the logic of scope? With natural language in mind, here is my answer: when an expression takes scope, it combines with one of its delimited continuations. The substructural logics given here, NL_λ and NL_{CL}, illustrate two equivalent ways to implement a concrete continuation-based grammar. These grammars are perfectly kosher substructural logics. In particular, they are sound and complete with respect to the usual class of models, they are conservative with respect to NL, and they enjoy cut elimination. Finally, although I presented a promising proof search strategy, a full account of decidability will have to wait for a future occasion.

References

1. Barendregt, H.P.: The Lambda Calculus: Its Syntax and Semantics. Elsevier, North-Holland (1984)
2. Barker, C.: Parasitic scope. Linguist. Philoso. **30**, 407–444 (2007)
3. Barker, C.: Scopability and sluicing. Linguist. Philosop. **36**(3), 187–223 (2013)
4. Barker, C.: Scope. In: Lappin, S., Fox, C. (eds.) The Handbook of Contemporary Semantics, 2d edn. Wiley-Blackwell, Malden (2014)
5. Barker, C.: Shan, Chung-chieh: Continuations and Natural Language. Oxford University Press, New York (2014)
6. Charlow, S.: On the semantics of exceptional scope. NYU Ph.D. dissertation (2014)
7. de Groote, P.: Towards abstract categorial grammars. In: Proceedings of the 40th Annual Meeting of the Association for Computational Linguistics. Morgan Kaufmann (2002)
8. Heim, I., Kratzer, A.: Semantics in generative grammar. Oxford (1998)
9. Kennedy, C., Stanley, J.: On average. Mind **118**(471), 583–646 (2009)
10. Kiselyov, O., Shan, C.: Continuation hierarchy and quantifier scope. In: McCready, E., Yabushita, K., Yoshimoto, K. (eds.) Formal Approaches to Semantics and Pragmatics. Springer, Netherlands (2014)
11. Kubota, Y.: Nonconstituent coordination in Japanese as constituent coordination: An analysis in Hybrid Type-Logical Categorial Grammar. Linguistic Inquiry 46.1 (2015)
12. Kubota, Y., Levine, R.: Unifying local and nonlocal modelling of respective and symmetrical predicates. In: Morrill, G., Muskens, R., Osswald, R., Richter, F. (eds.) Formal Grammar. LNCS, vol. 8612, pp. 104–120. Springer, Heidelberg (2014)
13. Lambek, J.: The mathematics of sentence structure. Am. Math. Monthly **60**(3), 154–170 (1958)
14. Moortgat, M.: Categorial Investigations: Logical and Linguistic aspects of the Lambek calculus. Foris, Dordrecht (1988)
15. Moortgat, M.: In situ binding: a modal analysis. In: Dekker, P., Stokhof, M. (eds.) Proceedings of the 10th Amsterdam Colloquium, pp. 539–549. Institute for Logic, Language and Computation, Universiteit van Amsterdam (1996a)
16. Moortgat, M.: Generalized quantification and discontinuous type constructors. In: Bunt, H.C., van Horck, A. (eds.) Discontinuous Constituency, pp. 181–207. Mouton de Gruyter, Berlin (1996b)
17. Moortgat, M.: Categorial Type Logics. In: van Benthem, J., ter Meulen, A.G.B. (eds.) The Handbook of Logic and Language. Elsevier, Amsterdam (1997)
18. Morrill, G., Valentn, O., Fadda, M.: The displacement calculus. J. Logic Lang. Inform. **20**(1), 1–48 (2011)
19. Muskens, R.: λ-grammars and the syntax-semantics interface. In: van Rooij, R., Stokhof, M. (eds) Proceedings of the 13th Amsterdam Colloquium, ILLC (2001)
20. Restall, G.: An introduction to substructural logics. Routledge, London (2000)
21. Valentín, O.: Theory of discontinuous lambek calculus. Ph.D. thesis, Universitat Autonma de Barcelona (2012)

Focus and Givenness Across the Grammar

Christopher Tancredi[✉]

Keio Institute of Cultural and Linguistic Studies, Minato-ku, Japan
cdtancredi@gmail.com

Abstract. This paper takes seriously the idea that a single expression can be simultaneously marked as given and as a focus, and works out some of the consequences of that assumption. I adopt Katz and Selkirk's (2011) suggestion that givenness is the flip side of newness rather than of focus, and argue that neither Rooth's semantics of focus nor Schwarzschild's analysis of givenness is by itself sufficient to account for a range of novel observations. I then show how both analyses can be maintained provided that the syntactic and phonological assumptions about focus/givenness marking and pitch accent assignment are appropriately revised.

Keywords: Focus · Givenness · Newness · Pitch accents

1 Introduction

In English, accentuation can affect the acceptability of a sentence in context. The standard approach to explaining these effects is to relate accent placement to identification of expressions as focused or given, provide a semantics for focus and for operators that are sensitive to focus, and place restrictions on discourse that are sensitive to the focal/given status of an expression. On the semantic/pragmatic side of grammar, three phenomena are standardly used for diagnosing focus: Contrast, Question-Answer Congruence (**QAC**), and Association With Focus (**AWF**).

(1) Contrast
    A: John talked with Mary. Then,
    B1: [BILL] *talked with Mary*.
    B2: *John talked with* [SUE].
    B3: *John* KISSED *Mary*.

(2) QAC
    Q1: Who talked with Mary?   A1: [BILL] *talked with Mary*.
    Q2: Who did Bill talk with?   A2: *Bill talked with* [MARY].
    Q3: What did Bill do?   A3: *Bill* [*talked with* MARY]

(3) AWF
    a. John talked with Sue. *Only* [BILL] *talked with* MARY.
    b. John talked with Sue. BILL *only talked with* [MARY].
    c. John asked Mary out to dinner. BILL *only* [TALKED *with*] *Mary*.

I would like to thank Daniel Büring, Makoto Kanazawa, and Michael Wagner for comments on an earlier draft of this paper that have led to improvements. Remaining errors are of course my own.

T. Murata et al. (Eds.): JSAI-isAI 2014 Workshops, LNAI 9067, pp. 200–222, 2015.
DOI: 10.1007/978-3-662-48119-6_15

In these examples, expressions identified as foci are placed in square brackets, capitals mark the location of a pitch accent, and italics indicate a lack of pitch accent. In the contrast examples, the foci are expressions that contrast semantically with something from the context sentence – *Bill* with *John* in (1a), *Sue* with *Mary* in (1b), and *kissed* with *talked with* in (1c). In the QAC examples, the focus is that part of the answer that corresponds to the *wh*-expression in the question. In the association examples, the focus is what gets substituted in generating a comparison class – *Bill* as opposed to *John* in (3a), *Mary* as opposed to *Sue* in (3b), and *talking with* as opposed to *asking out to dinner* in (3c).

On the phonological/phonetic side of grammar, a focus is typically pronounced with phonetic prominence: a pitch accent when new, lengthening when given. As can be seen by comparing A2 and A3 in the QAC examples, however, phonetic prominence is not in general sufficient to identify a semantic/pragmatic focus, as already noted in Chomsky (1971). Furthermore, while there is a tendency for prominence to be toward the right edge of a focus, as in A3, this tendency can be overridden by other considerations, as in the AWF example in (3c) where phonetic prominence shows up on the verb *talked* rather than on the equally in-focus *with*.

Givenness, like focus, affects how an expression relates to a discourse context. Schwarzschild (1999) argues that an expression is given iff it is entailed[1] by or coreferent with an antecedent. In the examples above, the italicized expressions outside of the foci all satisfy this requirement w.r.t. the context sentence/question that precedes them. On the phonological/phonetic side of grammar, givenness typically shows up as a lack of pitch accent. As can be seen in (2.A3), however, a lack of pitch accent does not by itself identify an expression as given. *Talked with*, in this example, bears no pitch accent and yet does not count as given in the context in which that example occurs.

The main challenge posed by examples like these is to provide a formal analysis that encompasses both the semantic/pragmatic side and the phonological/phonetic side of grammar and that predicts felicitous and infelicitous patterns of pitch accent assignment in different discourse contexts. In this paper I will examine in detail two analyses that aim to solve portions of this problem, those of Rooth (1992, 1995) and of Schwarzschild (1999). I will show that neither analysis on its own accounts for the full extent of what it sets out to explain, let alone what the other analysis does best. I show this by setting out four empirical challenges that any complete theory of focus and givenness needs to be able to account for, and showing that neither analysis can account for all four. I then propose to maintain Rooth's analysis of focus unmodified while making changes in the syntax and phonology of Schwarzschild's analysis to explain all four challenges. The solution to the problems will rely crucially on Katz and Selkirk's (2011) proposal that givenness is the complement not of focus but of newness.

In setting out the problems and the proposed solutions below, I will be implicitly assuming the following organization of grammar:

Pragmatics $\leftrightarrow$ Semantics $\leftrightarrow$ Syntax $\leftrightarrow$ Phonology $\leftrightarrow$ Phonetics

---

[1] See Schwarzschild (1999) for details and below for a somewhat simplified discussion.

The significance of this organization lies in the accessibility relations it licenses. A pragmatic restriction on appropriateness of a sentence in a given context, for example, can directly access the semantic interpretation of the sentence, but does not have direct access to the syntax, phonology or phonetics of that sentence under this assumption. If it is found that a certain phonetic aspect of a sentence correlates in some way with discourse appropriateness, under this organization of grammar that connection can only be explained by way of a chain of connections linking the phonetics to the phonology, the phonology to the syntax, the syntax to the semantics, and the semantics to the pragmatics. While not always made explicit, the analysis pursued in this paper obeys the restrictions implicit in this organization of grammar.

2 Previous Analyses

2.1 Schwarzschild (1999)

Schwarzschild adopts from Selkirk (1984, 1996) the idea that pitch accents in the phonology affect discourse felicity indirectly through their relation to F-marking in the syntax. While F-marking is given a direct phonological interpretation, however, in the semantics it receives no interpretation. Rather, only absence of F-marking is directly interpreted, as givenness. Summarizing and slightly simplifying,

Schwarzschild's Analysis of Givenness
If x is of type e, x is given iff it has a coreferential antecedent.
If x is of a conjoinable type (i.e. $<..., t>$), x is given iff its existential F-closure is entailed by an antecedent modulo existential type shifting.
Givenness: Non-F-marked expressions must be given.
AvoidF: F-mark as little as possible within the bounds of Givenness.
Basic F-Rule: An accented word is F-marked.
Foc-Rule: Foc-marked material must contain an accent.
    (A Foc-marked node is an F-marked node that is not immediately dominated by another F-marked node.)

The existential type shift of an expression of conjoinable type closes any open arguments through existential closure, creating an interpretation of type t. Existential F-closure applies to the result, replacing F-marked expressions with existentially bound variables of the same type.[2] The result is of type t and hence something that can be entailed. To see how this applies, consider the contrast example (1A-B2), repeated here with F-marking made explicit:[3]

(1) A: John talked with Mary. Then, B2: [*John* [*talked with* [SUE$_F$]]]

Non-F-marked expressions	Existential type shift	Existential F-closure
John talked with SUE$_F$	John talked with Sue$_F$	$\exists x$ (John talked with x)
talked with SUE$_F$	$\exists y$ (y talked with Sue$_F$)	$\exists x \exists y$ (y talked with x)
talked with	$\exists x \exists y$ (y talked with x)	$\exists x \exists y$ (y talked with x)
John	----	----

Since there is only a single F-mark in the sentence, that on *SUE*, every other constituent in the sentence is required to be given. With the exception of *John*, the result

---

[2] The ordering of existential type shifting before existential F-closure is unnecessary in the official formalization proposed in Schwarzschild (1999), which makes no explicit use of variables in interpreting F-marked expressions. The simplification used here (and by Schwarzschild himself) does not affect any of the arguments in this paper.

[3] I treat *talk with* here as a single lexical item for simplification.

of applying existential type shift and existential F-closure is in all cases entailed by the context sentence *John talked with Mary*. These expressions thus all count as given in the context, as required. In addition, *John* counts as given since it has a coreferent antecedent in the context sentence. In addition to the requirements of givenness, AvoidF also requires that the F-marking assigned be minimal. This requirement too is clearly met. F-marking cannot be removed from *Sue* since there is no expression in the context that is coreferent with *Sue*, and any additional F-marking would violate AvoidF.

Schwarzschild shows how the above analysis can apply to cases of QAC like those in (2). The key to making the account work is to associate *wh*-questions like *What did Bill do?* with existential formulas like $\exists P(P(Bill))$ (or perhaps $\exists P(P(Bill)$ & $action(P)))$ for the purpose of licensing givenness. This makes it possible to analyze the answer in (2.Q3-A3), for example, as follows:

(2) Q3: What did Bill do?   A3: [*Bill* [*talked-with$_F$* MARY$_F$]$_F$]
     Non-F-marked expressions   **Existential type shift**   **Existential F-closure**
     Bill [talked-with$_F$ MARY$_F$]$_F$ Bill [talked-with$_F$ Mary$_F$]$_F$ $\exists$P (P(Bill))
     Bill   ----   ----

Note that existential F-closure here only substitutes a variable for the highest F-marked expression, i.e. the Foc, not for the F-marked sub-constituents contained inside that expression. This means that for givenness to be satisfied the context only needs to contain an antecedent that entails that some property holds of Bill, not that some relation holds between Bill and some individual. Given Schwarzschild's analysis of questions this requirement is met. Since *Bill* also has a coreferent antecedent in the context, givenness is satisfied. It can further be seen that removing any of the F-marks would result in a violation of givenness, since there is no coreferring antecedent for *Mary*, and the context does not entail either that there was any talking or that Bill is related to something. Thus this analysis also satisfies AvoidF.

While Schwarzschild accounts well for the examples in (1) and (2), he does not account for the AWF examples in (3) for the simple reason that he does not provide an analysis of *only*. To see what is at issue, consider (3b) (= John talked with Sue. BILL *only talked with* [MARY].) The pitch accent on *Mary* is required here because it lacks a coreferential antecedent in the context. However, nothing in Schwarzschild's analysis predicts that this should result in *Mary* appearing to associate with *only*. F-marking is not given a direct interpretation, only lack of F-marking is, but even that is only related to discourse felicity and not to the semantics of associative particles like *only*. At the very least, then, Schwarzschild's analysis will need to be supplemented with an analysis of *only* that can account for its apparent association with F-marked expressions. I will argue below that such an analysis can be given, but only by allowing F-marking to play a role in the semantics, and hence by giving up Schwarzschild's assumption that only lack of F-marking is relevant to the semantics/pragmatics.

2.2   Rooth (1992, 1995)

Like Schwarzschild, Rooth assumes that phonological/phonetic focus is given a syntactic representation. Unlike Schwarzschild, Rooth takes the representation of focus to have a direct interpretation in the semantics. Formally, focus on an expression gives rise to a set of type-identical alternatives to that expression, the focus semantic value (FSV) of the expression. The FSV of a non-focused expression is the set that results from pointwise composition of expressions in the FSVs of its daughters. In the case of an expression not containing any focus, its FSV is the unit set of its normal semantic value.

Rooth's Analysis of Focus Semantic Values
Terminal nodes:
$[\![x_F]\!]^f$ = the focus semantic value of x_F
= the set of all alternatives to the normal semantic value of x (i.e. to $[\![x_F]\!]^o$)
$[\![x]\!]^f$ = $\{[\![x]\!]^o\}$
Non-terminal nodes:
$[\![x_{(F)} y_{(F)}]\!]^f$ = $\{x'(y'): x' \in [\![x_{(F)}]\!]^f \& y' \in [\![y_{(F)}]\!]^f\}$
(where x is of type $<\sigma,\tau>$ and y of type σ, order irrelevant)

While FSVs are properly semantic, they only affect truth conditions and discourse appropriateness for Rooth through their interaction with the $\sim$ operator. This operator uses the normal and focus semantic values of an expression to place restrictions on a discourse variable. For an expression of the form [[X] $\sim$ C], the variable C is presupposed to either be a member of $[\![X]\!]^f$ that is distinct from $[\![X]\!]^o$, or a subset of $[\![X]\!]^f$ that contains both $[\![X]\!]^o$ and one meaning distinct from $[\![X]\!]^o$. Since C is taken to be anaphoric, the $\sim$ operator can account for the interaction between focus in a sentence and the discourse context it occurs in. To see how, consider again (1.A-B2), repeated with the required focus marking (F), indexing and $\sim$ operator made explicit.

(1) A: [John talked with Mary]$_1$ Then, B2: [[*John talked with* SUE$_F$] $\sim$C$_1$]

The $\sim$ operator operates over the FSV of the expression it attaches to, making it necessary to calculate this value. The relevant calculations are given below.

x	$[\![x]\!]^o$	$[\![x]\!]^f$
SUE$_F$	Sue	$\{x: x \in D_e\}$
talked with	$\lambda x \lambda y.$ talked-with(y,x)	$\{\lambda x \lambda y.$ talked-with(y,x)$\}$
talked with SUE$_F$	$\lambda y.$ talked-with(y,Sue)	$\{\lambda y.$ talked-with(y,x): $x \in D_e\}$
John	John	$\{$John$\}$
John talked with SUE$_F$	talked-with(John,Sue)	$\{$talked-with(John,x): $x \in D_e\}$
[[*John talked with* SUE$_F$] $\sim$C$_1$]	talked-with(John,Sue)	$\{$talked-with(John,Sue)$\}$

The value of C_1 is presupposed in this case to be a member of the set *{talked-with (John,x): $x \in D_e$}* that is distinct from *talked-with(John,Sue)*. Taking C_1 to be anaphoric on A in (1) (indicated by co-indexing) satisfies this presupposition.

While the analysis just given for (1.A-B2) does not mention givenness, it does indirectly impose givenness on *John talked with*. Every formula in the FSV of *John talked with* SUE$_F$ will be of the form *talked-with(John,x)*, with some individual substituted for x, and the antecedent for C needs to be a member of this FSV. The antecedent will thus have to contain *John talked with* under this analysis of (1.B2).[4] However, Rooth does not impose any minimization conditions on focus identification, leaving open different possible identifications of focus. Adopting standard assumptions about the relation between pitch accents and foci, focus in this example could be on the PP *with SUE*, the VP *talked with SUE* or on the S *John talked with SUE*. Each of these

---

[4] As we will see below, the focus antecedent can follow rather than precede the interpretation of focus, something that is not possible for givenness. This is problematic for an analysis like Rooth's that reduces givenness to focus.

alternatives effectively results in a weaker constraint on what must be given, the last requiring nothing to be given at all. This means Rooth's overall analysis of focus does not account for the intuition that *John talked with* in (1.B2) is felt to be given in the context in which it occurs. This shortcoming could be overcome by adding a preference for narrow focus over broad whenever appropriate, a charitable assumption I will make both here and below. (See Truckenbrodt 1995 and Wagner 2012 for related discussion.)

Rooth's analysis applies straightforwardly to the QAC examples as well. The analysis differs from that given above in that the variable introduced by $\sim$ is taken to have a set of propositions as its value rather than a single proposition. The relevant analysis of the example in (2.Q3-A3) will be as follows:

(2) Q3: [What did Bill do?]$_2$ A3: [[*Bill* [*talked-with* MARY]$_F$] $\sim C_2$]

The value of C_2 here is presupposed to be a subset of the FSV of the sentence that $\sim C_2$ attaches to, i.e. $\{P(Bill): P \in D_{et}\}$. Identifying the question in Q3 as the antecedent of C_2 satisfies this presupposition, rendering the sentence felicitous.

Unlike Schwarzschild's analysis, Rooth's analysis is specifically designed to be able to handle AWF examples. *Only* operates semantically over both the normal semantic value of its sister and a discourse variable, requiring that the normal semantic value of the sister be the only value in the value of the discourse variable that will make the sentence true. By identifying this discourse variable with the variable introduced by a $\sim$ operator attached to the sister, the appearance of association follows, as illustrated below.

(4) John [kissed Mary and Sue]$_3$. Bill only(C$_3$) [[kissed MARY$_F$]$\sim$C$_3$]
   = [[only]] (C$_3$) ([[kissed MARY$_F$]]^o) ([[Bill]]): presupposition: C$_3 \subseteq$ [[kissed MARY$_F$]]^f
       Set C3 = {λy.kissed(y,m), λy.kissed(y,s)}
   = [[only]] ({λy.kissed(y,m), λy.kissed(y,s)}) (λy.kissed(y,m)) (bill)

The final line is true iff the only member of $\{\lambda y.kissed(y,m), \lambda y.kissed(y,s)\}$ that is true of Bill is $\lambda y.kissed(y,m)$, deriving the appearance of association with *Mary*.

3 Empirical Challenges

In this Sect. 1 examine four empirical challenges that any analysis of focus and givenness needs to account for. I will show that neither the analysis of Rooth nor that of Schwarzschild can account for all four of the challenges.

3.1 Given Foci

The first challenge is the existence of expressions that simultaneously qualify as given and as focused. Two subcases need to be distinguished: those in which the relevant expression is marked phonologically/phonetically only as focused, as in the case of (5), and those in which the expression is phonologically/phonetically identified as both focused and given, as in (6).

(5) Contrast:
Mary's father saw Bill. Then *he* HEARD MARY/#*Mary*.

QAC:
Q: Who did Mary's father see? A: *He saw* MARY/#*Mary*.

AWF:
Mary's father saw many people. However, *he only* HEARD MARY/#*Mary* (& nobody else).

In all three cases in (5), the final occurrence of *Mary* counts as given, having a coreferent antecedent in the first sentence. However, it also qualifies as focused, contrasting with *Bill*, answering to *who*, and associating with *only*. In all cases it surfaces phonetically with an obligatory pitch accent. These cases contrast with those in (6). (Here SMALL CAPS indicate phonetic prominence in length and intensity but not pitch accent.)

(6) Contrast:
John saw Mary and Sue saw Bill. Then,
ALICE saw #*'er/HER/*#HER/*MARY/*#MARY and TOM saw #*'im/HIM/*#HIM/*BILL/*#BILL

QAC:
John saw Mary. Who did Bill see?
BILL *saw* #*'er/HER/*#HER/*MARY/*#MARY

AWF:
John saw Mary. In fact, *he* ONLY *saw* #*'er/HER/*#HER/*MARY/*#MARY.

In these latter cases, the object *Mary/her* once again qualifies semantically/pragmatically as both given and a focus. In contrast to the cases in (5), however, in these examples *Mary/her* cannot surface with a pitch accent, suggesting that it is obligatorily marked as given.[5] Unlike other occurrences of given pronouns, however, *her* in these examples cannot be fully reduced, suggesting that it is also focused. (Cf. Rooth 1996, Beaver 2004, and Selkirk 2008 for further evidence of expressions that are both focused and given.)

These cases are problematic for both Rooth and Schwarzschild. Rooth can account straightforwardly for the pitch accents in (5). These pitch accents identify the accented expressions as focused, and in all cases they act as focused in the discourse. If Rooth were to take focus to be marked obligatorily in the syntax on an expression that qualifies as a focus in the semantics/pragmatics, he could furthermore account for the obligatory nature of these pitch accents. However, such an extension would land him in hot water with respect to (6). It would lead to a prediction of obligatory accenting in these examples as well, but accenting is infelicitous here. Since givenness for Rooth is simply a side effect of focus interpretation, he has no way of analyzing *Mary/her* in (6) as both focused and given.

Schwarzschild is in no better shape than Rooth. While Schwarzschild explicitly claims that being given does not preclude being F-marked and hence potentially accented,

---

[5] For the QAC example, pronouncing *Mary* with a pitch accent is possible in the context given, though doing so gives the impression that the answerer is ignoring the first sentence and relating the answer exclusively to the question. Since *Mary* is not given with respect to the question, accentuation would be expected in this case. This same accenting option is not available for an anaphorically interpreted *her* in this context, presumably because the anaphora makes it impossible to exclude the first sentence from the relevant discourse context.

he has no way of distinguishing between the cases in (5) where F-marking wins out over givenness in the phonetics and the cases in (6) where both givenness and focus are given equal weight in the phonetics. Particularly problematic for Schwarzschild is the impossibility of deaccenting *Mary* in the Contrast and AWF examples in (5). Since *Mary* is given in these contexts and its givenness is compatible with all other accentless expressions being identified as given, lacking an F-mark should be a possibility, and so by AvoidF it should be the only possibility. Without an F-mark, however, *Mary* should at least be allowed to not be accented, but the pitch accent on *Mary* is obligatory.

3.2   Non-constituent Foci

The second empirical challenge comes from expressions that together appear to function as a single focus for the purposes of semantics/pragmatics but which do not form a syntactic constituent.

(7) Contrast:
  a.   A: John fell. B: [MARY TRIPPED] *him*.
  b.   A: John saw pictures of Mary. B: BILL *saw* [SUE'S SCULPTURE] *of Mary*.

(8) QAC:
  a.   Q: What happened to John? A: [MARY TRIPPED] *him*.
  b.   Q: What did John see relating to Mary? A: *He saw* [SUE'S SCULPTURE] *of her*.

(9) AWF:
  a.   Sue doesn't know that John cried. She knows only that [MARY TRIPPED] *him*.
  b.   John didn't see Mary. *He saw* only [SUE'S SCULPTURE] *of Mary*.

Schwarzschild can give a straightforward account of the examples in (7) and (8) by F-marking not only the pitch accented words but also several of the constituents dominating these words. For example, the (a) cases can be analyzed as containing the structure $[MARY_F [TRIPPED_F him]_F]$. The F-marking on *Mary* and on *tripped* is needed since these are not given in the discourse context, while the absence of F-marking on *him* is justified by its having a coreferent antecedent. That the F-marking of the VP is necessary can be seen from the fact that eliminating this F would require the sentence to have an antecedent that entailed $\exists x, R\ (R(x,j))$. That is, the context would need to contain a two place predicate, one of whose arguments is *John*. This requirement is patently not met in these examples, making F-marking on VP a necessity, but also sufficient for accounting for the examples. Once again, since Schwarzschild does not analyze the phenomenon of AWF he has no account for the examples in (9).

The success of Schwarzschild's analysis in accounting for (7) and (8) comes from its ignoring F-marked expressions in determining the givenness of a non-F-marked type *e* expression. Rooth, on the other hand, assigns a central role to focus marking in calculating the FSV of an expression, and this causes problems. If we take pitch accent location to indicate F-marking, the (a) cases in (7)–(9) will have the following syntactic analysis: $[[MARY_F [TRIPPED_F him]] \sim C]$. This analysis will lead to the presupposition that C is a member of $\{R(x,j): R \in D_{<e,et>}\ \&\ x \in D_e\}$, i.e. that C is a proposition constructed from a 2-place predicate and two arguments, one of which is *John*. There is no antecedent in any of the (a) examples of (7)–(9) that satisfies this presupposition, however, leading to the incorrect prediction that these examples should be infelicitous. Adding F-marking to the VP and/or the S will potentially avoid this problem, but only at a cost. Adding F-marking to the VP will lead to C having to be a member of $\{P(x): P \in D_{et} \& x \in D_e\}$. Adding F-marking to S will lead to C having to be a member of $\{p: p \in D_t\}$. These presuppositions are easily satisfied in the contexts of (7)–(9). However, such an analysis would wrongly predict that *him* need not be given, since givenness for Rooth is an epiphenomenon that results from an expression being part of every alternative in an FSV introduced by a $\sim$ operator. This would wrongly predict discourses like (10) to be perfectly felicitous, with the analysis given.

(10) [Sue fell.]₄ [[MARY_F [TRIPPED_{(F)} John]_F]~C₄]

3.3 Connectedness of Givenness

As observed in Tancredi (1992), when a predicate and one of its arguments are both deaccented, it is not sufficient for the two words to have independent antecedents in the discourse context. Rather, there must be a single antecedent consisting of an entailing predicate standing in an identical thematic relation to an entailing/coreferential argument. I call this phenomenon *connectedness of givenness*: given expressions in a sentence act as if they are thematically connected, and require an antecedent that contains the same connectedness.[6] This can be seen clearly in the examples below.

(11) a.   A: John saw Mary. B: Then, BILL *saw Mary*.
     b.   A: John saw Mary. B: #Then, BILL *saw John*.

Rooth does not account for this observation. His failure to do so stems from the flexibility in where focus gets interpreted. Interpretation at the sentence level leads to both the acceptability of (11a) and the unacceptability of (11b), since in that case the VP will make a constant contribution to every member of the FSV of the sentence, one that occurs in the A sentence of (11a) but not of (11b). However, Rooth does not require focus to be interpreted at the sentence level, and if we interpret focus directly on the subject as in *[[BILL_F] $\sim$ C] saw Mary/John*, then the VP plays no role in determining any FSV relevant to the interpretation of the sentence in its context. The possibility of such local interpretation of focus means that the only requirement imposed by focus is for there to be a type *e* antecedent in the discourse context that is distinct from Bill, a requirement that is met in both (11a) and (11b). Under this analysis, then, both discourses are predicted to be felicitous, contrary to fact. We can overcome this problem by following Truckenbrodt (1995) and Wagner (2012) in requiring the domain/scope of focus interpretation to be as broad as possible, an assumption that I will again charitably adopt.

Like Rooth, Schwarzschild too correctly predicts the acceptability of (11a). Since the verb and object are each separately given, neither will be F-marked under his analysis. The subject, on the other hand, is not given, and so will have to be F-marked.

---

[6] Below I will analyze connectedness of givenness in terms of Givenness Semantic Values. I thus use this term to refer to the semantic properties of a sentence being interpreted, not to the properties of an appropriate antecedent.

Since F-marking is not needed on VP or S, AvoidF requires its absence, as in (12a) below. This means that not only will the verb and object need to satisfy givenness separately, but the VP and S will need to as well. Since all these expressions do in fact satisfy givenness under Schwarzschild's analysis – there is an antecedent entailing $\exists x$ *(x saw Mary)* – the analysis correctly predicts the acceptability of (11a). Schwarzschild also accounts for the unacceptability of (11b). The formulation of AvoidF makes it impossible to impose the same F-marking in (11b) since AvoidF requires minimal assignment of F-marking *within the bounds of Givenness*, and such F-marking would not satisfy Givenness However, (12b,c) below both satisfy Givenness, making them both competing alternatives.

(12) a.　[[BILL$_F$] [*saw Mary*]]
　　b.　[[BILL$_F$] [*saw* JOHN$_F$]]
　　c.　[[BILL$_F$] [*saw John*]$_F$]$_F$

(12b) contains less F-marking than (12c), making (12b) alone satisfy AvoidF. This in turn makes (12b) the only acceptable representation under Schwarzschild's analysis. However, the F-marking in (12b) leads to an obligatory accent on *John* since *John* counts as a Foc and every Foc needs to bear an accent, accounting for the unacceptability of (11b) where *John* fails to bear an accent.

3.4　Optional Accents

The fourth empirical challenge comes from the observation that, in cases like the (a) and (c) examples in (13)–(15), some pitch accents are optional.[7]

(13) Contrast:
　　a.　John went dancing.　　　THEN *he* [*drank* BEER] / [DRANK BEER] / #[DRANK *beer*]
　　b.　Mary's singing next door.　ALSO, [JOHN's *dancing*] / [JOHN's DANCING] / #[*John's* DANCING]

(14) QAC:
　　a.　Q: What is John doing?　　A: *He's* [*drinking* BEER] / [DRINKING BEER] / #[DRINKING *beer*]
　　b.　Q: What's happening next door?　A: [JOHN's *dancing*] / [JOHN's DANCING] / #[*John's* DANCING]

(15) AWF:
　　a.　Bill went dancing. John *only* [*drank* BEER] / [DRANK BEER] / #[DRANK *beer*]
　　b.　Mary said there's a party. Sue *only said* [JOHN's dancing] / [JOHN's DANCING] / #[*John's* DANCING]

The generalization that characterizes this phenomenon is that within a non-given expression, any word can optionally bear a pitch accent in addition to those words that do so obligatorily. This phenomenon is not explained by Schwarzschild's analysis. Schwarzschild takes pitch accents to correlate with F-marking by the Basic F-Rule and the Foc-Rule: every accented expression is F-marked, and every Foc-marked expression contains an accent. This means that a sentence with two pitch accents has to have

---

[7] There are additional accenting possibilities here and in examples throughout the paper. In the discourse: *A: John went dancing. B: Then, HE drank BEER(, though everyone else drank wine)*, for example, the accent on *he* in the second sentence is perfectly acceptable. Since it does not contrast with anything that precedes, does not answer a wh-question and is not the associate of a particle like *only*, it is plausible to analyze it not as a focus but as a topic. Consideration of topics is not possible within the length limitations of the current paper, and so these possibilities are systematically set aside.

at least two F-marks in the syntax. This by itself is not problematic, since in all of the cases in (13)–(15) above, both of the accentable words need to be F-marked independently since neither qualifies as given. This cannot be the full extent of the F-marking, however, since this F-marking would violate Givenness. In (13a), for example, such F-marking would require there to be an antecedent entailing $\exists R\exists x(R(x, john))$, a requirement not satisfied. At a bare minimum, F-marking is also required on the VP. Givenness in such a case will then only require a coreferent antecedent for *John* and an antecedent entailing $\exists P(P(john))$. AvoidF will block any additional F-marking. The problem now is one of distinguishing obligatory accents from optional ones. In the (a) examples, *beer* is obligatorily accented and *drinking* only optionally so. In the (b) examples it is *John* that is obligatorily accented and *dancing* whose accent is optional. Nothing in Schwarzschild's analysis, however, predicts this pattern. In particular, the third accent pattern in each example satisfies all of Schwarzschild's requirements and yet is unacceptable.

An additional problem arises for Schwarzschild when these examples are embedded in a context in which F-marking is not required on anything beyond the lexical items in question, as in (16).

(16) John ate pizza. THEN *he* [*drank* BEER] / [DRANK BEER] / #[DRANK *beer*]

Here unlike in (13a), F-marking on *drank* and on *beer* is sufficient to satisfy Givenness since the context sentence entails $\exists R\exists x(R(x,john))$. AvoidF then blocks additional F-marking on the VP. For Schwarzschild this means that both words will be independently identified as foci by the Foc-Rule and will therefore have to be assigned a pitch accent. While this accounts for the dual accent possibility, however, it fails to allow for the single accent option or to distinguish the good single-accent pattern from the bad.

The phenomenon of optional accents is only somewhat less problematic for Rooth, but only because Rooth does not give an independent characterization of the relation between focus in the semantics and pitch accent assignment in the phonology/phonetics. If each pitch accent is taken to identify a separate focus, then the problems that arise from cases of non-constituent foci will all arise here as well. In particular, *DRANK BEER* will then be treated as two separate foci, and each will need a type-identical antecedent. We see in (13)-(15) that this requirement is not met, and yet all of the examples are acceptable. Nothing in Rooth's analysis prevents analyzing the VP as focus in these examples, of course. However, doing so leaves us without an explanation for why the pitch accent on *beer* is obligatory while that on *drank* is merely optional. The phenomenon of optional accents thus shows that Rooth's analysis is at the very least incomplete.

3.5 Summary

In this Sect. 1 presented four empirical challenges to a theory of focus and givenness: given foci, non-constituent foci, connectedness of givenness, and optional accents. I showed that Schwarzschild's analysis can handle non-constituent foci and connectedness of givenness, but that it does not explain given foci and is incompatible with optional accents. Rooth's analysis, on the other hand, was seen to be compatible with given foci and with optional accents, but it does not explain either phenomenon, nor does it explain the connectedness of givenness, and it is furthermore incompatible with non-constituent foci.

The very existence of expressions that are phonetically explicitly identified both as given and as focused shows that givenness and focus cannot be two sides of the same coin. Since both Rooth and Schwarzschild treat the phenomena of givenness and focus as complementary, it follows that simply combining their analyses, e.g. by adopting Schwarzschild's analysis for givenness and Rooth's analysis for focus, will not be sufficient. I propose instead to build on the insight of Katz and Selkirk (2011) that focus needs to be distinguished from discourse newness, and that discourse newness is the complement of givenness. I then modify the non-semantic parts of Schwarzschild's analysis and combine it with Rooth's analysis to account for all of the phenomena examined in this section.

4   Phonological Phrasing: A Possible Solution to Optional Accents for Schwarzschild?

The problem of optional accents was seen to be devastating to Schwarzschild's analysis. It is worth considering whether a minimal modification to his analysis that makes such optional accents possible would be viable. One potential place to look to make such a modification is to phonological phrasing. Truckenbrodt (1995) argues that pitch accent location is determined at the level of the phonological phrase (**P-phrase**), with one accent assigned per P-phrase. The relevance of P-phrasing to pitch accents can be illustrated with the following example.

(17) What happened?
  a. #*Mary kissed* BILL.
  b. MARY *kissed* BILL.
  c. *MARY (and this I know first hand) *kissed* (and it shocked me) BILL
  d. MARY (and this I know first hand) KISSED (and it shocked me) BILL

In a response to the discourse initial question *what happened?*, every expression in an answer will, under Schwarzschild's analysis, be F-marked. This should lead to the sentence containing a single Foc – the sentence itself – and hence a single pitch accent, presumably on the right-most expression *Bill*.[8] What we find, however, is something more complicated. First, the predicted pattern given in (17a) is unacceptable: an extra accent is minimally required on the subject *Mary*, as in (17b). Second, we see that even this accent pattern can be made unacceptable through phonological phrasing as seen in (17c). Here the addition of parentheticals forces each word to constitute a separate intonation phrase (**I-phrase**) and hence an independent P-phrase as well. As we can see by comparing (17c) to (17d), it is not permitted in English for such a P-phrase/I-phrase to lack a pitch accent.

---

[8] Schwarzschild does not give rules for how to locate a pitch accent within a focus, so in principle it would be possible under his analysis for the single pitch accent to surface on the subject as in #*MARY kissed Bill*, or on the verb as in #*Mary KISSED Bill*. The fact that both of these variants are unacceptable in the context of (17) shows that the problem with (17a) is not merely one of accent location.

If we add to Schwarzschild's analysis Truckenbrodt's proposal that pitch accents stand in a one-to-one relation with P-phrases, then the facts in (17c, d) follow directly. As a Foc, the sentence as a whole has to contain a pitch accent by the Foc-Rule, and it does. Each pitch accented expression is also F-marked as required by the Basic F-Rule. We can further account for the contrast in (17a, b) if we put a restriction on the size of a P-phrase, allowing it to contain no more than two prosodic words. Since the answers in (17a, b) contain three prosodic words, such a restriction will force them to be broken into two P-phrases and hence to contain a minimum of two pitch accents.

Can such an analysis of the relation between P-phrases and pitch accents be the solution to optional accents under Schwarzschild's analysis? The obvious way to analyze (13a)/(16) on that analysis would be as in (18), where set brackets are used to delimit P-phrases.

(18) a.   {He drank$_F$ BEER$_F$}
     b.   {He DRANK$_F$} {BEER$_F$}

Unfortunately, this analysis does not mesh with AvoidF. We saw earlier that AvoidF made it impossible to F-mark an expression $[X_F\ Y_F]$ when the F-marks on X and Y by themselves are sufficient for satisfying Givenness. This is exactly the situation we have in (16), so AvoidF makes it impossible to F-mark the VP. Without such an additional F-mark, however, each of *drinking* and *beer* is a Foc, and so by the Foc-Rule needs to contain an accent. AvoidF and the Foc-Rule thus together rule out any representation like (18a) that contains only a single pitch accent.

5   Analysis

To handle given foci, I adopt the suggestion from Katz and Selkirk (2011) that givenness is the complement of newness. In principle this allows for either newness or givenness to be marked in the grammar, though Occam's Razor dictates that they not both be marked simultaneously. I opt for marking givenness, via syntactic G-marking. Assuming the structure of grammar outlined in Sect. 1, a full analysis then needs to do the following:

General Requirements

Assign F-marking in the syntax that identifies semantic foci.
Assign G-marking in the syntax that identifies discourse given and discourse new expressions in the semantics.
Give rules for appropriate use of semantically identified foci, new and given expressions in the discourse.
Use F- and/or G-marking to determine pitch accent distribution and relative prominence.

My proposal in outline for how to accomplish this is the following:

Focus

All and only semantic foci are F-marked in the syntax.
Semantic foci must contrast with an antecedent.
F-marking increases phonetic prominence, but does not affect pitch accent location.

Givenness/Newness

Given and new expressions are complementary to each other w.r.t. a selected discourse context.

Expressions semantically interpreted as given must be discourse given.

Semantically given lexical expressions are G-marked in the syntax.

G-marking projects from a head to its syntactic projections.

G-marked lexical items lack a pitch accent.

The details of implementation are of course crucial to explaining all of the phenomena examined above. In the remainder of this Sect. 1 will spell out these details, with additional applications of the analysis given in Sect. 6.

5.1   Semantics

Givenness. I propose to maintain the semantic core of Schwarzschild's analysis. Schwarzschild's analysis of givenness, however, could not semantically identify expressions simultaneously as given and as focussed because givenness for Schwarzschild derived from the absence of F-marking and all foci are F-marked. By employing G-marking for givenness and F-marking for focus we can straightforwardly overcome this shortcoming: given foci can be marked with both F and G in the syntax, with F- and G-marking interpreted independently in the semantics. In order to maintain a strict separation of components in the grammar, I analyze every expression as having a Givenness Semantic Value (**GSV**) in addition to its normal semantic value and its FSV.

Givenness Semantic Values (GSVs):

For a non-G-marked expression, its GSV is a type-identical variable.

For a G-marked terminal expression, its GSV is its normal semantic value.

For a G-marked non-terminal expression, its GSV is the result of composing the GSVs of its daughters.

GSVs play two important roles in the grammar: they correlate with G-marking in the syntax, and they impose antecedence requirements on the discourse context. The antecedence requirement I call Givenness, following Schwarzschild. Givenness will be satisfied by an expression if its GSV counts as Discourse Given. Adding a rule that maximizes G-marking results in an analysis that is roughly equivalent to Schwarzschild's:

Givenness: A GSV must be Discourse Given.
Discourse Givenness:

A semantic value is Discourse Given iff:

It is a variable; or

It is a type e expression and has a coreferring antecedent; or

The existential closure of its existential type shift is entailed by (the existential closure of) an antecedent.[9]

MaximizeG: G-marking is maximized within the limits of Givenness.

Note that Discourse Givenness is a general property that can hold of any kind of semantic value. The normal semantic value of *Mary* in the answer to the QAC example in (5), for instance, will be Discourse Given despite not being analyzed as Given. The Discourse Givenness of *Mary* in this case plays no role, however, in the sentence it's contained in satisfying Givenness.

To account for the connectedness of givenness, I propose to supplement the semantic core of Schwarzschild's analysis with the following syntactic constraint on G-marking:

ProjectG : A lexical head is G - marked iff its syntactic projections are.

To see how this constraint works, consider once again (11b), the example showing connectedness of givenness. Here I will consider four separate options for lexical G-marking: on both *saw* and *John*, only on *saw*, only on *John*, and on neither. In all four cases I take the verb to be the head of both VP and S and so by ProjectG these expressions will have the same G-marking as the verb.[10]

(11b)   A: John saw Mary. Then,
  i.   B: [BILL [$saw_G John_G$]$_G$]$_G$
  ii.   B: [BILL [$saw_G John$]$_G$]$_G$
  iii.   B: [BILL [$saw John_G$]]
  iv.   B: [BILL [$saw John$]]

Each of *saw* and *John* are Discourse Given and so can, but need not, be analyzed semantically as given as well, i.e. as having non-variable GSVs. They will be G-marked if and only if so analyzed, and then by ProjectG this G-marking will obligatorily project. However, (i) fails to satisfy Givenness (and so also violates MaximizeG) since there is no antecedent entailing the GSV $\exists x(x\ saw\ John)$ of the VP and S. (ii) satisfies both MaximizeG and Givenness. However, *John* is not G-marked, and under plausible phonological assumptions, this will require it to bear a pitch accent, which it does not do (see Sect. 5.3 for details). (iii) satisfies Givenness. Whether it also satisfies MaximizeG depends on how we interpret maximization. Since G-marking on *saw* projects to VP and S and that on *John* does not project, the G-marking in (ii) could conceivably be taken to be greater than that in (iii) for the purposes of MaximizeG making (iii) fail to satisfy MaximizeG. Formalizing such a notion of MaximizeG, however, would be far from straightforward. Alternatively and more plausibly, (iii) could be taken to satisfy MaximizeG but be ruled out because of the lack of an accent on the verb *saw*. MaximizeG on such an approach would only compare two

---

[9] Parallel to Schwarzschild's analysis, existential closure (Schwarzschild's existential F-closure) binds variables substituted for non-G-marked (Schwarzschild's F-marked) expressions, whereas existential type shifting binds unsaturated argument positions.

[10] A more plausible assumption would be that the verb only heads the VP, with the subject generated within the VP and raised to its surface position. Adopting this assumption would require relating Givenness to traces/copies. Though I do not see any inherent problems with doing so, I put off consideration of movement effects for a separate occasion.

representations if the G-marking of one completely subsumes the G-marking of the other, and not if their G-marking only partially overlaps or fails to overlap at all. Finally, (iv) satisfies Givenness trivially, but clearly fails to satisfy MaximizeG. This leaves (ii) as the only possible representation of (B). If (ii) is as hypothesized incompatible with a lack of pitch accent on *John*, then the unacceptability of (11b) follows. Note crucially that ProjectG blocks a representation containing G-marking only on *saw, John* and S, but not on VP. Such a representation would satisfy Givenness and plausibly MaximizeG as well, and so in the absence of ProjectG would be wrongly predicted to be acceptable.

Focus. Rooth gave a compositional semantics for focus based on the assumption that focus can be identified in the syntax. Givenness on this analysis was seen to be a mere side effect of focus. Non-constituent foci were seen to be problematic on these assumptions. On that analysis, the only way to account for the givenness of the object of a sentence when both the subject and verb are accented is to take the subject and verb to each be independent foci. In some cases, though, these two foci act as one, but treating them as one would require a non-compositional step in the interpretation. We have seen, however, that focus and givenness are not in fact complementary. This opens up an alternative solution to the non-constituent focus problem: take the focus in the apparent non-constituent foci cases to be some single constituent that contains both pitch accented expressions. In the case of (7a) this would lead to the following syntactic representation.

(7a) A: [John fell]₃ B: [[MARY TRIPPED *him*$_G$]$_F$~C₃]

This representation satisfies Rooth's semantics as well as our revised semantics of givenness. The sole G-marked expression *him* has a coreferring antecedent in *John* as required by Givenness, and there is an antecedent for C_3 that is a member of the FSV of the sentence (= the set of all propositions) as required by the ~ operator. Note, however, that givenness of *him* under this analysis does not follow from the interpretation of focus alone.

If focus and givenness come apart as suggested, it should in principle be possible for a focused expression in a sentence to take a different antecedent than a given expression in that same sentence, regardless of what syntactic relation holds between the two expressions. In (19), where I have included Rooth's focus interpretation operator and the requisite accompanying variables and indices, we see that just such a possibility exists.

(19) A: Many people know that John₁ fell.
   B: Most of them, however, only(C₅) [[know that [MARY TRIPPED *him*$_{1,G}$]$_F$]~C₅]
   C: They don't also [know that that was an accident]₅.

Here the G-marking of *him* in (B) is licensed by *John* in (A). The VP sister of *only*, on the other hand, cannot be taken to contrast with the matrix VP of (A) on pain of contradiction. On the non-contradictory interpretation of (B), the VP sister of *only* contrasts instead with the matrix VP of (C). Since (C) contains no potential antecedent for *him*, this example clearly shows the separability of focus and givenness. The important consequence of these observations for the semantics is that accounting for apparent non-constituent foci no longer requires a revision to Rooth's semantics of focus. It just needs to be accepted that focus semantics does not also account for givenness.[11]

Given foci pose a greater challenge to Rooth's analysis. In the contrast case in (5), both *he* and *Mary* have coreferent antecedents, and *heard* is not discourse given. MaximizeG thus requires that both *he* and *Mary* be G-marked. In this regard, the contrast case in (5) is parallel to the non-contrast example in (20). The fact that (20) is acceptable shows that the problem with the contrast case in (5) is not (or at least not exclusively) a problem of Givenness.

(20) Mary's father is a good person. *He* LOVES *Mary*.

This suggests that the problem with (5) derives from contrast, which by assumption relates to the semantics of focus. At a minimum on this view, it needs to be shown that absence of any focus as in (21a) is acceptable, while presence of focus as in (21b) is not.

(21) a. [*he*$_G$ [LOVES *Mary*$_G$]]
b. [[*he*$_G$ [HEARD$_F$ *Mary*$_G$]]~C]

This much is relatively straightforward. The variable C in (21b) requires an antecedent whose semantic value is a member of the set *{R(Mary's father, Mary): R $\in D_{<e, et>}$}*. While the NP *Mary's father* does presuppose a proposition of the requisite form (namely that Mary's father is the father of Mary), the NP itself does not have such a proposition as a value and so plausibly cannot be the antecedent to C. The only potential antecedent of the right semantic type is the entire first sentence *Mary's father saw Bill*, but the interpretation of this sentence is not of the required form.

While the impossibility of analyzing (5) as in (21b) can easily be accounted for under Rooth's semantics, more challenging is eliminating the possibility of representing (5) as (22a) or (22b).

(22) a. [[*he*$_G$ [HEARD$_F$ *Mary*$_{F,G}$]]~C]
b. [[*he*$_G$ [HEARD *Mary*$_G$]$_F$]~C]

(22a) differs from (21b) only in analyzing *Mary* simultaneously as a focus and as given, while (22b) differs from (21b) only in having focus on the VP rather than on the V. Neither difference in focus has an effect on Givenness, with both representations in (22) satisfying Givenness just like (21b) does. However, they do make a difference for satisfying the presupposition of the ~ operator. By analyzing *heard* and *Mary* as two separate foci as in (22a), Rooth's analysis requires an antecedent for C that is a member of the set *{R(Mary's father, x): R $\in D_{<e, et>}$ & x $\in D_e$}*, and this requirement is clearly satisfied in the context in which the sentence occurs. Similarly, by analyzing the VP as

---

[11] If focus is interpreted at a higher constituent than where it is marked, then the semantics of focus will in effect still give rise to a kind of givenness effect since the focus antecedent will still have to contain the non-focused parts of the higher constituent. However, this effect differs from that derived from G-marking in that it can in principle be cataphoric and need not result in deaccenting, as in the second sentence in: *John came to my party. He only MET MARY there, though. He didn't meet TOM.* To get *only* to associate intuitively with *Mary*, *Mary* has to be analyzed as the focus, with *met* being new and focus interpreted at the level of the VP. Such an analysis requires an antecedent for the VP that includes *meeting*, though *met* in the second sentence does not thereby count as given. Only the third sentence satisfies the antecedent requirement for the focus.

a focus as in (22b), Rooth's analysis requires an antecedent for C that is a member of the set $\{P(Mary's\ father):\ P \in D_{et}\}$, and again the context sentence satisfies this requirement. If G-marked expressions surface as unaccented, then with either of these representations the contrast example in (5) is predicted to be acceptable without an accent on *Mary*, contrary to observation.

I do not see a way of formally blocking the representations in (22) by adjusting either the semantics of givenness or that of focus. I instead propose an ad hoc solution of requiring maximum parallelism between focus antecedence and givenness antecedence when a single expression is subject to both requirements. This forces adoption of the analysis in (22a) over that in (22b) since only (22a) requires a focal antecedent for *Mary*, which independently requires a givenness antecedent. It also forces *Mary* in (22a) to have a givenness antecedent that is the same as its focus antecedent *Bill*, however, which imposes the contradictory requirements on *Bill* of being coreferent with *Mary* and semantically distinct from *Mary* at the same time.

That some extra-semantic explanation is needed to account for the contrast example in (5) is independently suggested by the difference between that example and (20): in (5) we seem pushed toward interpreting the second sentence as contrasting with the first, while in (20) we do not, and yet intuitively it is this need to contrast that causes the problems in (5). The fact that there is a formal analysis of (5) that does not impose contrast and that satisfies all givenness and focus requirements – (21a) with *heard* in place of *loves* – makes no difference. The context seems to lead us down a garden path requiring contrast and resulting in unacceptability rather than allowing a non-contrasting understanding that would be acceptable.

If the contrast example in (5) can be explained by appeal to maximizing parallelism, then this example does not require us to make any changes to Rooth's semantics of focus. While acknowledging that the ad hoc analysis proposed needs further investigation and deeper justification, I will thus accept Rooth's semantics unmodified.

5.2 Syntax

The syntax of F-marking I take to be trivial: F-marking is assigned to all and only those expressions identified in the semantics as foci. The syntax of G-marking cannot be made trivial in the same way, however. If it were, we would have no way of explaining the connectedness of givenness illustrated in (11b), repeated here.

(11b)   A: John saw Mary. B: #Then, BILL *saw John*.

In this example, assignment of G-marking based on Discourse Givenness would dictate G-marking on *saw* and on *John*, but would not predict any further connectedness between these expressions, leading to the incorrect prediction that the discourse should be felicitous. To account for the connectedness found, minimally the VP must be required to be Given as a whole. Since this requirement cannot come from Discourse Givenness, the only other plausible source is syntactic restrictions on G-marking: G-marking must be required to project to the level of the VP.

The examples of non-constituent foci show that G-marking cannot be taken to project automatically from just any G-marked expression. In (7a), projection of

G-marking from *him* to the VP would lead to a requirement that the discourse context contain a 2-place predicate with *John* filling one of the argument positions.

(7a) A: John fell. B: [MARY TRIPPED] *him*.

Since this requirement is not met in (7a) and yet the discourse is felicitous, it follows that such projection must not be imposed. The obvious way to navigate the opposing requirements of (11b) and (7a) is to take G-marking to project obligatorily from a syntactic head to its projections, but never from a non-head, as codified in ProjectG.

5.3   Phonology[12]

Truckenbrodt (1995) accounts for the location of pitch accents based on the assumption that every P-phrase bears a unique accent, located on the head of that P-phrase. He adopts the following constraints on P-phrases:

(23) Truckenbrodt's Constraints on P-phrases
  i. Every P-phrase has a unique head, x_φ. (inviolable)
  ii. x_φThe head of a P-phrase is the rightmost expression bearing an asterisk on the ω level. (= Align(φ,R,x_φ,R) or Align φ, violable)[13]
  iii. Every lexically headed XP must be contained in a P-phrase. (= Wrap-XP, violable)
  iv. Every lexically headed XP must contain a phrasal stress x_φ. (= Stress-XP, violable)

While these constraints adequately generate what could be considered the default pronunciations of all-new sentences, they do not account for optional accents. This can be seen by considering examples like (13), modified slightly below to eliminate G-marked expressions and with P-phrasing made explicit.

(24)   (*) (*)   (*) (*) (*)   (*)   (*)   (*   )
        JOHN *likes* BEER   JOHN LIKES BEER   JOHN's DANCING   JOHN's *dancing*

Truckenbrodt's analysis generates the first and third structures in (24), but not the second or fourth. The second structure violates Truckenbrodt's Wrap-XP (= (iii)) since the VP is not contained in a single P-phrase, while the fourth violates his Align φ (= (ii)) and Stress-XP (= (iv)) since the head of the P-phrase is on the left and the VP fails to contain a P-phrase level asterisk. While these violations are in principle allowable if all other candidate representations have either equally severe or more severe violations, in the present case this situation does not obtain. The first and third representations satisfy all of the constraints in (i)–(iv), and there are no obvious additional constraints to propose whose violation would balance the first and third cases out with the second and fourth.

---

[12] While adequate to the task of explaining the examples in this paper, the phonological analysis given here is insufficient for handling other problems of accent location. Addressing the inadequacies, however, is not possible within the length limitations of this paper, so I address them instead in a companion paper, Tancredi (2015), where I give a more comprehensive phonological analysis of accent location.

[13] Here and below, ω is the level of prosodic words and φ the level of P-phrases, and $x\varphi$ is the head of a P-phrase.

To overcome the challenges proposed by the sentences in (24), I propose that P-phrasing is based primarily on a lexical difference in metrical phonology between verbs and names. I analyze both names and verbs as prosodic words, i.e. as having an inherent ω-level asterisk. I analyze names, however, as having a lexically specified φ-level asterisk as well. Default pitch accent assignment is the result of constructing P-phrases without any modification to lexically determined asterisks, subject to the following constraints:

(25) Proposed Constraints on P-phrases
    i. Every P-phrase has a unique head. (inviolable)
    ii. P-phrases are at most binary, i.e. they contain at most two asterisks on the ω level. (inviolable)
    iii. A non-head prosodic word is to the left of the head in its P-phrase when possible within the constraints in (i) and (ii).

In addition to these constraints, I assume that the distribution of asterisks can also be modified in two ways. First, any word can be promoted to the head of a P-phrase, i.e. it can have asterisks added at the φ level and, if necessary, at the ω level as well. Second, G-marked phrases get demoted, i.e. they have all φ-level and ω-level asterisks removed.

To see how these constraints apply, consider the four representations in (24). All the words in (24) come with ω-level asterisks (not shown), and the names all have φ-level asterisks as well, as specified in the lexicon. Each P-phrase (indicated with parentheses) has a unique head (indicated by an asterisk) as required by (25.i). Furthermore, no P-phrase has more than two prosodic words, satisfying (25.ii). If no words are promoted to P-phrase heads, (25.iii) dictates the P-phrasing in the first and fourth examples, and blocks placing the subject and verb in a single P-phrase in the first example. The second and third examples result from promoting the verb to a P-phrase head.

Optional accents under the analysis proposed come from the optionality of supplementing lexically determined metrical asterisks with additional φ-level (and if need be ω-level) asterisks. The deaccenting associated with Givenness derives from removal of all ω-level and φ-level asterisks. Since G-marked expressions cannot be optionally accented, under a rule-based phonology the removal of asterisks will have to follow supplementation of asterisks so that any supplementation gets undone. Under an Optimality Theoretic approach, the same effect can be had by ranking deaccenting of G-marked expressions higher than faithfulness for lexically specified metrical asterisks, with faithfulness requiring presence of all such asterisks but not prohibiting addition of extra asterisks. Under either approach, optional accents will be correctly limited to non-G-marked expressions.

6 Application

In this Sect. 1 apply the analysis from Sect. 5 to select data from Sect. 3 not yet covered.

6.1 Given Foci

The remaining cases of given foci are straightforward. The obligatory accent on *Mary* in the QAC case in (5), repeated as (26a), comes from the impossibility of simultaneously analyzing the verb and the object as G-marked, while the AWF case repeated in (26b) is given the same treatment as the Contrast case.

(26) a. Q: Who did Mary's father see? A: *He saw* MARY/#*Mary*.
   b. Mary's father saw many people. However, *he only* HEARD MARY/#*Mary* (& nobody else).

In the examples in (6), G-marking *Mary* is compatible with all other G-marking required by MaximizeG and so is necessary, and its F-marking is compatible with constraints on focus. If we additionally assume that focus is marked when possible, we also account for the residual prominence found in these examples. Their analysis is given below.

(27) a. [John saw Mary and Sue saw Bill]$_1$. Then, [ALICE$_F$ [*saw*$_G$ *HER*$_{F,G}$]$_2$~C$_3$ and$_G$ TOM$_F$ [*saw*$_G$ *HIM*$_{F,G}$]$_3$~C$_2$]~C$_1$

b. John saw Mary. [Who did Bill see?]$_2$ [[BILL *saw*$_G$ *HER*$_{F,G}$] ~ C$_2$]^{14}

c. John saw Mary. In fact, *he* ONLY [*saw HER*$_{F,G}$]~C

6.2 Non-constituent Foci

The phenomenon of non-constituent foci has already been accounted for in its essentials in Sect. 5.1. The phenomenon itself was argued to not be real – what appeared to be multiple foci turned out to be multiple pitch accents assigned within a single focus. We have not yet seen how these multiple pitch accents get assigned in the phonology, however. To do so, consider once again the example from (7a), repeated here.

(7a) A: John fell. B: [MARY TRIPPED him$_G$]$_F$.

Lexically, *Mary* comes with both ω-level and φ-level asterisks, while *tripped* has only an ω-level asterisk. The G-marking on *him* results in its having no asterisks on either prosodic level. If we make no additions to the metrical structure, we predict that the sentence should surface with a single P-phrase and hence a single pitch accent, on the subject *Mary*. This pronunciation is indeed possible, though perhaps not preferred. Recall, though, that we also allow any non-G-marked word to be promoted to a P-phrase head. Promoting the verb *tripped* will result in *tripped* heading a P-phrase including the object *him*, and hence receiving a pitch accent along with the subject *Mary*. We thus predict the accenting in (7a) to be one of two possible pronunciations of this example, which is exactly what we find.

---

[14] The accent on *Bill* in (27b) comes from its not being able to be marked as given when both *saw* and *her* are so marked. It need not be analyzed as a focus or as a topic. Formally, absence of either F- or G-marking identifies *Bill* as discourse new. I take that here to mean that it is new with respect to its givenness antecedent (i.e. the first sentence), not necessarily with respect to the focus antecedent (the second sentence).

7 Conclusion

In this paper, I have examined the analyses of Schwarzschild (1999) and Rooth (1992, 1995) and shown both to be inadequate to the task of accounting for a range of focus and givenness effects. Importantly, not only was each analysis found to be inadequate on its own, but the combination of the two analyses was also shown to be inadequate, since neither analysis accounted for the behavior of given foci or for the range of optional accents observed. Rooth's analysis additionally could only account for connectedness of givenness if focus was required to be as narrow as possible and its domain of interpretation as wide as possible. Finally, Rooth's analysis failed to account for (apparent) non-constituent foci. I then showed that it is possible to account for all of these phenomena by adopting Rooth's semantics of focus and a slightly modified version of Schwarzschild's semantics of givenness while making major adjustments to the syntactic and phonological analyses of givenness and of pitch accent assignment.

The most important adjustment made was to separate the representation of givenness from that of focus in both the syntax and the semantics, a separation argued for independently by Katz and Selkirk (2011). Given foci are analyzed as being simultaneously F-marked and G-marked in the syntax, and as having non-trivial FSVs and GSVs in the semantics. This separation makes it possible to maintain Rooth's analysis of focus, though that analysis is thereby limited to accounting only for focus effects, not givenness effects. Schwarzschild's analysis, on the other hand, required substantive changes. Trivially, the semantics of givenness had to be re-cast to relate to syntactic G-marking rather than F-marking. To highlight the semantic nature of givenness, I accomplished this by proposing Givenness Semantic Values that relate simultaneously to the syntax (through G-marking) and the discourse (through Givenness). Though superficially different from Schwarzschild's analysis, this aspect of the proposal is in essence only a different technical implementation of the same core semantic idea. The syntactic and phonological parts of the proposed analysis, on the other hand, differ from those in Schwarzschild's analysis in ways that are not merely superficial. ProjectG imposes identity of syntactic G-marking between a head and its projections, and plays a central role in accounting for the connectedness of givenness under the proposal presented. Also, pitch accent location is analyzed as related to newness (non-Givenness), not to focus, and the rules generating pitch accents are unrelated to those proposed by Schwarzschild for connecting F-marking with pitch accents. The proposed rules build on lexically specified metrical structure, and allow for optional accents on any discourse new expression. This same analysis also accounts for the appearance of non-constituent foci. That appearance derived from the presence of independent pitch accents on expressions that do not by themselves form a syntactic constituent, a subject and a transitive verb in the case examined in most detail. This pattern of pitch accent location was seen to derive from promotion of the verb to a P-phrase head, an operation that is independent of focus. The apparent non-constituent foci examples could then be analyzed as containing only a single broad focus at the sentence level, with the contained object identified as Given and accents assigned to the subject and verb as one option in the phonology.

To get the analysis to account for certain subcases of given foci, it was necessary to supplement the core analyses of focus and givenness with a separate analysis of contrast. The ad hoc nature of the added proposal clearly constitutes a weakness in the overall analysis that needs to be addressed. Additionally, while the proposal made determines what expressions can and cannot be analyzed as given, and it also determines which expressions must, which can and which cannot bear a pitch accent, it does not determine which expressions get analyzed as foci or what pragmatic effects follow from analyzing an expression as a focus, and neither does it say anything about the pragmatic effects of optional accenting. Filling in these gaps I leave as a task for future research.

References

Beaver, D.: 'Sense and sensitivity: explorations on the interface between focus and meaning', handout, Stanford University (2004)

Chomsky, N.: Deep structure, surface structure, and semantic interpretation. In: Steinberg, D.D., Jakobovits, L.A. (eds.) Semantics: An Interdisciplinary Reader in Philosophy, Linguistics, and Psychology. Cambridge University Press, Cambridge (1971)

Katz, J., Selkirk, E.: Contrastive focus vs. discourse-new: evidence from phonetic prominence in english. Language **87**(4), 771–816 (2011)

Rooth, M.: A theory of focus interpretation. Nat. Lang. Semant. **1**, 75–116 (1992)

Rooth, M.: Focus. In: Lappin, S. (ed.) The Handbook of Contemporary Semantic Theory. Blackwell, London (1995)

Rooth, M.: On the interface principles for intonational focus. In: Galloway, T., Spence, J. (eds.) SALT VI, pp. 202–226. Cornell University, Ithaca (1996)

Schwarzschild, R.: Givenness AVOIDF and other constraints on the placement of accent. Nat. Lang. Semant. **7**, 141–177 (1999)

Selkirk, E.: Phonology and Syntax: The Relation between Sound and Structure. MIT Press, Cambridge (1984)

Selkirk, E.: Sentence prosody: intonation, stress and phrasing. In: Goldsmith, J.A. (ed.) The Handbook of Phonological Theory. Blackwell, London (1996)

Selkirk, E.: Contrastive focus, givenness and the unmarked status of 'discourse-new'. Acta Linguistica Hung. **55**, 331–346 (2008)

Tancredi, C.: Deletion, deaccenting and presupposition, Ph.D. dissertation, MIT (1992)

Tancredi, C.: The phonology of accent. In: Reports of the Keio Institute of Cultural and Linguistic Studies 46 (2015)

Truckenbrodt, H.: Phonological phrases: their relation to syntax, focus and prominence, Ph.D. dissertation, MIT (1995)

Wagner, M.: Focus and givenness: a unified approach. In: Kučerová, I., Neeleman, A. (eds.) Contrasts and Positions in Information Structure, pp. 102–148. Cambridge University Press, Cambridge (2012)

JURISIN 2014

Eighth International Workshop on Juris-Informatics (JURISIN 2014)

Satoshi Tojo[✉]

Japan Advanced Institute of Science and Technology,
1-1 Asahidai, Nomi, Ishikawa 923-1292, Japan
tojo@jaist.ac.jp

Juris-informatics is a new research area which studies legal issues from the perspective of informatics. The purpose of this workshop is to discuss both the fundamental and practical issues among people from the various backgrounds such as law, social science, information and intelligent technology, logic and philosophy, including the conventional "AI and law" area. JURISIN 2014 was held on November 23 and 24, 2014, in association with the Sixth JSAI International Symposia on AI (JSAI-isAI 2014) supported by the Japanese Society for Artificial Intelligence (JSAI).

From submitted papers, we accepted thirteen papers. They cover various topics such as legal reasoning systems, formal argumentation theory, legal text processing, and so on.

As guest speakers, we invited Professor Bart Verheij, University of Groningen, who has been a leading scientist of this field, and he gave a talk on *The Future of Argumentation Technology as guided by the needs of the law*. Also, we have invited Professor Riichiro Mizoguchi, Japan Advanced Institute of Science and Technology, who is a specialist of ontology in AI, and he gave a talk on *Ontology engineering - Theory and practice -*. As this topic concerns the general issue of AI, we have shared this talk with LENLS workshop. In return, LENLS workshop kindly opened a talk of Professor Matthew Stone, Rutgers University, to us on *Logic and probability in grounded semantics*.

A new event in this workshop is Bar Exam Competition. This time, JURISIN invited participation in a legal information extraction and entailment competition. Previous conferences/workshops have not conducted such a shared task on a large legal data collection, so we hoped that the 2014 workshop would establish a major experimental effort in the legal information extraction/retrieval field.

According to discussions of JURISIN 2014, authors revised their papers. The program committee reviewed these revised papers, and selected four papers.

- Tetsuji Goto and Satoshi Tojo: *Classification of Precedents by Modeling Tool for Action and Epistemic State: DEMO*
- Mi-Young Kim, Ying Xu and Randy Goebel: *Legal Question Answering Using Ranking SVM and Syntactic/Semantic Similarity*
- Shruti Gaur, Nguyen Vo, Kazuaki Kashihara and Chitta Baral: *Translating Simple Legal Text to Formal Representations*
- Pimolluck Jirakunkanok, Katsuhiko Sano and Satoshi Tojo: *Analyzing Reliability Change in Legal Case*

© Springer-Verlag Berlin Heidelberg 2015
T. Murata et al. (Eds.): JSAI-isAI 2014 Workshops, LNAI 9067, pp. 225–226, 2015.
DOI: 10.1007/978-3-662-48119-6_16

Acknowledgement. JURISIN 2014 was held in conjunction with JSAI-isAI 2014 supported by JSAI. We thank all staffs of JSAI-isAI 2014 and JSAI for their supports. And also JURISIN 2014 was supported by members of the steering committee, the program committee and advisory committee. We really appreciate their support.

Classification of Precedents by Modeling Tool for Action and Epistemic State: DEMO

Tetsuji Goto[✉] and Satoshi Tojo

School of Information Science, JAIST,
19th Floor, Shinagawa Inter-City Tower A, Konan 2, Minato-ku, Tokyo, Japan
{goto.tetsuji,tojo}@jaist.ac.jp
http://www.jaist.ac.jp/

Abstract. To determine whether the crime is really caused by the defendant, the judge examines the causal relation of each action in the case to an external factor in the Penal Code. In this process, the judgement is greatly influenced by the predictability of results and the awareness about actions. In this paper, we model these predictability or awareness by Dynamic Epistemic Logic (DEL), and thereafter we describe the change of knowledge of the judge by Action Model. For this purpose, we pick up several typical precedents, and classify them from the viewpoints of predictability and awareness. We implement the process of these precedents in the trial on DEMO (Dynamic Epistemic MOdeling) which can specify epistemic models and action models, and we observe the change of the judge's epistemic states during the trial. Based on this observation, we categorize the outputs of DEMO into several patterns.

Keywords: Dynamic epistemic logic · Action model · Model checking · Penal Code

1 Introduction

To determine a criminality specified in the Penal Code, the followings are examined by a judge [15].

1. The defendant's action comes under the external factors defined by the Penal Code (*Actus reus*)[1].
2. There is no justifiable reason to dismiss the illegality[2].
3. There is no justifiable reason to dismiss the responsibility[3].

If these conditions are matched, the criminality is decided. Note that, in this paper, we deal mainly with the process of verifying the correspondence. (We take the position that the external factor includes the intent and the lapse. [17])

---

[1] The guilty acts or typified criminal acts, sometimes called as the objective element of a crime.

[2] A reason that there is no illegality about the act that illegality is usually estimated.

[3] A reason to deny the responsibility of the act that responsibility is accepted as a general rule.

© Springer-Verlag Berlin Heidelberg 2015
T. Murata et al. (Eds.): JSAI-isAI 2014 Workshops, LNAI 9067, pp. 227–243, 2015.
DOI: 10.1007/978-3-662-48119-6_17

In general, as external factos of crime, "Action", "Result" and "Causality" are required. The evaluation of the defendant's action by intent (*Mens rea*)[4] or by lapse[5] is greatly influenced by the awareness about the action by the defendant and the predictability of results. For example, the intent is determined based on the awareness about a fact and the prediction of a result. For lapse, the predictability and the duty to prevent the result are the issue. Of actual crimes, there are so many cases [16] in which a fortuitous event happens between the action by the defendant and the result, or the action based on an uncertain awareness by the defendant causes the criminal result.

It is the commonly acknowledged that each case should be considered separately from other cases, and this attitude makes us difficult to classify the cases systematically. It should be more easy for us to handle the case, if the judge's epistemic states through the trial could be categorized. For example, argumentation frameworks have been studied and applied to judicial reasoning (recent examples are [1, 13, 20].) and these models can compute diagrams.

On the other hand, Dynamic Logic have been applied to describe belief revision ([3, 9]). We have focused on predictability or awareness and need to represent the epistemic states of the judge or the defendant individually. So we try to classify some typical precedents, to represent them by using the DEL (Dynamic Epistemic Logic) [7], because Action Model in DEL can represent the local epistemic states and can update the states by various epistemic actions.

This paper is organized as follows. In Sect. 2, we briefly introduce the DEL and Action Model which are used to describe the precedents later sections. In the following Sect. 3, we define the usage of this language in the context of the judgement of crimes and actually pick up some typical precedents classifying them into 6 cases according to the judgement process. In Sect. 4, these precedents are modeled by using DEL and implemented on DEMO (Dynamic Epistemic MOdeling) software [8] and we observe these outputs and categorize them. Finally, we summarize our contribution.

2   DEL and Action Model

2.1   DEL and Action Model

Knowledge and belief are not static because of the communication between agents. Dynamic Epistemic Logic is an extension of epistemic logic [10] with dynamic operators '[ ]', and $[\pi]\varphi$ is read as "successfully executing program π yields a φ state" [7]. Namely, given a model M and a possible world s,

$$M, s \models [\pi]\varphi$$

iff M is properly changed by the execution of π and as a result φ holds. Public Announcement Logic [18] is an example of DEL where the epistemic action is

---

[4] A guilty mind or an intention to commit a crime.

[5] A failure to take reasonable care when they act by taking account of the potential harm to other people.

only restricted to public announcement. Action Models [2] are used to describe epistemic actions.

Definition 1 (Action model). Let $\mathcal{L}$ be any logical language for given parameters agents A and atoms P. An $S5$ action m kodel M is a structure $\langle S, \sim, \text{pre} \rangle$ where S is a domain of action points and for each agent a, $\sim_a$ is an equivalent relation in S, stating that two states are indistinguishable for a. pre: $S \rightarrow \mathcal{L}$ is a preconditions function that assigns a formula in $\mathcal{L}$ to each $s \in S$. A pointed $S5$ action model is a structure (M, s) with $s \in S$

An epistemic state can be changed by an epistemic action, so the new state after updating is described as a pair of an old world with an action that has taken place in that state. The expression (s, s) indicates that action s is executable in the state s.

$$M, s \models \text{pre(s)}$$

The two factual states are indistinguishable, if the following relation exists, where index a indicates for agent a.

$$(s, \text{s}) \sim_a (t, \text{t}) \text{ iff } s \sim_a t \text{ and } s \sim_a t \text{ (in S5 action model)}$$

Example 1: Read [7]. There are two epistemic states (0/1) where the proposition P is true (p) or false ($\neg p$) respectively and P is true actually. At first Agent a and b didn't know whether the value of P was true or false, so there is a link between 0 and 1 for a and b. This means that they cannot distinguish these states. A letter came to a that told p and a read it and knew that but b couldn't distinguish an action 'p' (a reads a letter which tells p) from an action 'np' (a reads a letter which tells $\neg p$), but b knew that a knew p or $\neg p$. In action model defined as below, this can be interpreted as a relation between these epistemic action points. The new epistemic state after updating by the epistemic action (Read, p) is expressed as the right figure of Fig. 1 where there is no link for agent a between state0 and state1 and the link for agent b remains.

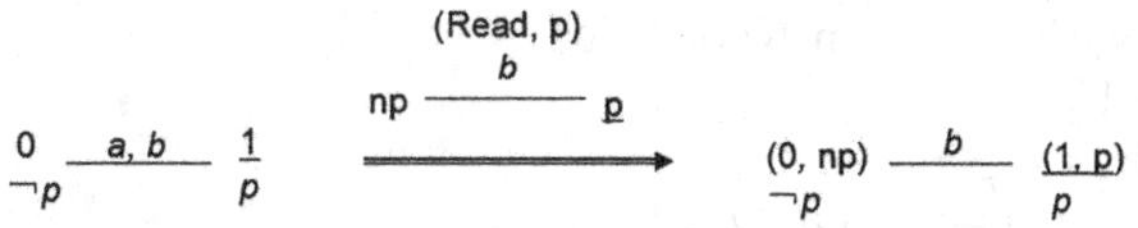

Fig. 1. State transition by an action read

Example 2: MayRead [7]. Agent b has left the table for a while, and when back, suspects a of having read the letter. There are two epistemic states (0/1) where the proposition P is true (p) or false ($\neg p$) respectively and P is true actually. In fact, agent a did not read the letter which tells P is true and doesn't know whether p is or is not. Agent b cannot know the agent a read or did not

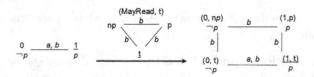

Fig. 2. State transition by an action MayRead

read it so he cannot distinguish three action points, i.e., a reads the letter and it contains p (p), a reads the letter and it contains $\neg p$ ($\neg$p) and a does not read (t), in addition to that he does not know whether p is.

The new epistemic state after updating by the epistemic action (MayRead, t) is expressed as the right figure of Fig. 2 where there is no link for agent a between upper two states which represents the a's action "read the letter", and there is a link for both agent a and b between lower two states which represents that agent a did nothing. The left vertical link represents that agent b cannot distinguish the state where $\neg p$ is and agent a did not read the letter from the state where $\neg p$ is and agent a read the letter containing $\neg p$, so there are two states where $\neg p$ is. Similarly the right vertical link represents that agent b cannot distinguish the state where p is and agent a did not read the letter from the state where p is and agent a read the letter containing p. There are also two p states. There are only four states according to the precondition of each action, i.e., the precondition of action $\neg$p is $\neg p$, p is for p and p or $\neg p$ for t.

Definition 2 (Syntax of Action Model Language). The language of action model logic is the union of the formulas of static epistemic logic and that of epistemic actions.

$$\varphi ::= p \mid \neg\varphi \mid (\varphi \wedge \varphi) \mid K_a\varphi \mid C_B\varphi \mid [\alpha]\varphi$$
$$\alpha ::= (\mathrm{M, s}) \mid (\alpha \cup \alpha)$$

Definition 3 (Semantics of Action Model). The semantics of Action Model can be defined as follows. The first 5 definitions are the same as the logic of Public Announcement with Common Knowledge.

$M, s \models p$ iff $s \in V_p$

$M, s \models \neg\varphi$ iff $M, s \nvDash \varphi$

$M, s \models \varphi \wedge \psi$ iff $M, s \models \varphi$ and $M, s \models \psi$

$M, s \models K_a\varphi$ iff for all $s' \in S : s \sim_a s'$ implies $M, s' \models \varphi$

$M, s \models C_B\varphi$ iff for all $s' \in S : s \sim_B s'$ implies $M, s' \models \varphi$

$M, s \models [\alpha]\varphi$ iff for all $M', s' : (M, s)[\![\alpha]\!](M', s')$ implies $M', s' \models \varphi$

where $[\![\alpha]\!]$ is the subset of domain where the precondition of α is true.

$(M, s)[\![\mathrm{M, s}]\!](M', s')$ iff $M, s \models \mathrm{pre(s)}$ and $(M', s') = (M \otimes \mathrm{M}, (s, \mathrm{s}))$

The updated model $M'(= M \otimes \mathrm{M})$ is a restricted modal product ($\otimes$) of an epistemic model and an action model, which is defined as an structure $\langle S', \sim', V' \rangle$ where $S' = \{(s, \mathrm{s}) \mid s \in S, \mathrm{s} \in \mathrm{S}$ and $M, s \models pre(\mathrm{s})\}$

2.2 DEMO

DEMO [8] is a modeling tool for Dynamic Epistemic Logic and it allows modeling epistemic updates, display of action models, formula evaluation in epistemic models, so DEMO can be used to check semantic intuitions about what goes on in epistemic update situations.

DEMO is programmed in Haskell [14] and imports three modules, List, Char and DPLL. Here List and Char are standard Haskell modules and used to describe the data structure (model). DPLL is a module for propositional reasoning with the Davis, Putnam, Logemann, Loveland procedure [5,6]. And in it's main file, DEMO defines Action Model and Epistemic state as a pointed model and defines the relation between these points.

It receives an input of model definitions (Episteimc Action and Epistemic State) of individual case and it's updatings. It outputs the updated models or the evaluation of propositional formulas. It can also output the files which corresponds to the dot form [12] to represent the graphical images of the updated models.

3   Classification of Precedents

3.1   Handling of Awareness and Predictability in the Penal Code

External Factors Defined in the Penal Code. In general, external factors are roughly divided into subjective and objective ones [17]. (There are also several opposite theories [21] against this, such that both intent and lapse are regarded as the responsibility and should not be included in the external factors.) The objective factor contains action, result and causality between an action and a result. The subjective factors are comprised of intent and lapse.

Intent in the Penal Code. Intent is "an intention to commit a crime". (The Penal Code Article 38 paragraph 1 [19]). At least, awareness of an objective external factor such as an action, a result and prediction for causality are needed. Further, in general, the probability of occurrence or the admittance of the results by the defendant [4] are taken into account by the judge. As a kind of intent, there is an uncertain intent, for example the willful negligence (*dolus eventualis*)[6] is classified as this type.

Lapse in the Penal Code. The lapse is defined in the Penal Code Article 38 paragraph 1 as "The action without awareness to commit a crime, and it is not punishable. However, if there is a special provision in the code, this shall not be applied to." [19]. The lapse is applied in the case where the defendant did not foresee the result which might have been able to predict. Recently the duty of the defendant to avoid criminal results is more emphasized [11].

---

[6] The defendant is uncertain about the realization of crime but knowing that crimes may be implemented and he has accepted it.

Causality. A relationship between an action and a result is called causality. In order to affirm the causality, there should be not only a conditional relationship (without that there should not be this) between an action and a result, but also it is required to be regarded reasonable from the experience of the social life of ordinary people. (Legally sufficient cause [21]).

Evaluation Process of Correspondence to External Factors. To determine whether the defendant's action conforms to the external factors, the objective and the subjective factors are examined [17] (Fig. 3).

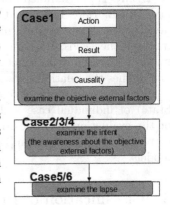

1. Awareness about the objective facts constituting the offense. At first, the defendant's recognition about the objective external factors such as an action, a result and a causality between them is verified. If the recognition is different from the actual fact, it is considered that a mistake in interpretation of facts [21] has occurred.

2. Awareness about subjective factors. Secondly, the intent (the awareness, the prediction and

Fig. 3. Evaluation process

the possibility of occurrence of the result, etc.) and the lapse (the breach of the duty of predicting the result, the breach of the duty of avoiding the result) are examined.

3.2 Description of the Issues in the Precedents by DEL

From the above, in order to represent a process of deciding a judgement of precedents dealt in this paper, the followings descriptions are needed.

- Description of facts (action, result, causality) constituting an offense
- Description of awareness about the fact and intention
- Description of predictability

So we define issues in the process of a trial as follows.

Action α_a: Fan action point of Action Model by agent a.
Result φ: Fa proposition of a possible world (epistemic state).
Causality $[\alpha]\varphi$: Fa relation between an action and a state in DEL, if there is a causality between an action and a result.
Predictability: The predictability is described as a possibility of link cut between epistemic states. We describe that agent a cannot predicate the state at the state s as $a's$ link between state s and state t.
Possibility of avoidance: This is represented as a link cut between the states of preconditions for the alternative action in Action Model of DEL.

Table 1. Classification of precedents

Precedents	Outline	Classified Cases
Accused of injury resulting in death case (No. A35, 2003)[a]	Four people assaulted the victim repeatedly. He ran away into the highway nearby and was run over by a car and died. The judge admitted a causal relationship between the assault and the death.	**Case1** The intervention of the unexpected action by the victim or the third person.
Accused of indecent document sale (No. A1713, 1953)	The defendant who translated and published the "Lady Chatterley's Lover" without knowing the legal meaning of "obscenity", were charged with selling obscene document	**Case2** (Normative) The recognition of meaning (legal concept)
Accused of murder case (No. RE 517, 1923)[b]	The defendant tried to murder the victim by strangulation and then made an attempt to conceal the crime by burying him in the sand of the coast, but he died because of sand absorption. The judge determined there was an intention because of the causal relationship.	**Case3** The mistake of a process or causality
Dealing with stolen goods case (No. RE238, 1947)[c]	The defendant bought stolen clothes without knowing that they were originally stolen. The judge applied the crime of illegal acquisition of the stolen goods.	**Case4** Willful negligence
Hokkaido University electric scalpel case (No. U219, 1974)	The nurse had mistakenly connected the cable of the scalpel, and the patient's right foot was damaged. The judge ruled professional negligence resulting in bodily injury.	**Case5** (lapse) Predictability (delinquency of duty of care)
The use of HIV contaminated blood in Teikyo University hospital (No. WA1879, 1996)	The doctor used unheated blood and the patient becomes HIV positive. The judge applied the innocence to the doctor.	**Case6** (lapse) a delinquency of duty of avoiding the result

[a]Similar Case: Accused of unlawful arrest and illegal confinement resulting in death case (No. A2901, 2005).The defendant imprisoned the victim in a rear trunk of a passenger car and a car driven by a third person bumped into the rear trunk and the victim died.

[b]Similar Case: Accused of murder and fraud (No. A1625, 2003) The defendant took the consciousness of the victim with chloroform and tried to murder him by drowning. The cause of his death is not clear either chloroform or drowning.

[c]Similar Case: The Stimulant Drug Control Law violations (No. A1038, 1998) The defendant carried stimulant drug without knowing the fact.

Intent: This is represented as an awareness about the prediction, so there is no link for the agent between an actual state and states where the precondition is false.

Lapse: Lapse is described as no intent by the defendant, the predictability and the possibility of avoidance from the view point of an usual person or a judge.

3.3 Classification of Precedents

According to the evaluation process written in the previous section, there are three main points where predictability or awareness is the issue in the judgement. The first point is the awareness of the objective facts which includes the problem of intervention of unexpected actions and a mistake in recognizing causality. The second is the intent of the defendant and the problem of willful negligence occurs at this point. The third is the lapse which includes the problem of predictability and possibility of avoiding results. From precedents often cited [16], some typical examples are listed below and they are classified into 6 cases from the point of view of awareness and predictability (Table 1) and these case are mapped to the three points above (Fig. 3).

4   Implementation and Result

The model checking for these cases is implemented by using DEMO [8]. In this implementation, the accessibility relations are restricted to S5 relation.

Updating the states is executed in two steps. The first step is by the defendant's actions during a crime (or the start of a trial where the judge has no prejudice.) and the second step is by the judge's actions from the start of a trial to the final judgement. In updating, the epistemic actions as follows are used.

message which notifies propositions to particular persons and the others may or may not know whether the message has reached. This corresponds to the situation that a prosecutor gives new evidence and the judge examines this evidence and the others don't know the determination of the judge's mind.

public which is the same as public announcement. This is used to describe the common sense which influence the criminal actions.

For example, (message b p) indicates that the judge knows p but the defendant may or may not know what is going on. (Fig. 4) The product of two states and two action points consists four states after updating, but there are only three states according to the preconditions of each action.

When we define the propositions, we take "p" as the primary (or the external factor) and "$q, r, ...$" as the subsidiary (the extraneous factor).

4.1   Case1: The Intervention of the Unexpected Action

The defendant committed the assault repeatedly. So the victim ran away into the highway nearby, was run over by a car and died. In this case, the defendant is on the crime of inflicting injury and the cause of the victim's death is an issue.

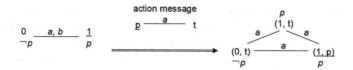

Fig. 4. Update by message

- Proposition p: The victim is injured.
- Proposition q: The victim is driven to the emotional corner.
- Agent a: The defendant (in the following all cases).
- Agent b: The judge (in the following all cases).

We set the proposition p (external factor) as the injury and q (substantial element) as the cause of a successive action. Four epistemic states are set where the valuation depends on these two propositions and it's true/false binary values respectively.

As the crime proceeds, the defendant takes actions. We interprets these non-epistemic actions as corresponding (whose preconditions are equivalent) epistemic actions which are points of Action Model. In this case there are two non-epistemic actions.

- "injure the victim" whose precondition is $\neg p$ (for injuring)
- "run into highway" whose precondition is q (for running into the highway)

It is necessary to update the epistemic states by two corresponding epistemic actions concerning these two non-epistemic actions described above. The defendant can recognize his own action of injuring the victim, so we update the states by the action of sending a message to himself whose precondition is the same as of the precondition of his non-epistemic action "injure". Figure 5 is an image of updating by this action "message $a\ \neg p$", where the $\neg p$ states (the states agent a believes $\neg p$) are copied for agent b because he cannot distinguish these states. These added states and links are represented as shaded states and doted lines in Fig. 5.

On the other hand, the victim's action "run into the highway" cannot be predicted by the defendant, so no additional information is sent to himself and the links between states where q is true or false($\neg q$) remain unchanged. That is the state at the time of the crime has happened. A part of the program code is as follows.

```
intervent = initE[P 0, Q 0] --defines initial epistemic states
initInt = upds intervent [message a (Neg p)] --updating
```

Then we updates this epistemic state by the action of the judge. This process is regarded as the proceeding of the trial at the court. The actions for the final state (Fig. 7) are as follows.

message b (**Neg** p) The judge knows the defendant injured the victim.
The result of updating by this action is shown in Fig. 6. The shaded states and doted lines describe the added ones by this updating.

message b (**Disj** $[p, q]$) The judge is informed that the victim was cornered by the defendant's action. (If there is a causal relation, the precondition of the action (assault) implies the result state under the occurrence of this defendant's action. And an implication can be represented by the disjunction. $\neg p \rightarrow q \Leftrightarrow p \vee q$ In the following cases, an implication is translated to a disjunction.)

message b (**K** a (**Neg** p)) The judge is informed that the defendant knows he injured the victim.

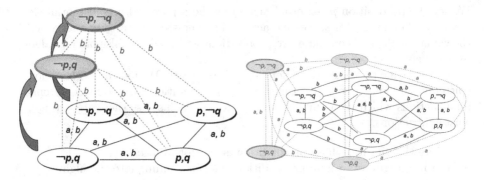

Fig. 5. Case1 updating by message $a \ \neg p$

Fig. 6. Case1 updating by message $b \ \neg p$

4.2   Case2: Recognition of the Meaning (Normative Case)

The defendant who translated and published the novel "Lady Chatterley's Lover" were charged with obscene document sale.
The propositions and the states are as follows.

– Proposition p: The novel is obscene (legal meaning). q: The public order is violated.
– Four initial states (0..3), 0:$\neg p, \neg q$, 1:$p, \neg q$, 2:$\neg p, q$, 3:p, q:

The defendant's actions which update the epistemic states through the crime

public (**Disj** [**Neg** q, p]) "It is known that the violation of public order is a crime."

message a q "The defendant knows publishing this novel violates the public order."

The judge's actions which update the epistemic states during the trial

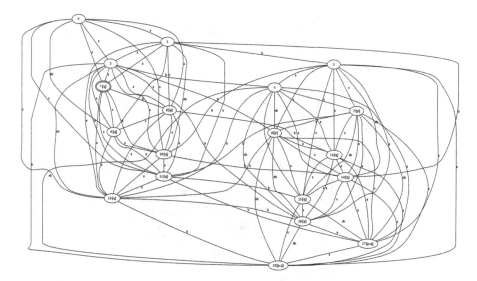

Fig. 7. Case1 the final state after the trial

message b p "The judge is informed that the novel is obscene in the legal
   meaning."
message b (**K** a q) "The judge is informed that the defendant knows the violation of the public order."

After these updates, the judge knows that the defendant knows the illegality
and he can inflict the punishment based on the defendant's intention. This can
be checked in DEMO as follows.

```
*DEMO> isTrue (upds initMeaning [message b p, message b (K a q)])
(K b (K a p))
True
```

The final state becomes like Fig. 8. The shaded circles (states) indicates the
actual state and states which can be reached from the actual state by the agents'
links. These are examined for classification later (in Subsect. 4.7).

4.3   Case3: Mistake of the Causality

The defendant tried to murder the victim by strangulation and then made an
attempt to conceal the crime by burying him in the sand of the coast, but he
did not die at that time and died because of sand absorption.
   The propositions and the states are as follows.

- Proposition p: The victim is dead. q: The victim survives. the defendant's
   strangulation.

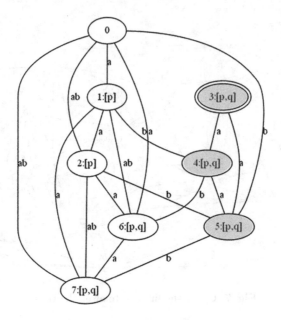

Fig. 8. Case2 the final state after the trial

The defendant's actions which update the epistemic states through the crime

message a (**Neg** p) "The defendant is conscious about his own action (strangulation)." (But the defendant does not know the victim survives.)

The judge's actions which update the epistemic states during the trial

message b (**Neg** p) "The judge knows the defendant strangulated the victim."
message b (**Disj** $[p, q]$) "The judge is informed that the concealment is a part of the process of strangulation."
message b (**K** a (**Neg** p)) "The judge is informed that the defendant knows he strangulated the victim."

The final state is the same as Case1.

4.4 Case4: The Willful Negligence

The defendant bought stolen clothes without knowing that they were originally stolen. He is accused of paid acquisition of stolen goods.
    The propositions and the states are as follows.

- Proposition p: The goods are stolen.   q: The probability of being stolen is high.
- State four states (0..3) $0:\neg p, \neg q$, $1:p, \neg q$, $2:\neg p, q$, $3:p, q$ Agent a, b can not distinguish these states.

The defendant's actions which update the epistemic states through the crime

public (Disj [Neg q, p**])** "It is known that the probability is high then the goods are perhaps stolen."

message a q "The defendant is informed that the probability of being stolen is high."

The judge's actions which update the epistemic states during the trial

message b p "The judge is informed that the goods are stolen."

message b (**K** a q) "The judge is informed that the defendant knows the probability is high."

The result is the same as case2. After these updates, the judge knows that the defendant knows the illegality, which deserves a punishment for the defendant's intention. This can be checked by DEMO, too.

```
*DEMO> isTrue (upds initWilneg [message b p, message b (K a q)])
(K b (K a p))
True
```

4.5   Case5: The Delinquency of Duty of Care

The doctor and the nurse made a mistake of connecting the cable of the scalpel incorrectly, and the patient's right foot below knee was damaged and resulted in an amputation. The propositions and the states are as follows.

- Proposition p: The patient is injured. q: The wrong connection highly tends to result in injury. r: The cables are connected wrongly.
- State eight states (0..7): Respectively binary values of P, Q, R.

The defendant's actions which update the epistemic states through the crime

public (Disj [Neg q, **Neg** r, p**])** "It is common sense that if the cable is connected wrongly and the wrong connection tends to result in injury, then the patient is damaged."

The judge's actions which update the epistemic states during the trial.

message b p "The judge knows the patient is injured."
message b q "The judge is informed that the probability is high."
message b r "The judge knows the cable is connected wrongly."

4.6   Case6: The Delinquency of Duty of Avoidance

The doctor used an unheated blood and the patient becomes HIV positive. The doctor is innocent.

The propositions and the states are as follows.

- Proposition p: The medicine is infected with HIV. q: The probability of infection with HIV is high. r: Cryoprecipitate is better than unheated blood for the patient.
- State eight states (0..7), binary values for P, Q R respectively.

The defendant's actions which update the epistemic states through the crime

public (Disj [Neg q, r**])** "It is common sense that if the probability is high, the doctor should give the patient cryoprecipitateis."

message a **(Neg** q**)** "The defendant is informed that the probability of infection isn't high."

message a **(Neg (Neg** r**))** "The defendant doesn't know it isn't better and easy to give cryoprecipitate than unheated blood." (The opposite of the precedent, because the defendant become innocent in the precedent.)

The judge's actions which update the epistemic states during the trial

message b p "The judge knows the patient is infected by HIV."

message b **(Neg** q**)** "The judge knows the probability isn't high."

message b **(Neg** r**)** "The judge is informed that it isn't better and easy to give cryoprecipitate than unheated blood."

The final state after the trial. Model is consists of 32 states and the actual state is one where the defendant cannot distinguish p from $\neg p$ and r from $\neg r$. (he has the possibility of taking the other action.). The judge can distinguish p, q, r.

4.7   Patterns of the Final States

The final states of six cases after updating can be categorized into three graphical patterns according to the actual state and the links from this state.

- Intentional case (Case1,2,3,4)

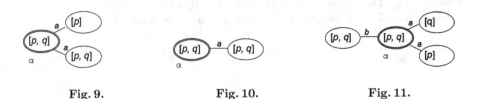

Fig. 9.   Fig. 10.   Fig. 11.

- The agent b can distinguish an actual world and there is no link between the value of primary proposition p and $\neg p$ for the agent a. For the subsidiary proposition q, there is a link between states where q or $\neg q$ for agent a. (Case1,3) (Fig. 9)
- The agent b can distinguish an actual world and for the agent a, there is no link between the states where both values of main proposition p and the subsidiary proposition q respectively. (Case2,4) (Fig. 10)
- Negligent (Lapse) case (Case5,6)
  - The agent b can access to worlds where the primary proposition is unique. But p is not decided for the agent a (the defendant) from $b's$ states. The subsidiary propositions q/r are not decided for a too. (Case5,6) (Fig. 11)

4.8   Legal Interpretation of Patterns

For the cases where the crime is committed by the defendant intentionally, the final state can be summarized to the first two patterns mentioned in the previous subsection where p can be decided (distinguishable from the $\neg p$ states).

- The judge knows the defendant's knowledge about his intention (p)
- The judge can distinguish an actual world.
- The defendant can distinguish (knows) the intention about p.

The first pattern is concerning the judge's awareness about facts (Case1,3) where the judge can know the defendant's intention about p directly from him and the subsidiary elements are examined by the judge.

The second is concerning the defendant's awareness about the meaning of his actions (Case2,4) where the judge can find the defendant's intention based on the common sense.

On the other hand, the cases where the crime is committed by the defendant's lapse, the final state can be summarized to the third pattern in which the judge knows that the defendant cannot distinguish p from $\neg p$. (Both worlds (p or $\neg p$) are reachable by the defendant's relation.) The subsidiary element (q or r) is not distinguishable for the defendant too. But the judge thinks p is distinguishable from $\neg p$ based on q or r(Case5,6).

5   Conclusion and Further Directions

5.1   Conclusion

We have examined the process of making judgement according to the Penal Code and found that awareness and predictability are the main factors to decide the correspondence of the defendant's action to the external factors. We picked up the typical six cases and verified these claims in Action Model.

In Action Model the predictability and the awareness in the precedents can be regarded as the link between epistemic states. If the states or the action is predictable from a state, there is no link between the two states.

These process of precedents can be reproduced by using DEMO. In this model, the epistemic states are updated by epistemic actions which simulates the change of the judge's epistemic state. Finally these result states after updating can be categorized into three patterns focusing on the actual state and it's links of the defendant and the judge from this actual state. For some typical and well cited precedents, the output of DEMO can be interpreted by adopting these patterns mentioned previously and these judgements can be checked.

5.2 Further Directions

To describe lapse or willful negligence, it is needed to employ the probability regarding to the predictability, and in this paper the propositions about probability are temporarily used. But the border of intent and lapse should be continuous. To describe that, the link between states should be directional and states are prioritized. And related to this, a model would require the defeasible reasoning.

In implementing these precedents, by the number of the states and the points of actions, the computational complexity increased rapidly, as we can find in Fig. 7. We can easily guess that if we deal with actual cases, as the number of factors is much larger, the link between epistemic states would be too complicated to be visible. Some of those actual cases, however, may be the combination of entangled predictability and awareness.Thus, if we could unravel this entanglement by human hand, we can reduce the complexity of actual cases to the tractable size and thus our classification may become feasible. Now, our contribution to real cases is summarized as follows; we may reduce the complication of an actual case to a tractable size in finding similarity to typical fundamental cases, or in other words, our results of six cases may serve as target analyses from the viewpoint of actual cases.

References

1. Araszkiewicz, M., Savelka, J.: Refined coherence as constraint satisfaction framework for representing judicial reasoning. In: Legal Knowledge and Information Systems, vol. 250, pp. 1–10 (2012)
2. Baltag, A., Moss, L.S.: Logics for epistemic program. Synthese **139**(2), 165–224 (2004)
3. van Benthem, J., van Eijck, J., Kooi, B.: Logics of communication and change. In: Information and Computation, vol. 204, pp. 1620–1662. Elsevier (2006)
4. Dandou, S.: The criminal law essentials, general remarks, 3rd edit. Soubunsya (1990) (in Japanese)
5. Davis, M., Logemann, G., Loveland, D.: A machine program for theorem proving. Commun. ACM **5**(7), 394–397 (1962)
6. Davis, M., Putnam, H.: A computing procedure for quantification theory. J. ACM **3**(7), 201–215 (1960)
7. van Ditmarsch, H., van der Hoek, W., Kooi, B.: Dynamic Epistemic Logic. Springer, Amsterdam (2008)

8. van Eijck, J.: Dynamic episteimc modeling: Technical Report. Centrum voor Wiskunde en Informatica, Amsterdam (2004)
9. van Eijck, J., Wang, Y.: Propositional dynamic logic as a logic of belief revision. In: Hodges, W., de Queiroz, R. (eds.) Logic, Language, Information and Computation. LNCS (LNAI), vol. 5110, pp. 136–148. Springer, Heidelberg (2008)
10. Fagin, R., Halpern, J.Y., Vardi, M.Y.: Reasoning About Knowledge. MIT Press, Cambridge (1995)
11. Fujiki, H. (ed.): Criminal lapse, a dispute on the old and new lapse theory. Gakuyoushobou (1975) (in Japanese)
12. Koutsofios, E., North, S.: Drawing graphs with dot. http://www.research.att.com/north/graphviz/
13. Lynch, C., Ashley, K.D., Pinkwart, N., Aleven, V.: Argument diagramming and diagnostic reliability. In: Proceedings of the JURIX 2009, pp. 106–115 (2009)
14. Jones, S.P. (ed.): Haskell 98 Language and Libraries (2002). http://www.haskell.org/onlinereport/
15. Mayer, M.E.: Der allgemeine teil des deutsch strafrechts, 2. C. Winter, Heidelberg (1923)
16. Nishida, N., Yamaguchi, A., Saeki, H.: The 100 selected precedents of the Penal Code 1, 6th edit. Yuhikaku (2008) (in Japanese)
17. Oya, M.: Lecture note on criminal law: general part. Seibundo (2007) (in Japanese)
18. Plaza, J.A.: Logics of public communications. In: Proceedings of the 4th International Symposium on Methodologies for Intelligent Systems, pp. 201–216 (1989)
19. The Penal Code. http://www.japaneselawtranslation.go.jp/law/detail/?id=1960
20. Wyner, A.Z., Bench-Capon, T.J.M., Atkinson, K.M.: Towards formalising argumentation about legal cases. In: Proceedings of the 13th International Conference on Artificial Intelligence and Law, pp. 1–10 (2011)
21. Yamaguchi, A.: Criminal law: General part. Yuhikaku (2006) (in Japanese)

Legal Question Answering
Using Ranking SVM
and Syntactic/Semantic Similarity

Mi-Young Kim$^{(\boxtimes)}$, Ying Xu, and Randy Goebel

Department of Computing Science, University of Alberta, Edmonton, Canada
{miyoung2,yx2,rgoebel}@ualberta.ca

Abstract. We describe a legal question answering system which combines legal information retrieval and textual entailment. We have evaluated our system using the data from the first competition on legal information extraction/entailment (COLIEE) 2014. The competition focuses on two aspects of legal information processing related to answering yes/no questions from Japanese legal bar exams. The shared task consists of two phases: legal ad hoc information retrieval and textual entailment. The first phase requires the identification of Japan civil law articles relevant to a legal bar exam query. We have implemented two unsupervised baseline models (tf-idf and Latent Dirichlet Allocation (LDA)-based Information Retrieval (IR)), and a supervised model, Ranking SVM, for the task. The features of the model are a set of words, and scores of an article based on the corresponding baseline models. The results show that the Ranking SVM model nearly doubles the Mean Average Precision compared with both baseline models. The second phase is to answer "Yes" or "No" to previously unseen queries, by comparing the meanings of queries with relevant articles. The features used for phase two are syntactic/semantic similarities and identification of negation/antonym relations. The results show that our method, combined with rule-based model and the unsupervised model, outperforms the SVM-based supervised model.

Keywords: Legal text mining · Question answering · Recognizing textual entailment · Information retrieval · Ranking SVM · Latent dirichlet allocation (LDA)

1 Task Description

The Competition on Legal Information Extraction/Entailment (COLIEE) 2014 focuses on two aspects of legal information processing related to answering yes/no questions from legal bar exams: Legal document retrieval (phase 1) and Yes/No Question answering for legal queries (phase 2).

Phase 1 is an ad hoc information retrieval (IR) task. The goal is to retrieve relevant Japan civil law articles that are related to a question in legal bar exams. Here **relevant** means, based on the articles, lawyers are able to infer the question's answer.

We approach this problem with two unsupervised models based on statistical information. One is the tf-idf model [1], i.e. term frequency-inverse document frequency. The relevance between a query and a document depends on their intersection

© Springer-Verlag Berlin Heidelberg 2015
T. Murata et al. (Eds.): JSAI-isAI 2014 Workshops, LNAI 9067, pp. 244–258, 2015.
DOI: 10.1007/978-3-662-48119-6_18

word set. The importance of words is measured with a function of term frequency and document frequency as parameters.

The other unsupervised model is a topic model-based IR system. The topic model we employ is the Latent Dirichlet Allocation (LDA) model [2]. The LDA model assumes that there is a set of latent topics for a corpus. A document can have several topics, and words in the document are assumed to be generated based on the topic distribution of such a topic model. The more similar two documents are, the more similar their topic distributions will be. While the tf-idf model considers only words that are both in the query and the document, the LDA based IR model also considers words that are in the query but not in the document. For example, if "seller" is only in the query and the document is about trading and contains words such as "buy," "customer," etc., then the probability that the document generates "seller" will be high.

In addition to implementing both models for the task, we also incorporate them into a Ranking SVM model [3]. That model is used to re-rank documents that are retrieved by the tf-idf model. The model's features include lexical words, and both tf-idf score and LDA model score of a document for a given query. The intuition is that, by including lexical words, the SVM model can re-assign weights to lexical words based on the training instances instead of solely on statistical information. By including both statistical models' scores, the SVM model can take advantage of both models.

Our experiments for phase 1 show that all features contribute to the improvement of IR results. The Ranking SVM model nearly doubles the Mean Average Precision (MAP) compared with both baseline models.

The goal of phase 2 is to construct Yes/No question answering systems for legal queries, by entailment from the relevant articles. The task investigates the performance of systems that answer "Y" or "N" to previously unseen queries, by comparing approximations to the semantics of queries and relevant articles.

Our system uses features that depend on the syntactic structure, the presence of negation, and the semantic similarities of the words. The combined model with rules and unsupervised learning model shows reasonable performance, which improves the baseline system, and outperforms an SVM-based supervised machine learning model.

2 Phase 1: Legal Information Retrieval

2.1 IR Models

In this section, we introduce our tf-idf model, our LDA-based IR model, and our Ranking SVM model. Queries and articles are all tokenized and parsed by the Stanford NLP tool. For the IR task, the similarity of a query and an article is based on the terms within them. Our terms can be a word or a dependency linked word bigram.

2.1.1 Tf-Idf
One of our baseline models is a tf-idf model implemented in Lucene, an open source IR system.[1]

---

[1] Lucene can be downloaded from http://lucene.apache.org/core/.

The simplified version of Lucene's similarity score of an article to a query is:

$$tf\text{-}idf(Q,A) = \sum_t \left[\sqrt{tf(t,A)} \times (1 + \log(idf(t)))^2 \right] \tag{1}$$

The score $tf\text{-}idf(Q,A)$ is a measure which computes the relevance between a query Q and an article A. First, for every term t in the query A, we compute $tf(t,A)$, and $idf(t)$. The score $tf(t,A)$ is the term frequency of t in the article A, and $idf(t)$ is the inverse document frequency of the term t, which is the number of articles that contain t. After some normalization process in the Lucene package, we multiply $tf(t,A)$ and $idf(t)$, and then we compute the sum of these multiplication scores for all terms t in the query A. This summation result is $tf\text{-}idf(Q,A)$. The bigger $tf\text{-}idf(Q,A)$ is, the more relevant between the query Q and the article A. The real version has some normalized parameters in terms of an article's length to alleviate the functions biased towards long documents.

2.1.2 LDA-Based IR

Our LDA-based IR model was first proposed in [4]. For more details about LDA model please refer to [2]. The score function is:

$$LDA(Q,A) = \prod_{t \in Q} P(t|A) \tag{2}$$

$P(t|A)$ is the probability of the term t given the article A, specified as:

$$P(t|A) = \sum_{z \in K} P(t|z) \times P(z|A) \tag{3}$$

where z is a latent topic in topic set K, $P(t|z)$ is the probability of the term t given topic z, and $P(z|A)$ is the probability of the topic z given the article A. The two probabilities are from an LDA-model trained with the Mallet package.[2]

2.1.3 Ranking SVM

The Ranking SVM model was proposed by [3]. The model ranks the documents based on user's click through data. Given the feature vector of a training instance, i.e. a retrieved article set given a query, denoted by $\Phi(Q,Ai)$, the model tries to find a ranking that satisfies constraints:

$$\Phi(Q,A_i) \rangle \Phi(Q,A_j) \tag{4}$$

where A_i is a relevant article for the query Q, while A_j is less relevant.

We incorporate four types of features.

– Lexical words: the lemmatized normal form of surface structure of words in both the retrieved article and the query. In the conversion to the SVM's instance

---

[2] Mallet is a machine learning software package. It can be downloaded from http://mallet.cs.umass. edu/.

representation, this type is converted into binary features whose values are one or zero, i.e. if a word exists in the intersection word set or not.
- Dependency pairs: word pairs that are linked by a dependency link. The intuition is that, compared with the bag of words information, syntactic information should improve the capture of salient semantic content. Dependency parse features have been used in many NLP tasks, and improved IR performance [5]. This feature type is also converted into binary values.
- TF-IDF score (Sect. 2.1.1).
- LDA-based IR score (Sect. 2.1.2).

2.2   Experiments

The legal IR task has several sets of queries with the Japan civil law articles as documents (755 articles in total). Here follows one example of the query and a corresponding relevant article.

Question: A person who made a manifestation of intention which was induced by duress emanated from a third party may rescind such manifestation of intention on the basis of duress, only if the other party knew or was negligent of such fact.

Related Article: (Fraud or Duress) Article 96 (1) Manifestation of intention which is induced by any fraud or duress may be rescinded. (2) In cases any third party commits any fraud inducing any person to make a manifestation of intention to the other party, such manifestation of intention may be rescinded only if the other party knew such fact. (3) The rescission of the manifestation of intention induced by the fraud pursuant to the provision of the preceding two paragraphs may not be asserted against a third party without knowledge.

Before the final test set was released, we received 4 sets of queries for a dry run. The 4 sets of data include 179 queries, and 223 relevant articles (average 1.25 articles per query). We used the corresponding 4-fold leave-one-out cross validation evaluation. The metrics for measuring our IR models is Mean Average Precision (MAP):

$$MAP(Q) = \frac{1}{|Q|} \sum_{q \in Q} \frac{1}{m} \sum_{k \in (1,m)} precision(R_k) \tag{5}$$

where Q is the set of queries, and m is the number of retrieved articles. R_k is the set of ranked retrieval results from the top until the k-th article. In the following experiments, we set two values of m, which are 5 and 10, for all queries, corresponding to the column MAP@5 and MAP@10 in Table 1.

The following experiments compare different IR models on the legal IR task, verifying whether the ensemble SVM-Ranking model improves the IR performance.

The LDA model is trained on the whole Japan civil law dataset. The parameters such as number of training iterations and number of topics are set according to the 4-fold cross validation IR performance. SVM's parameters are in a similar manner. Given the top 20 articles returned by the tf-idf model, the SVM model extracts features for every article and trains according to the order that correct articles are ranked higher than incorrect ones. The LDA values used as a feature in the SVM models are from the best LDA-based IR system. Table 1 presents the results of different models. The result shows that including dependency information slightly improves IR. The ranking SVM almost doubles the MAP. It does learn a better term weight for IR systems. Including the baseline tf-idf scores as features slightly improves the performance, and including LDA-based score further improves the performance.

Table 2 shows the IR result on test data. This result was obtained by using the model Id 6. We did not use the model Id 8 because of the relatively low performance gain after adding LDA information. The test data consists of 47 questions and 51 corresponding articles (average 1.09 relevant articles per question). As shown in Table 2, our system produced 61 relevant articles, and precision of 60.66 %, recall of 72.55 %, and F-measure of 66.07 %. This performance was ranked No. 1 for the phase 1 of the competition.

Table 1. IR results on dry run data with different models.

Id	Models	MAP@10	MAP@5
1	tf-idf with lemma	0.108	0.154
2	LDA-based	0.105	0.147
3	tf-idf with lemma and dependency pair	0.111	0.158
4	SVM-ranking with lemma	0.199	0.287
5	SVM-ranking with lemma and dependency pair	0.204	0.294
6	model 5 plus tf-idf score	0.211	0.306
7	model 5 plus LDA-based score	0.207	0.298
8	model 5 plus LDA-based score plus tf-idf score	**0.214**	**0.312**

Table 2. IR results on test data using the model ID. 6

Participant ID	Performance on phase one	
Alberta-KXG	* The number of submitted articles: 61	
	* The number of correctly submitted articles: 37	
	Precision	0.6066
	Recall	0.7255
	F-measure	0.6607

2.3   Discussion

The baseline tf-idf model's performance confirms that legal case information retrieval is a difficult task. One problem is language variability, i.e., different expressions convey the same meaning. For explanation purposes, consider the following example:

Question: In a demand for compensation based on delayed performance of loan claim, if person Y gives proof that the delayed performance is not based on the reasons attributable to him or herself, person Y shall be relieved of the liability.

Related Article: (Special Provisions for Monetary Debt) Article 419 (1) The amount of the damages for failure to perform any obligation for the delivery of any money shall be determined with reference to the statutory interest rate; provided, however, that, in cases the agreed interest rate exceeds the statutory interest rate, the agreed interest rate shall prevail. (2) The obligee shall not be required to prove his/her damages with respect to the damages set forth in the preceding paragraph. (3) The obligor may not raise the defense of force majeure with respect to the damages referred to in paragraph 1.

There is no common content word between the question and the article. An expert can consider several alignments to help decide that the two paragraphs are related: "force majeure'" can be aligned to the negative of "reasons attributable to him or herself;" "agreed interest rate shall prevail" can be aligned to "person Y shall be relieved of the liability;" and "failure to perform" can be aligned to "delay;" etc.

From this example, we can see the difficulty of the task. It cannot solely depend on lexical information, but also take semantics into consideration. Any approach to address this problem will require at least a larger corpus from which to build improved semantic models.

3   Phase 2: Answering 'Yes'/'No' Questions Based on Textual Entailment

Our system combines different kinds of syntactic and semantic information to predict textual entailment. We exploit syntactic/semantic similarity features, and negation. For negation and antonym patterns, we use Kim et al. [6] 's approach.

3.1   Our System

3.1.1   Condition and Conclusion Detection

Previous Textual Entailment systems compute lexical or syntactic similarities between two whole sentences [7–10]. However, the sentences in this task are more complex; for example they often contain components like condition (premise), conclusion, and exceptional case description. Even when two sentences have similar word occurrences, if condition and conclusion are placed in a different order, or if a condition part matches an exceptional case, then the result of textual entailment should be 'no.' Therefore, we first need to identify condition, conclusion, and exceptional sentences in civil law articles, before computing similarity.

We also exploit keywords indicating a condition within dry run data. The keywords are as follows: "in case(s)," "if," "unless," "with respect to," "when," and comma. The sentence which explain exceptional cases also have keywords such as "provided," and "not apply," and those sentences typically appear in the last position in an article.

We can split the articles into three parts using these keywords. Condition and conclusion are in the first sentence, but an exceptional case is explained in the second sentence. There also exist articles that do not mention exceptional cases.

$$conclusion := segment_{last}(first\ sentence,\ keyword_{condition}),$$

$$condition := Concatenate(segment_i(first\ sentence,\ keyword_{condition})),$$

$$exception := second\ sentence\ which\ includes\ keyword_{exception}$$

Condition is the concatenation of segments that include the keywords for conditions in the first sentence. Conclusion is the last segment in the first sentence. An exceptional case is the second sentence that includes the keywords for exceptional cases.

For example,

<Civil Law Article 295-1 > : If a possessor of a Thing belonging to another person has a claim that has arisen with respect to that Thing, he/she may retain that thing until that claim is satisfied. Provided, however, that this shall not apply if such claim has not yet fallen due.

(1) Condition=> If a possessor of a Thing belonging to another person has a claim that has arisen with respect to that Thing,

(2) Conclusion=> he/she may retain that thing until that claim is satisfied.

(3)Exception=> Provided, however, that this shall not apply if such claim has not yet fallen due.

3.1.2   Textual Entailment Between a Query Sentence and Several Articles

Previous textual entailment systems compare two sentences and deduce a result [7–10]. However, in our task, a query sentence can have multiple relevant articles, and one article consists of several sentences. So, in order to propose the appropriate textual entailment result, we have to compare the meanings of one query sentence and multiple article sentences (one sentence vs. multiple articles).

Our approach assumes that there exists one most relevant article amongst relevant articles which can answer 'yes' or 'no' for a query sentence. For simplicity, we select the most relevant article and then try to deduce a textual entailment result by comparing between a query sentence and most relevant article (one sentence vs. one article). Here, we will describe how we choose the most relevant article with a query, from previously identified relevant articles:

$$most\ relevant\ article\ sentence := \underset{article_{n,j} \in\ articles\ relevant\ with\ query_n}{\arg\max} \{overlap(condition_{article_{n,j}}, condition_{query_n})$$

$$+overlap(conclusion_{article_{n,j}}, conclusion_{query_n})\}$$

The basic idea is that sentence $article_{n,j}$ is the most relevant article if it has maximum word overlap with the $query_n$ sentence. $condition_s$ is the condition part of the sentence s, and $conclusion_s$ is the conclusion of the sentence s. When we compute the word overlap, we separately compute the word overlap in condition parts and conclusion parts, and then use their sum.

For this approach to textual entailment, we have tried three methods: applying hand-constructed rules, creating a simple model with unsupervised learning, and then using an SVM-based supervised learning method.

3.1.3   Applying Rules

Because our language domain is restricted for both the input questions and law articles, there are some questions that can be answered easily using only negation and antonym information. If the question and article share the same word as the root in each syntactic tree, we consider the question as easy, which means it can be answered using only negation/antonym detection. Here is an example:

Question: If person A sells owned land X to person B, but soon after, sells the same land X to person C then if the registration title is transferred to B, then person B can assert against C in the acquisition of ownership of land X.

Article 177: Acquisitions of, losses of and changes in real rights concerning immovable properties may not be asserted against third parties, unless the same are registered pursuant to the applicable provisions of the Real Estate Registration Act and other laws regarding registration.

The conclusion and condition segments obtained from our system are as follows:

Conclusion of the question: then person B can assert against C in the acquisition of ownership of land X.
Condition of the question: If person A sells owned land X to person B, but soon after, sells the same land X to person C then if the registration title is transferred to B,
*Conclusion of the article: Acquisitions of, losses of and changes in real rights concerning immovable properties may **not** be asserted against third parties,*
*Condition of the article: **unless** the same are registered pursuant to the applicable provisions of the Real Estate Registration Act and other laws regarding registration.*

In the above example, the conclusions in both the question and the article use the root word "assert" of the corresponding syntactic trees. Therefore, this example can be answered using only the confirming negation and antonym information. If the sum of the negation levels of a question is the same with that of the corresponding article, then we determine the answer is "yes," and otherwise "no."

The negation level is computed as following: if [negation + antonym] occurs an odd number of times in a condition (conclusion), its negation level is "1." Otherwise if the [negation + antonym] occurs an even number of times, its negation level is "0." In the above example, the negation level of the condition of the question is zero, and that of the conclusion of the question is also zero. The negation level of a condition of the article is one, and that of a conclusion of the article is also one. Since the sum of the negation levels of the question is the same with that of the corresponding article, we determine the answer of the question is "yes."

A more concise description for this rule is shown in Fig. 1. In this Figure, $article_n$ is the most relevant article of the query $query_n$. The output of our rule-based system is also used below in an unsupervised learning model for assigning labels of condition (conclusion) clusters.

> if $(neg\_level(condition_{article\,n})+neg\_level(conclusion_{article\,n})$
> $= neg\_level(condition_{query\,n})+neg\_level(conclusion_{query\,n}))$,
> $Answer_n := yes$,
> otherwise, $Answer_n := no$,
> where $neg\_level() := 1$ if negation and antonym occur odd number of
>   times.
> $neg\_level() := 0$ otherwise.

Fig. 1. Answering rule for easy questions

3.1.4 Unsupervised Learning

We also try to construct a deeper representation for complex sentences. Fully general solutions are extremely difficult, if not impossible; for our first approximation for the non-easy cases, we have developed a method using unsupervised learning with more detailed linguistic information. Since we do not know the impact each linguistic attribute has on our task, we first run a machine learning algorithm that learns what information is relevant in the text to achieve our goal.

The types of features we use are as follows:

Word matching Having the same lemma.

Tree structure features Considering only the dependents of a root.

Lexical semantic features Having the same Kadokawa [11] thesaurus concept code.

We use our learning method on linguistic features to confirm the following semantic entailment features:

Feature 1 : if $w_{root}(condition_{query_n}) = w_{root}(condition_{article_n})$

Feature 2 : if $w_{root}(conclusion_{query_n}) = w_{root}(conclusion_{article_n})$

Feature 3 : if $W_{dep}(conclusion_{query_n}) \cap W_{dep}(conclusion_{article_n}) \neq \phi$

Feature 4 : if $c_{root}(condition_{query_n}) = c_{root}(condition_{article_n})$

Feature 5 : if $c_{root}(conclusion_{query_n}) = c_{root}(conclusion_{article_n})$

Feature 6 : if $neg\_level(condition_{query_n}) = neg\_level(condition_{article_n})$

Feature 7 : if $neg\_level(conclusion_{query_n}) = neg\_level(conclusion_{article_n})$

In the features above, $article_n$ is the most relevant article of the query $query_n$. $w_{root}(s)$ means the root word in the syntactic tree of the sentence s, and $W_{dep}(s)$ is the set of all the dependents of the root word in the syntactic tree of the sentence s. $c_{root}(s)$ is the Kadokawa concept code of the root word in the syntactic tree of the sentence s. For Features 1 and 2, we check if the root word in the syntactic tree of the $condition_{query\_n}(conclusion_{query\_n})$ is the same with that of the $condition_{article\_n}$ ($conclusion_{article\_n}$). For Feature 3, we examine if one of the dependents of the root word in the syntactic tree of the $conclusion_{query\_n}$ is also a dependent of the root word in the syntactic tree of the $conclusion_{article\_n}$. For Features 4 and 5, we check if the Kadokawa concept code of the root word in the syntactic tree of the $condition_{query\_n}(conclusion_{query\_n})$ is the same with that of the $condition_{article\_n}(conclusion_{article\_n})$.

For Features 6 and 7, we compare *neg_level()* between *condition $_{query\_n}$ (conclusion$_{query\_n}$)* and *condition$_{article\_n}$(conclusion$_{article\_n}$)*. Features 1, 2, 3 consider both lexical and syntactic information, and Features 4 and 5 consider semantic information. Features 6 and 7 incorporate negation and antonym information. Features 1 and 2 are used to check if conditions (conclusions) of a question and corresponding article share the same root word in the syntactic tree. Feature 3 is used to determine if each dependent of a root in the conclusion of a question appears in the article. We heuristically limit the number of dependents as those three nearest to the root. Features 4 and 5 confirm if the root words of conditions (conclusions) of the question and corresponding article share the same concept code. We use some morphological and syntactic analysis to extract lemma and dependency information. Details of the morphological and syntactic analyzer are given in Sect. 3.2.

The inputs for our unsupervised learning model are all the questions and corresponding articles. The outputs are two clusters of the questions. The yes/no outputs based on rules described in Sect. 3.1.3 are used as a key for assigning a yes/no label of each cluster. The cluster which includes higher portion of "yes" of the easy questions is assigned the label "yes," and the other cluster is assigned "no." We determine their yes/no answers using their clustering labels.

3.1.5 Supervised Learning with SVM

We compare our method with SVM, as a kind of supervised learning model. Using the SVM tool included in the Weka [12] software, we performed cross-validation for the 179 questions using 7 features explained in Sect. 3.1.4. We used a linear kernel SVM because it is popular for real-time applications as they enjoy both faster training and classification speeds, with significantly less memory requirements than non-linear kernels because of the compact representation of the decision function.

3.2 Experimental Setup for Phase 2

In the general formulation of the textual entailment problem, given an input text sentence and a hypothesis sentence, the task is to make predictions about whether or not the hypothesis is entailed by the input sentence. We report the accuracy of our method in answering yes/no questions of legal bar exams by predicting whether the questions can be entailed by the corresponding civil law articles.

There is a balanced positive-negative sample distribution in the dataset (55.87 % yes, and 44.13 % no) for dry run, so we consider the baseline for true/false evaluation is the accuracy when returning always "yes," which is 55.87 %. Our data for dry run has 179 questions, with total 1044 civil law articles.

The original examinations are provided in Japanese and English, and our initial implementation used a Korean translation, provided by the Excite translation tool (http://excite.translation.jp/world/). The reason that we chose Korean is that we have a team member whose native language is Korean, and the characteristics of Korean and Japanese language are similar. In addition, the translation quality between two languages ensures relatively stable performance. Because our study team includes a Korean researcher, we can easily analyze the errors and intermediate rules in Korean.

We used a Korean morphological analyzer and dependency parser [13], which extracts enriched information including the use of the Kadokawa thesaurus for lexical semantic information. We use a simple unsupervised learning method, since the data size is not big enough to separate it into training and test data.

3.3   Experimental Results

Evaluation of question answering systems is in general almost as complex as the construction of the question-answering itself. So one must make the choice to consider several features of QA systems in the evaluation process, e.g., query language difficulty, content language difficulty, question difficulty, usability, accuracy, confidence, speed and breadth of domain [14].

Table 3 shows our results on the dry run data. A rule-based model showed accuracy of 53.77 %. We also use a K-means clustering algorithm with K = 2 for unsupervised learning for the rest of the questions, and it showed accuracy of 60.85 %. The overall performance when combining the use of rules and unsupervised learning showed 61.96 % of accuracy which outperformed unsupervised learning for all questions, and even SVM (59.43 %), the supervised learning model we use with a linear kernel. According to p-value measures (p = 0.01) between the baseline and the combined model in the true/false determination, the combined model with rule-based model for easy questions and unsupervised learning model for non-easy questions significantly outperformed the baseline. Since previous methods use supervised learning with syntactic and lexical information, we consider the supervised learning experiment with SVM in Table 2 approximately represents the performance of previous methods.

Table 4 shows our performance of Phase 2 on test data. The test data consists of 41 questions, where 20 questions were 'Yes', and 21 questions were 'No'. The accuracy on test data is 63.41 %, and it ranked No. 1 on the COLIEE competition.

We also evaluated the performance combining our two systems for phase 1 and phase 2. Table 5 shows the accuracies of yes/no answers of two different systems. The first system is our textual entailment system, which receives the legal bar exam queries and relevant articles as input, and produces yes/no answers for the input queries by entailment between queries and relevant articles retrieved by legal experts. The second system is the combined system of legal IR and textual entailment. The combined system receives only legal bar exam queries as input. Then, our legal IR system

Table 3. Experimental results on dry run data for phase 2

Our method	Accuracy (%)
Baseline	55.87
Rule-based model	53.77
Unsupervised learning (K-means)	60.85
Cross-validation with Supervised learning (SVM)	59.43
Rule-based model for easy questions + unsupervised learning for non-easy questions (combined)	61.96

Table 4. Experimental results on test data for phase 2

Our model	Accuracy on phase 2 (%)
Rule-based model for easy questions + unsupervised learning for non-easy questions (combined)	63.41

Table 5. Performance after combining two systems for phase 1 and phase 2

Our system	Accuracy of 'Yes/No' answers (%)
Textual entailment system	67.39
(Input: legal bar exam queries and relevant articles)	
Combined system of legal IR and textual entailment	60.87
(Input: legal bar exam queries)	

retrieves relevant articles for the queries, and our textual entailment system determines "yes" or "no" for each query through textual entailment. It is natural that the combined system shows poorer performance than that using only textual entailment system: this is because, in the combined system, the relevant articles were retrieved by our legal IR system, while in the textual entailment system, the relevant articles were confirmed by human experts.

We used the test data which was released for phase 1 excluding the last query (we've found that the last query was omitted in the translation process), and the accuracy of 'Yes/No' answers for our textual entailment system was 67.39 % as shown in Table 5. The combined system showed 6.52 % decrease of accuracy in answering "Yes/No." But we found some questions which were correctly answered, even though the retrieved relevant articles were incorrect. There is an obvious need for deeper and detailed analysis, which will require at least the integration of information extraction techniques to identify and exploit legal relationships.

3.4   Discussion

From unsuccessful instances, we classified the error types as shown in Table 6. We can see that the challenge of paraphrasing causes most of the errors of our system. As just mentioned, the broad challenges of accurate Question Answering are contingent on legal relationship identification and exploitation. The development of that knowledge could be addressed by exploiting expert knowledge and much larger corpora, in companion with existing automatic information extraction methods. One of the most obvious places to focus on information extraction is the need to do more extensive temporal analysis.

Table 6 also shows that the error rate attributed to problems of negation/antonym was 9.68 %. This error mostly arose from the weakness of our antonym dictionary, as there were no errors of misinterpreted negations in the translation process.

The rate of the incorrectly translated sentences was 15.22 % in the test data. Because the original sentences are relatively clear and the syntactic characteristics of

Table 6. Error types in the result of the textual entailment system

Error type	Accuracy (%)	Error type	Accuracy (%)
Negation/antonym	9.68	Paraphrasing	54.84
Exceptional case	12.90	Constraints in condition	9.68
Condition, conclusion mismatch	6.45	Other errors	6.45

Japanese and Korean are so similar, there were only a 4.35 % error rate on syntactic structures in the translated sentences. Most of the errors resulted from the unnatural translation of a word in a sentence. It will be interesting to compare our performance using Korean-translated sentences with that using original sentences, in which case we would expect that using original sentences will show better performance.

In the legal bar exam, the human passing score is usually around 63 % and our system achieved 60.87 %. We think our performance in the first COLIEE competition is promising. For more accurate comparison with the human scores, we need to test the whole legal bar exam of a specific year including all difficult cases (e.g., a query identifies a legal article that refers to another legal article).

4 Related Work

SemEval 2014 Task 1 [15] evaluates system predictions of semantic relatedness (SR) and textual entailment (TE) relations on sentence pairs from the SICK dataset, which consist of 750 images and two descriptions for each image. The top ranked system (UIUC) [10] in this task uses distributional constituent similarity and denotational constituent similarity features. Their method of comparing constituents between the whole two sentences are not useful in our task, because our task consists of one legal query and multiple corresponding article sentences.

There was a textual entailment method from Bdour et al. [16] which provided the basis for a Yes/No Arabic Question Answering System. They used a kind of logical representation, which bridges the distinct representations of the functional structure obtained for questions and passages. This method is not appropriate for our task. If a false question sentence is constructed by replacing named entities with terms of different meaning in the legal article, a logic representation can be helpful. However, false questions are not simply constructed by substituting specific named entities, and any logical representation can make the problem more complex. Kouylekov and Magnini [17] experimented with various cost functions and found a combination scheme to work the best for RTE. Vanderwende et al. [18] used syntactic heuristic matching rules with a lexical-similarity back-off model. Nielsen et al. [19] extracted features from dependency paths, and combined them with word-alignment features in a mixture of an expert-based classifier. Zanzotto et al. [20] proposed a syntactic cross-pair similarity measure for RTE. Harmeling [21] took a similar classification-based approach with transformation sequence features. Marsi et al. [22] described a system using dependency-based paraphrasing techniques. All previous systems uniformly conclude that syntactic information is helpful in RTE: we also use syntactic information combined with lexical semantic information. As further research, we can enrich our knowledge base with deeper analysis of data, and add paraphrasing dictionary getting help from experts.

5 Conclusion

We have described our implementation for the Competition on Legal Information Extraction/Entailment (COLIEE) 2014 Task.

For phase 1, legal information retrieval, we implemented a Ranking-SVM model for the legal information retrieval task. By incorporating features such as lexical words, dependency links, tf-idf score, and LDA-based IR score, our model doubles the mean average precision.

For phase 2, we have proposed a method to answer yes/no questions from legal bar exams related to civil law. We used the knowledge base of Kim et al. [6] by analyzing negation patterns and antonyms in the civil law articles. To make the alignment easy, we first segment questions and articles into condition, conclusion, and exception. We then extract deep linguistic features with lexical, syntactic information based on morphological analysis and dependency trees, and lexical semantic information using the Kadokawa thesaurus. Our method uses a hybrid model that combines a rule-based model for easy questions and unsupervised learning model for non-easy questions. This achieved quite encouraging results in both true and false determination. We also show the performance combining the models for phase 1 and phase 2. To improve our approach in future work, we need to create deeper representations (e.g., to deal with paraphrase), and analyze the temporal aspects of legal sentences. In addition, we will complement our knowledge base with paraphrasing dictionary with the help of experts.

Acknowledgements. This research was supported by the Alberta Innovates Centre for Machine Learning (AICML) and the iCORE division of Alberta Innovates Technology Futures.

References

1. Jones, K.S.: A statistical interpretation of term specificity and its application in retrieval. In: Willett, P. (ed.) Document Retrieval Systems, pp. 132–142. Taylor Graham Publishing, London (1988)
2. Blei, D.M., Ng, A.Y., Jordan, M.I.: Latent dirichlet allocation. J. Mach. Learn. Res. 3, 993–1022 (2003)
3. Joachims, T.: Optimizing search engines using clickthrough data. In: Proceedings of the Eighth ACM SIGKDD International Conference on Knowledge Discovery and Data Mining, KDD 2002, pp. 133–142. ACM, New York (2002)
4. Wei, X., Croft, W.B.: LDA-based document models for ad-hoc retrieval. In: Proceeding of the 29th Annual International ACM SIGIR Conference on Research and Development in Information Retrieval, SIGIR 2006, pp. 178–185. ACM, New York (2006)
5. Maxwell, K.T., Oberlander, J., Croft, W.B.: Feature-based selection of dependency paths in ad hoc information retrieval. In: Proceedings of the 51st Annual Meeting of the Association for Computational Linguistics, vol. 1, Long Papers, pp. 507–516. Association for Computational Linguistics, Sofia, August 2013

6. Kim, M-Y., Xu, Y., Goebel, R., Satoh, K.: Answering yes/no questions in legal bar exams. In: JURISIN (2013)
7. Jikoun, V., de Rijke, M.: Recognizing textual entailment using lexical similarity. In: Proceedings of the PASCAL Challenges Workshop on RTE (2005)
8. MacCartney, B., Grenager, T., de Marneffe, M.-C., Cer, D., Manning, C.D.: Learning to recognize features of valid textual entailments. In Proceedings of HLT-NAACL (2006)
9. Sno, R., Vanderwende, L., Menezes, A.: Effectively using syntax for recognizing false entailment. In: Proceedings of HLT-NAACL (2006)
10. Lai, A., Hockenmaier, J.: Illinois-LH: a denotational and distributional approach to semantics. In: Proceedings of SemEval 2014: International Workshop on Semantic Evaluation (2014)
11. Ohno, S., Hamanishi, M.: New Synonym Dictionary. Kadokawa Shoten, Tokyo (1981)
12. Hall, M., Frank, E., Holmes, G., Pfahringer, B., Reutemann, P., Witten, I.H.: The WEKA data mining software: an update. SIGKDD Explor. 11(1), 10–18 (2009)
13. Kim, M-Y., Kang, S-J., Lee, J-H.: Resolving ambiguity in inter-chunk dependency parsing. In: Proceedings of 6th Natural Language Processing Pacific Rim Symposium, pp. 263–270 (2001)
14. Walas, M.: How to answer yes/no spatial questions using qualitative reasoning? In: Gelbukh, A. (ed.) CICLing 2012, Part II. LNCS, vol. 7182, pp. 330–341. Springer, Heidelberg (2012)
15. Marelli, M., Bentivogli, L., Baroni, M., Bernardi, R., Menini, S., Zamparelli, R.: SemEval-2014 task 1: evaluation of compositional distributional semantic models on full sentences through semantic relatedness and textual entailment. In: Proceedings of SemEval 2014: International Workshop on Semantic Evaluation (2014)
16. Bdour, W.N., Gharaibeh, N.K.: Development of yes/no arabic question answering system. Int. J. Artif. Intell. Appl. 4(1), 51–63 (2013)
17. Kouylekov, M., Magnini, B.: Tree edit distance for recognizing textual entailment: estimating the cost of insertion. In: Proceedings of the second PASCAL Challenges Workshop on RTE (2006)
18. Vanderwende, L., Menezes, A., Snow, R.: Microsoft research at RTE-2: syntactic contributions in the entailment task: an implementation. In: Proceedings of the Second PASCAL Challenges Workshop on RTE (2006)
19. Nielsen, R.D., Ward, W., Martin, J.H.: Toward dependency path based entailment. In: Proceedings of the Second PASCAL Challenges Workshop on RTE (2006)
20. Zanzotto, F.M., Moschitti, A., Pennacchiotti, M., Pazienza, M.T.: Learning textual entailment from examples. In: Proceedings of the Second PASCAL Challenges Workshop on RTE (2006)
21. Harmeling, S.: An extensible probabilistic transformation-based approach to the third recognizing textual entailment challenge. In: Proceedings of ACL PASCAL Workshop on Textual Entailment and Paraphrasing (2007)
22. Marsi, E., Krahmer, E., Bosma, W.: Dependency-based paraphrasing for recognizing textual entailment. In: Proceedings of ACL PASCAL Workshop on Textual Entailment and Paraphrasing (2007)

Translating Simple Legal Text to Formal Representations

Shruti Gaur[✉], Nguyen H. Vo, Kazuaki Kashihara, and Chitta Baral

Arizona State University, Tempe, AZ, USA
{shruti.gaur,nguyen.h.vo,kkashiha,chitta}@asu.edu

Abstract. Various logical representations and frameworks have been proposed for reasoning with legal information. These approaches assume that the legal text has already been translated to the desired formal representation. However, the approaches for translating legal text into formal representations have mostly focused on inferring facts from text or translating it to a single representation. In this work, we use the NL2KR system to translate legal text into a wide variety of formal representations. This will enable the use of existing logical reasoning approaches on legal text (English), thus allowing reasoning with text.

Keywords: Natural language processing · Natural language understanding · Natural language translation

1 Introduction and Motivation

One of the tasks of the Competition on Legal Information Extraction and Entailment [1] consists of finding whether a given statement is entailed by the given legal article(s) or not. This is similar to the Recognition of Textual Entailment (RTE) challenge [3]. It has been observed by Bos and Markert [7] that classification based on shallow features alone performs better than theorem proving, for RTE. Androutsopoulos and Malakasiotis [3] state that most approaches for RTE do not focus on converting natural language to its formal representation. However, we believe that approaches using statistical or machine learning methods on shallow features do not offer much explanation about why a certain sentence is entailed or not, hence providing little insight into the cause of entailment. In this respect, we consider the approaches based on logical reasoning to be more promising.

There have been several works that propose logical representations and logics for representing and reasoning with legal information [11–14,21]. Reasoning rules and frameworks assume that the information given in the form of natural language can somehow be understood and represented in the required form. However, current methods to convert legal text to formal representations [4,8,17,18] either focus on extracting important facts or are not generalizable to a wide variety of representations. Currently, there is no consensus on a single representation to express legal information. Therefore, a system that can translate

© Springer-Verlag Berlin Heidelberg 2015
T. Murata et al. (Eds.): JSAI-isAI 2014 Workshops, LNAI 9067, pp. 259–273, 2015.
DOI: 10.1007/978-3-662-48119-6_19

natural language to a wide variety of formal languages, depending on the application, is desired. In this paper we show how our NL2KR system can be used for translation of simple legal sentences in English to various formal representations. This will facilitate reasoning with various frameworks.

2   Related Work

Some approaches to translate text into formal representations focus on extraction of specific facts from the text. For example, Lagos et al. [17] present a semi-automatic method to extract specific information such as events, characters, roles, etc. from legal text by using the Xerox Incremental Parser (XIP) [2]. The XIP performs preprocessing, named entity extraction, chunking and dependency extraction, and combination of dependencies to create new ones. Bajwa et al. [4] propose an approach to automatically translate specification of business rules in English to Semantic Business Vocabulary and Rules (SBVR). Their method is essentially a rule-based information-extraction approach, which identifies SBVR elements from text. The goal of other approaches like the work by McCarty [18] is to obtain a semantic interpretation of the complete sentence. This approach uses the output from the state-of-the-art statistical parser to obtain a semantic representation called Quasi-Logical Form (QLF). QLF is a rich knowledge representation structure which is considered an intermediate step towards a fully logical form.

There have been similar efforts in other languages. Nakamura et al. [20] present a rule-based approach to convert Japanese legal text into a logical representation conforming to Davidsonian style. They ascertain the structure of legal sentences and identify cue phrases that indicate this structure, by manually analyzing around 500 sentences. They also define transformation rules for some special occurrences of nouns and verbs. In their subsequent work [16], they propose a method to resolve references that point to other articles or itemized lists, by replacing them with the relevant content.

As mentioned in the previous section, different legal reasoning frameworks expect input in different logical representations. Even though Legal Knowledge Interchange Format (LKIF) [15] was an attempt to standardize the representation of legal knowledge in the semantic web, currently, no single representation has been unanimously considered the de-facto standard for legal text. Therefore, we need a system that can translate natural language to a particular representation depending on the application.

3   The NL2KR Framework

NL2KR is a framework to develop translation systems that translate natural language to a wide variety of formal language representations. It is easily adaptable to new domains according to the training data supplied. It is based on the algorithms presented in Baral et al. [5]. The workflow using the NL2KR systems consists of two phases: (1.) learning and (2.) translation, as shown in Fig. 1.

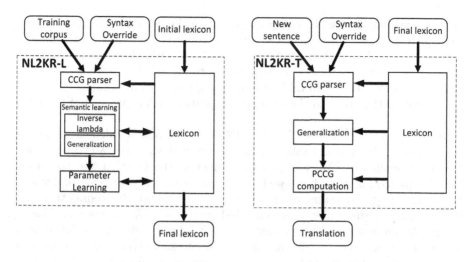

Fig. 1. The NL2KR system showing learning (left) and translation (right)

In the learning phase, the system takes training data, an initial dictionary and any optional syntax overrides, as inputs. The training data consists of a number of natural language sentences along with their formal representations in the desired target language. The initial dictionary (or lexicon) contains meanings of some words. The dictionary is manually supplied to the system. Using these inputs, NL2KR tries to learn the meanings of as many words as possible. Thus, the output of the learning phase is an updated dictionary which includes the meanings of all newly learned words. The translation phase uses the dictionary created by the learning phase to translate previously unseen sentences.

At the core of NL2KR are two very elegant algorithms, Inverse Lambda and Generalization, which are used to find meanings of unknown words in terms of lambda (λ) expressions. NL2KR is inspired[1] by Montague's approach [19]. Every word has a λ expression meaning. The meaning of a sentence is successively built from the combination of the λ expressions of words according to the rules of combination in Lambda (λ) calculus [9]. The order in which the words should be combined is given by the parse tree of the sentence according to a given Combinatory Categorial Grammar (CCG) [22]. As an example illustrating this approach, consider the sentence "John loves Mary" shown in Table 1. The CCG category of "loves" is (S\NP)/NP. This means that this word

Table 1. Example

John	loves	Mary
NP	$(S\backslash NP)/NP$	NP
$john$	$\#y.\#x.loves(x,y)$	$mary$

$$S\backslash NP$$
$$\#x.loves(x,mary)$$

$$S$$
$$loves(john,mary)$$

---

[1] NL2KR cannot be said to be based on Montague Semantics as it does not use intensional semantics. The translation of natural language to formal language with the use of lambda calculus, however, is in the same spirit as Montague's approach.

takes arguments of type NP (noun-phrase) from the left and the right, to form a complete sentence. From the CCG parse, we observe that "loves" and "Mary" combine first and then their combination combines with "John" to form a complete parse. The λ expression corresponding to "loves" is $\#y.\#x.loves(x,y)^2$, which means that this word takes two inputs, $\#x$ and $\#y$ as arguments and the application of this word to the arguments results in a λ expression of the form $loves(x,y)$.

The close correspondence between CCG syntax and λ calculus semantics is very helpful in applying this method. In the first step, the λ expression for "loves" is applied to "Mary", with the former as the function and the latter as the argument, in accordance with CCG categories. This application, denoted as $\#y.\#x.loves(x,y)@mary$ results in $\#x.loves(x,mary)$. Proceeding this way, the meaning of the sentence is generated in terms of λ expressions. This is a very elegant way to model semantics and has been widely used [5,6,10,23]. The problem, however, is that for longer sentences, λ expressions become too complex for even humans to figure out. This problem is addressed by NL2KR by employing the Inverse Lambda and Generalization algorithms to automatically formulate λ expressions from words whose semantics are known.

Learning Algorithms: The two algorithms used to learn λ semantics of new words are the Inverse Lambda and Generalization algorithms. When the λ expressions of a phrase and that of one of its sub-parts (children in the CCG parse tree) are known, we can use this knowledge to find the λ expression of the unknown sub-part. The Inverse Lambda operation computes a λ expression F such that $H = F@G$ or $H = G@F$ given H and G. These are called Inverse-L and Inverse-R algorithms, respectively. For example, if we know the meaning of the sentence "John loves Mary" (Table 1) as $loves(john,mary)$ and the meaning of John as $john$, we can find the meaning of "loves Mary" using Inverse Lambda, as $\#x.loves(x,mary)$. Going further, if we know the meaning of "Mary" as $mary$, we can find the meaning of "loves" using Inverse Lambda.

The Generalization algorithm is used to learn meanings of unknown words from syntactically similar words with known meanings. It is used when Inverse Lambda algorithms alone are not enough to learn new meanings of words or when we need to learn meanings of words that are not even present in the training data set. For example, we can generalize the meaning of the word "likes" with CCG category $(S\backslash NP)/NP)$, from the meaning of "loves", which we already know from the previous example. The meaning of "likes" thus generated will be $\#y.\#x.likes(x,y)$. We will illustrate learning in later sections with the help of examples.

For every sentence in the training set, we first use the CCG Parser to obtain all possible parse trees. Using the initial dictionary supplied by the user, the system assigns all known meanings to the words (at the leaf) in each parse tree. Moving bottom up, it combines as many words with each other as possible(in the

---

[2] $\#$ is used in place of λ to enable typing into a terminal.

order dictated by the parse tree) by performing λ applications. The meaning of each complete sentence is known from the training corpus. We need to traverse top-down from this known translation, while simultaneously traversing bottom up, by filling in missing word or phrase meanings. Meanings of unknown words and phrases are obtained using Inverse Lambda and Generalization, as applicable, until nothing new can be learned.

Dealing with Ambiguity: To deal with ambiguity of words, a parameter learning method [23] is used to estimate a weight for each word-meaning pair such that the joint probability of the training sentences getting translated to their given formal representation is maximized. However, this method might not work in all cases and more complex approaches, possibly involving word sense identification from context, might have to be used. Completely addressing this problem is a part of future work.

Translation Approach: Given a sentence, we consider all the possible parse trees, consisting of meanings of every word learned by the system or obtained from Generalization algorithm. Then we use Probabilistic CCG (PCCG) [23] to find the most probable tree, according to weights assigned to each word.

Availability: NL2KR is freely available for Windows, Linux and MacOSX systems at http://nl2kr.engineering.asu.edu. It is configurable for different domains and can be adapted to work with a large number of formal representations. A tutorial has also been provided.

4 Translating to Formal Legal Representations

NL2KR can be used to translate sentences into various logical representations, either directly or by using an intermediate language[3]. It can be customized to different domains based on the initial dictionary and training data provided. The quality of these inputs affects NL2KR's performance. A language class can be considered a good analogy of NL2KR. The effectiveness of learning depends on the richness of vocabulary imparted to the students beforehand (similar to initial lexicon) and the sentences chosen to teach the language (training data). In our experiments, we observed that learning simpler sentences before complex ones aided learning. We will also give some guidelines for creating the initial dictionary. Several logics have been proposed in the literature for representing legal information [11–14,21], from which we have selected a few. In this section, we will illustrate the method of creating a good initial lexicon and demonstrate how to use the system to learn new word meanings, with respect to these examples. We will start with simple examples and progress to more complicated ones.

---

[3] The choice of intermediate language depends on the domain and target languages. Once an intermediate language has been decided, the conversion can be automated.

4.1   Translating to First Order Logic Representations

In this section, we demonstrate translating a sentence from the Competition on Legal Information Extraction and Entailment [1] corpus to a first order logic representation.

Sentence: Possessory rights may be acquired by an agent.
Translation: $rights(X) \wedge type(X, possessory) \wedge agent(Y) >$
$acquirable(X, Y, may)$

Here $>$ is used to denote implication. The form of an action, for e.g., "acquirable" is $action(X, Y, Z)$. It denotes X(possessory rights) is being acquired by Y(agent) and the type of this action is Z. In the given example, *acquiring* is a possibility, not an obligation, which is why we use *may* as its type.

Once we provide this training data and other required inputs to the NL2KR learning interface (Fig. 2), we can start the Learning process. We will describe how to create inputs for learning in the next sub-section. The initial dictionary contains a list of words and their meanings in terms of λ expressions. Even if we do not know meanings of some words, we can use the system to figure them out on its own, using Inverse Lambda or Generalization algorithms. Figure 2 shows that the system learns the meaning of "rights" automatically using Inverse Lambda.

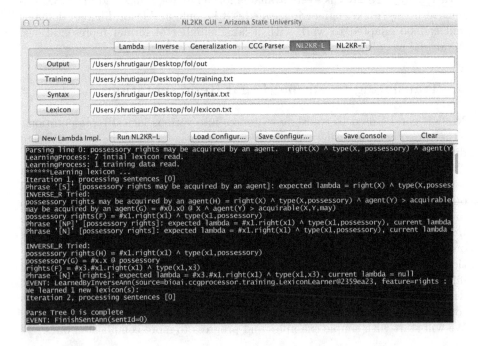

Fig. 2. NL2KR automatically learning the meaning of "rights" using Inverse Lambda Algorithm : feature = rights : [N] : #x3.#x1.right(x1) $\wedge$ type(x1,x3)

4.2   Translating Sentences with Temporal Information

Consider the following sentence and translation, which shows an example of temporal ordering.

Sentence: After the invoice is received the customer is obliged to pay.
Translation: $implies(receipt(invoice, T1) \wedge (T2 > T1),$
$obl(pay(customer, T2)))$

Here $implies(x, y)$ denotes $x \rightarrow y$. The predicate obl denotes that the action is an *obligation* (usually marked by words such as obliged to, shall, must, etc.) in contrast to a *possibility* (usually marked by words such as may). T1 and T2 are the instances of time at which the two events occurred.

Words that do not contribute significantly to the meaning of the sentence can be assigned the trivial meaning $\#x.x$ in the dictionary. It is a λ expression that does not affect the meaning of other λ expressions. We can assign it to words such as "is", "to" and "the" since these do not carry much meaning in this example. Next, we can start entering the meanings that are evident from looking at the target representation. Since "invoice" occurs as itself, we can give it the simple meaning *invoice* (similarly for "customer"). From the representation, we observe that "received" is a function called *receipt* with two arguments, hence we can give it the meaning $\#x.\#t.receipt(x, t)$ (similarly for "pay"). *Obliged* is a more complicated function because it takes another function (pay) as its argument and therefore uses @y@t to carry forward the variables in *pay* to the next higher level of the tree, where we obtain the real arguments (customer and T2). Once all these meanings (Table 2) have been supplied, the system can automatically find the meaning of the word "after" using Inverse Lambda algorithm (Fig. 3). This is remarkable from the perspective that the meaning of "after" looks complicated and it might be tedious for users to supply such meanings manually in the initial lexicon. This demonstrates one of the advantages of using NL2KR. The meaning of "after" makes intuitive sense. The λ expression $\#x12.\#x11.implies(x12$ @ $T1 \wedge T2 > T1, x11$ @ $T2)$ means that "after" is a λ function which takes two inputs: x11 and x12, where the first input event (x12) occurs at time T1, T2 > T1, the second input event (x11) occurs at time T2 and x12 implies (or leads to) x11. Hence, we were able to learn a significantly complicated meaning automatically by providing relatively simple λ expressions in the initial dictionary.

4.3   Translating to Temporal Deontic Action Laws

Giordano et al. [14] have defined a Temporal Deontic Action Language for defining temporal deontic action theories, by introducing a temporal deontic extension of Answer Set Programming (ASP) combined with Deontic Dynamic Linear Time Temporal Logic (DDLTL). This language is used for expressing domain description laws, for e.g., action laws, precondition laws, causal laws, etc., which describe the preconditions and effects of actions. It is also used for expressing obligations, for e.g., achievement obligations, maintenance obligations, contrary

Table 2. λ expressions and CCG categories in the initial dictionary for the sentence "After the invoice is received the customer is obliged to pay."

Word	Syntax	Meaning
invoice	N	invoice
is	(S\NP)/NP	#x.x
received	NP	#x.#t.receipt(x,t)
customer	N	customer
obliged	NP/NP	#x.#y.#t.obl(x@y@t)
to	NP/(S\NP)	#x.x
pay	S\NP	#x.#t.pay(x,t)
the	NP/N	#x.x

to duty obligations, etc. We will take examples of several domain description laws from the paper and demonstrate how to translate them automatically from natural language to the Deontic action language, using NL2KR.

Since NL2KR does not support some special symbols used in the Temporal Deontic Action Language, we first use NL2KR to convert the natural language sentences to an intermediate representation which is directly convertible to the Temporal Deontic Action Language. Then using the one-to-one correspondence between the intermediate language and the action language, we obtain the desired representation. In the Intermediate representation shown below, we have defined the predicate $creates(x, y)$, which means x creates y. We have also changed the representation of $until$ to have two parameters a and b denoting "a until b" (as defined by Giordano et al. [14]).

Action Law:
Sentence: The action $accept\_price$ creates an obligation to pay.
Translation: $[accept\_price]\mathbf{O}(\top U < pay > \top)$
Intermediate: creates(action(accept_price),O(until(a(T),b(pay,T))))

The NL2KR learning component can be used to make the system learn words and meanings from this sentence. The iterative learning process is depicted in the screenshot in Fig. 4. We start by giving meanings of simple words first. We give trivial meanings $#x.x$ to "the", "an" and "to", because they do not significantly affect the meaning of the sentence. Next, we guess the meanings of words from the target representation. Since "action" is a function that accepts a single argument, we give it the meaning $#x.action(x)$. Similarly, "accept_price" which occurs as itself is given the meaning $accept\_price$. Similarly, Obligation, O is also a function, but it contains more structure, which can be obtained from the target representation. We interpret an obligation to also have an implicit notion of time by having "until" embedded in its meaning. However, the verb "pay" should be replaceable, because there can be other sentences such as "The action $accept\_price$ creates an obligation to ship". Therefore, we leave it as a

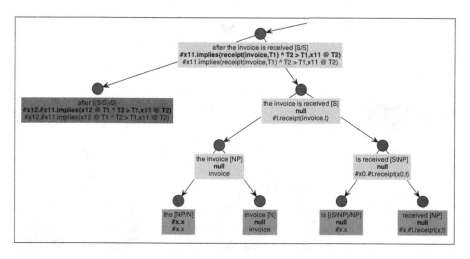

Fig. 3. NL2KR automatically learning the meaning of "after" using Inverse Lambda Algorithm : feature = after : [(S/S)/S] : #x12.#x11.implies(x12 @ T1 ∧ T2 >T1,x11 @ T2)

variable input. The word "pay" could have been simply *pay* but we assign it the meaning #x.x@pay. This is because the node, "an obligation", expects "to pay" to be an argument to it, but their CCG categories dictate otherwise. In cases where there is such inconsistency, we use meanings prefixed with #x.x@ for the function (according to CCG categories), so that their role is *flipped* to that of arguments[4]. The λ expressions and CCG categories of the constituent words are shown in Table 3. After giving these meanings, we find that the meaning of "creates" is obtained automatically by the system using the Inverse Lambda algorithm (Fig. 4).

Once the learning process is complete, we can use the Translation component of NL2KR to translate a new sentence. In this case, we use NL2KR to translate the following action law.

Action Law:
Sentence: The action *cancel_payment* cancels the obligation to pay.
Translation: [*cancel_payment*]¬**O**(⊤$U < pay > $⊤)
Intermediate: cancels(action(cancel_payment),O(until(a(T),b(pay,T))))

The screenshot of the translation process is shown in Fig. 5. We observe that the sentence was automatically translated by the system successfully. This was done by generating the meanings of unknown words ("cancel_payment"

---

[4] Let the required function be A and the required argument be B. Let the CCG-determined function be B and the CCG-determined argument be A. Recall that @ denotes λ application. By giving a meaning of the form #x.(x@b) to B, and performing application as determined by CCG, we obtain the result as (#x.(x@b))@a or a@b.

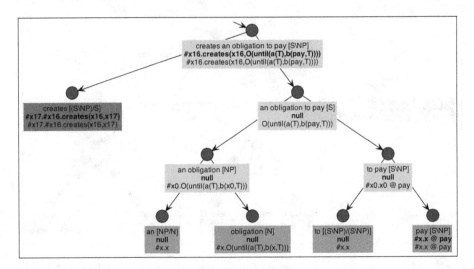

Fig. 4. Screenshot of the Learning Process in NL2KR for the sentence "The action *accept_price* creates an obligation to pay." The meaning of "creates" is obtained automatically by the system using the Inverse Lambda algorithm

and "cancels") using Generalization (Fig. 6) on the words learned from the first action law.

4.4 Translating to Temporal Object Logic in REALM

Regulations Expressed as Logical Models (REALM) [13] is a system that models regulatory rules in temporal object logic. The concepts and relationships occurring in this rule are mapped to predefined types and relationships in a Unified Modeling Language (UML) model. Using some examples from this paper, we will show how NL2KR can be used to translate rules specified in natural language to this temporal object logic representation.

Table 3. λ expressions and CCG categories for the words in the action law "The action *accept_price* creates an obligation to pay."

Word	Syntax	Meaning
the	NP/N	#x.x
action	N/N	#x.action(x)
accept_price	N	accept_price
an	NP/N	#x.x
obligation	N	#x.O(until(a(T),b(x,T)))
to	(S\NP)/(S\NP)	#x.x
pay	S\NP	#x.x@pay

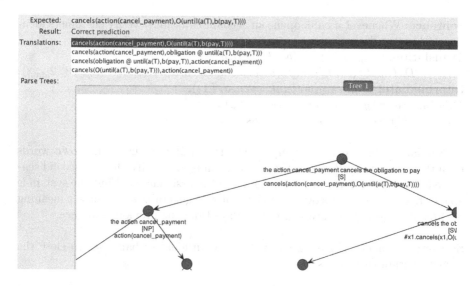

Fig. 5. Screenshot of the Translation Process in NL2KR for the sentence "The action *cancel_payment* cancels the obligation to pay."

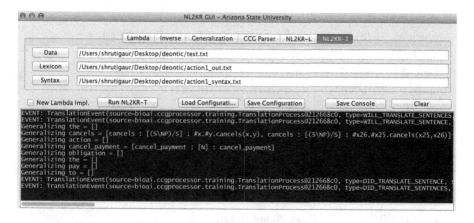

Fig. 6. Generating the meanings of unknown words (cancel_payment and cancels) using Generalization during the Translation Process in NL2KR for the sentence "The action *cancel_payment* cancels the obligation to pay."

As in the previous section, we have created an intermediate representation which can directly be converted to the desired temporal object logic representation. This is needed due to unavailability of certain symbols in NL2KR's vocabulary. We also assume that coreference in sentences has been resolved. For example, in the following sentence, the second occurrence of "bank" has replaced the pronoun "it".

Sentence: Whenever a bank opens an account *bank* must verify customers identity within two_days

Translation: $\Box t_{open}(DoOn_F(bank, open, a) \rightarrow$
$\Diamond t_{verify}(DoInput_F(bank, verify, a.customer.record) \wedge t_{verify} - t_{open} \leq 2_{[day]})$
Intermediate: $implies(g(do(bank, open, a, T1)),$
$f(do(bank, verify, a\_customer\_record, T2)$
$\wedge equals(difference(T2, T1), two\_days)))$

Similar to the previous examples, we use NL2KR to learn unknown words from these sentences. We do not give the meaning of "verify" for the second sentence (which is different from its meaning in the first sentence) but the system is able to figure it out on its own. Moreover, it also generalizes the correct meaning of three_days using the meaning of two_days from the previous sentence.

Sentence: Whenever a bank can not verify an identity bank has to close the account within three_days

Table 4. Initial Lexicon containing λ expressions and CCG categories for both REALM examples

Word[Syntax]	Meaning
whenever [(S/S)/S]	#y.#x.implies(g(y@T1),f((x@T1)@T2))
a [NP/N]	#x.x
bank [N]	bank
opens [(S\NP)/NP]	#y.#x.#t1.do(x,open,y,t1)
an [NP/N]	#x.x
account [N]	a
must [(S\NP)/(S\NP)]	#x.x
verify [(S\NP)/NP]	#x.x@#x1.#x2.#x3.#x4.#x5.(do(x3,verify,x1,x5)
	∧ x2@x4@x5)
customers [NP/N]	#x.x
identity [N]	a_customer_record
within [(NP\NP)/NP]	#z.#y.#x.x@y@#t1.#t2.equals(difference(t2,t1),z)
has [(S\NP)/(S\NP)]	#x.x
to [(S\NP)/(S\NP)]	#x.x
the [NP/N]	#x.x
can [(S\NP)/(S\NP)]	#x.x
close [(S\NP)/NP]	#x.x@#x1.#x2.#x3.#x4.#x5.(do(x3,close,x1,x5)
	∧ x2@x4@x5)
not [(S\NP)/(S\NP)]	#y.#x.#t1.(y @ x @ t1 ∧ isfalse)
two_days [N]	two_days

Translation: $\Box t_{open}(DoOn_F(bank, open, a) \rightarrow$
$\Diamond t_{verify}(DoInput_F(bank, verify, a.customer.record) \land t_{verify} - t_{open} \leq 2_{[day]})$
Intermediate: $implies(g(do(bank, verify, a\_customer\_record, T1) \land is false),$
$f(do(bank, close, a, T2) \land equals(difference(T2, T1), three\_days)))$

We observe that the initial dictionary for this case (Table 4) looks more complicated than the one in Sect. 4.3. This is because the target language in this case is such that the functions which would have been intuitive according to their natural language meanings, for e.g., "opens", "verify", etc. are not functions but arguments of an artificially created function, "do". It is obvious that the language of REALM was designed for different purposes than that of translation, which is why such a situation exists. Our motivation here is to give the reader an explanation of why some languages are easy for NL2KR to translate, while others are more difficult.

5 Conclusion and Future Work

Although legal text is written in natural language, one needs to do some kind of formal reasoning with it to draw conclusions. The first step to do that is to translate legal text to an appropriate logical language. At present there is no consensus on a single logical language to represent legal text. Therefore, one cannot develop a translation system targeted to a single language. Thus, a platform that can translate legal text to the desired logical language depending on the application, is needed. We have developed such a system called NL2KR. In this paper, we showed how NL2KR is useful in translating sentences from legal texts in English to various formal representations defined in various works, thereby bridging the gap from language to logical representation and enabling the use of various logical frameworks over the information contained in such texts.

So far we have experimented with a few small sentences picked from the literature on logical representation of legal texts. However, we need to expand this approach to capture nuances of legal texts used in real laws and statutes. Further enhancements are needed in NL2KR to equip it to deal with longer and more complicated sentences. One approach that can be used would involve breaking the sentence into smaller parts and subsequently dealing with each part separately. Such a parser, called L-Parser is available at http://bioai8core. fulton.asu.edu/lparser. We also plan to combine statistical and logical methods in the future. In particular, we are considering using a combination of distributional semantics and hand curated linguistic knowledge to characterize content words (especially, noun, verbs and adjectives) and use logical characterization for grammatical words (prepositions, articles, quantifiers, negation, etc.).

Acknowledgements. We thank Arindam Mitra and Somak Aditya for their work in developing the L-Parser. We thank NSF for the DataNet Federation Consortium grant OCI-0940841 and ONR for their grant N00014-13-1-0334 for partially supporting the development of NL2KR.

References

1. Jurisin legal information extraction and entailment competition (2014). http://webdocs.cs.ualberta.ca/miyoung2/jurisin_task/index.html
2. Aït-Mokhtar, S., Chanod, J.P., Roux, C.: Robustness beyond shallowness: incremental deep parsing. Nat. Lang. Eng. **8**(3), 121–144 (2002)
3. Androutsopoulos, I., Malakasiotis, P.: A survey of paraphrasing and textual entailment methods. J. Artif. Int. Res. **38**(1), 135–187 (2010)
4. Bajwa, I.B., Behzad, L.M.: SBVR business rules generation from natural language specification. In: AAAI 2011 Spring Symposium AI for Business Agility, San Francisco, USA, pp. 2–8 (2011)
5. Baral, C., Dzifcak, J., Gonzalez, M.A., Zhou, J.: Using Inverse lambda and Generalization to Translate English to Formal Languages. CoRR abs/1108.3843 (2011)
6. Blackburn, P., Bos, J.: Representation and Inference for Natural Language: A First Course in Computational Semantics. Center for the Study of Language and Information, Stanford (2005)
7. Bos, J., Markert, K.: Recognising textual entailment with logical inference. In: Proceedings of the Conference on Human Language Technology and Empirical Methods in Natural Language Processing, HLT 2005, pp. 628–635. Association for Computational Linguistics, Stroudsburg (2005)
8. Brüninghaus, S., Ashley, K.D.: Improving the representation of legal case texts with information extraction methods. In: Proceedings of the 8th International Conference on Artificial Intelligence and Law, ICAIL 2001, pp. 42–51. ACM, New York (2001)
9. Church, A.: An unsolvable problem of elementary number theory. Am. J. Math. **58**(2), 345–363 (1936)
10. Costantini, S., Paolucci, A.: Towards translating natural language sentences into ASP. In: Faber, W., Leone, N. (eds.) CILC, CEUR Workshop Proceedings, vol. 598. CEUR-WS.org (2010)
11. De Vos, M., Padget, J., Satoh, K.: Legal modelling and reasoning using institutions. In: Bekki, D. (ed.) JSAI-isAI 2010. LNCS, vol. 6797, pp. 129–140. Springer, Heidelberg (2011)
12. Distinto, I., Guarino, N., Masolo, C.: A well-founded ontological framework for modeling personal income tax. In: Proceedings of the Fourteenth International Conference on Artificial Intelligence and Law, ICAIL 2013, pp. 33–42. ACM, New York (2013)
13. Giblin, C., Liu, A.Y., Müller, S., Pfitzmann, B., Zhou, X.: Regulations expressed as logical models (REALM). In: Proceedings of the 2005 Conference on Legal Knowledge and Information Systems, JURIX 2005, The Eighteenth Annual Conference, pp. 37–48. IOS Press, Amsterdam (2005)
14. Giordano, L., Martelli, A., Dupré, D.T.: Temporal deontic action logic for the verification of compliance to norms in ASP. In: Proceedings of the Fourteenth International Conference on Artificial Intelligence and Law, ICAIL 2013, pp. 53–62. ACM, New York (2013)
15. Hoekstra, R., Breuker, J., Bello, M.D., Boer, E.: The LKIF core ontology of basic legal concepts. In: Proceedings of the Workshop on Legal Ontologies and Artificial Intelligence Techniques, LOAIT 2007 (2007)
16. Kimura, Y., Nakamura, M., Shimazu, A.: Treatment of legal sentences including itemized and referential expressions – towards translation into logical forms. In: Hattori, H., Kawamura, T., Idé, T., Yokoo, M., Murakami, Y. (eds.) JSAI 2008. LNCS, vol. 5447, pp. 242–253. Springer, Heidelberg (2009)

17. Lagos, N., Segond, F., Castellani, S., O'Neill, J.: Event extraction for legal case building and reasoning. In: Shi, Z., Vadera, S., Aamodt, A., Leake, D. (eds.) IIP 2010. IFIP AICT, vol. 340, pp. 92–101. Springer, Heidelberg (2010)
18. McCarty, L.T.: Deep semantic interpretations of legal texts. In: Proceedings of the 11th International Conference on Artificial Intelligence and Law, ICAIL 2007, pp. 217–224. ACM, New York (2007)
19. Montague, R.: English as a formal language. In: Thomason, R.H. (ed.) Formal Philosophy: Selected Papers of Richard Montague, pp. 188–222. Yale University Press, New Haven (1974)
20. Nakamura, M., Nobuoka, S., Shimazu, A.: Towards translation of legal sentences into logical forms. In: Satoh, K., Inokuchi, A., Nagao, K., Kawamura, T. (eds.) JSAI 2007. LNCS (LNAI), vol. 4914, pp. 349–362. Springer, Heidelberg (2008)
21. Riveret, R., Rotolo, A., Contissa, G., Sartor, G., Vasconcelos, W.: Temporal accommodation of legal argumentation. In: Proceedings of the 13th International Conference on Artificial Intelligence and Law, ICAIL 2011, pp. 71–80. ACM, New York (2011)
22. Steedman, M.: The Syntactic Process. MIT Press, Cambridge (2000)
23. Zettlemoyer, L.S., Collins, M.: Learning to map sentences to logical form: structured classification with probabilistic categorial grammars. In: UAI, pp. 658–666. AUAI Press (2005)

Analyzing Reliability Change in Legal Case

Pimolluck Jirakunkanok[✉], Katsuhiko Sano, and Satoshi Tojo

School of Information Science, Japan Advanced Institute of Science and Technology,
1-1 Asahidai, Nomi, Ishikawa 923-1292, Japan
{pimolluck.jira,v-sano,tojo}@jaist.ac.jp

Abstract. Reliability among agents plays a significant role in both human and agent communications. An agent may change her reliability for the other agents, when she receives a new piece of information from one of them. In order to analyze such reliability change, this paper proposes a logical formalization with two dynamic operators, i.e., downgrade and upgrade operators. The downgrade operator allows an agent to downgrade some specified agents to be less reliable in terms of the degree of reliability, while the upgrade operator allows the agent to upgrade them to be more reliable. Furthermore, we demonstrate our formalization by a legal case from Thailand.

Keywords: Reliability change · Belief · Legal case · Modal logic · Signed information

1 Introduction

In agent communication, an agent needs some criteria to decide which information she should believe. A common criterion is to consider the reliability of an information source. If the agent considers that a source of received information is reliable, she would accept and might believe the received information. On the other hand, the agent may reject the received information if she considers that the source is not reliable.

In legal proceedings, since a consideration of reliability has a strong influence on a judge's decision, the judge also needs a concept of the reliability of an information source. That is, when the judge receives a piece of information from a witness, the judge should consider if the witness is reliable or not. In addition, when the judge receives new information, she might change her reliability of the witness. This paper aims to investigate an effect of reliability change of the judge in legal judgment.

Recently, many studies [1–3] presented the use of logic-based approaches in the legal systems. Dynamic epistemic logic (DEL) [4,5] is a logical tool to study reasoning about information change due to communication between agents. Based on these frameworks, several works [6–9] proposed to formalize the notion of reliability. Among of them, Lorini et al. [8] introduced a modal framework for reasoning about signed information. In their framework, the agents can keep track of the information source by using the notion of signed statement. They

© Springer-Verlag Berlin Heidelberg 2015
T. Murata et al. (Eds.): JSAI-isAI 2014 Workshops, LNAI 9067, pp. 274–290, 2015.
DOI: 10.1007/978-3-662-48119-6_20

also considered the notion of reliability over the information sources. However, they did not deal with any dynamics of reliability relations among agents.

For this reason, we propose to formalize reliability change of an agent. First, we apply a concept of signed statement based on [8] to formalize the source of information. Then, we introduce two dynamic operators, i.e. downgrade and upgrade operators, in order to capture the change of reliability ordering between agents. The downgrade operator is used to downgrade some specified agents to be less reliable in terms of the degree of reliability, while the upgrade operator is used for upgrading. Finally, we reformulate a careful policy [8] in terms of DEL and employ it to consider which pieces of received signed information an agent should believe. Moreover, we demonstrate our formalization in an example of a legal case from Thailand.

The remainder of this paper is organized as follows. Section 2 describes the target legal case. Then, a formal tool for analyzing the legal case is presented in Sect. 3. In Sect. 4, we propose a dynamic logical analysis of the target legal case. Finally, our conclusion and future works are stated in Sect. 5.

2   Target Legal Case

Firstly, we summarize a story of our target legal case that occurred on 26th January 2003 in Trang province, Thailand[1] as follows:

> One day, a victim v had a drink with his friends f_1, f_2 and d at f_2's house. After that, v was punched and stabbed with a hand scraper in the back by an offender, and as a result, v had bleeding in the lung. However, v was still alive.

In the inquiry stage, a police po, who is an inquiry official, interviewed four witnesses v, f_1, f_2, mo that gave the following statements.

(I_1) v told that d was the offender who punched and stabbed v.
(I_2) f_1 also told that d was the offender who punched and stabbed v.
(I_3) f_2 stated that v and d had a dispute, but did not have any fighting.
(I_4) mo, who is v's mother, told that d was the offender according to v's saying.
   More details can be shown as follows:

> At night of the accident, mo visited v in the hospital. Then, v told her that v went to have a drink with d, f_1 and f_2 at f_2's house. During drinking, v and d had a dispute, then d punched v and stabbed with a hand scraper in the back of v.

From the interview, po accused d of attempting to kill v.

In the Civil Court, v and f_1 changed their statements as follows. First, v told that one of a group of unknown teenagers was the offender who punched and stabbed v with a knife. More details can be shown as follows:

---

[1] This legal case can be referred from http://deka2007.supremecourt.or.th/deka/web/ search.jsp (in Thai).

At 19 o'clock, v and f_1 were invited to drink by x who was their neighbor. After drinking, v and f_1 went to a market. While f_1 was riding a motorcycle from x's house, a group of unknown teenagers came to punch v. Then, one of them stabbed with a knife in the back of v.

Second, f_1 only stated that v was punched by d, but could not state that v was stabbed by d or not. More details can be shown as follows:

At 18 o'clock, v and f_1 were invited to drink by d. Then, v and f_1 went to f_2's house by a motorcycle (v was a rider), and d also followed them. Next, v had a drink with f_1, f_2, d and two other friends at f_2's house. Around 21 o'clock, v and d had a dispute and then d punched v. f_2 came to forbid them from fighting, while f_1 went to bring the motorcycle. After that, v came to sit behind f_1's motorcycle and said that he was stabbed.

Moreover, po was called to be a witness for testifying all statements in the inquiry stage.

Thus, there are six testimonies in the Civil Court as follows:

(T_1) v told that one of a group of unknown teenagers was the offender who punched and stabbed v with a knife.

(T_2) f_1 only stated that v was punched by d, but could not state that v was stabbed by d or not.

(T_3) po stated that v told that d was the offender who punched and stabbed v.

(T_4) po stated that f_1 told that d was the offender who punched and stabbed v.

(T_5) po stated that f_2 stated that v and d had a dispute, but did not have any fighting.

(T_6) po stated that mo told that d was the offender according to v's saying.

From the above testimonies, testimonies of v and f_1 in the inquiry stage (T_3 and T_4) are more reliable than that in the Civil Court (T_1 and T_2) because of the following reasons. First, the judge believed that po and f_2 had never had any arguments against d. So, there is no reason that they will allege or testify against d to be punished. Second, according to T_1 and T_2, the judge believed that v and f_1 tried to distort the facts in order to prevent d who is their friend from the punishment. Therefore, the judge decided that d was the offender and intended to kill v by the following reasons.

- Since the hand scrapper was a dangerous weapon, d used it in a possibly lethal attack. This shows that d intended to kill v.
- d stabbed v while v was turning back. At that time, d could choose other alternative positions for attacking. Nevertheless, d strongly stabbed v in the lung that is a vital organ. It is obvious that d intended to kill v.
- From the statement of the doctor, v was seriously injured, i.e., there was air leaking and bleeding in the chest cavity and the lung, and would be dead unless v got the treatment in time. This shows that the attack of d was possibly lethal.

For this reason, the Civil Court judged d to be sentenced to ten years' imprisonment by Article 288 and Article 80 of Penal Code: [2]

Article 288 (offence causing death): Whoever, murdering the other person, shall be punished by death or imprisoned as from fifteen years to twenty years.

Article 80 (commitment): Whoever commences to commit an offence, but does not carry it through, or carries it through, but does not achieve its end, is said to attempt to commit an offence. Whoever attempts to commit an offence shall be liable to two-thirds of the punishment as provided by the law for such offence.

In the Appeal Court and the Supreme Court, d appealed that he did not intend to kill v; in fact, he only intended to attack v. However, the judge agreed with the decision of the Civil Court and adopted the result, i.e., d was imprisoned for ten years by Articles 288 and Article 80 of Penal Code.

3   Formal Tool for Analyzing Target Legal Case

3.1   Static Logic of Agents' Beliefs for Signed Information

To analyze the previous legal case from a logical point of view, we introduce a modal language, based on previous work [8], which enables us to formalize each agent's belief, the reliability of information sources, and signed information.

Let G be a fixed *finite* set of agents. Our syntax $\mathcal{L}$ consists of the following vocabulary: (i) a countably infinite set $\mathsf{Prop} = \{p, q, r, ...\}$ of propositional letters, (ii) Boolean connectives: $\neg$, $\wedge$, (iii) the belief operators $\mathsf{Bel}(a, \cdot)$ $(a \in G)$, (iv) the signature operators $\mathsf{Sign}(a, \cdot)$ $(a \in G)$, and (v) the constants for reliability ordering $b \leqslant_a c$ $(a, b, c \in G)$. A set of formulas of $\mathcal{L}$ is inductively defined as follows:

$$\varphi ::= p \mid \neg\varphi \mid \varphi \wedge \varphi \mid \mathsf{Bel}(a, \varphi) \mid \mathsf{Sign}(a, \varphi) \mid b \leqslant_a c,$$

where $p \in \mathsf{Prop}$ and $a, b, c \in G$. For intuitive readings of formulas, the reader can be referred to Table 1. Note that $b <_a c$ stands for b is strictly more reliable than c, i.e., $(b \leqslant_a c) \wedge \neg(c \leqslant_a b)$, and $b \approx_a c$ which stands for b and c are equally reliable can be defined as $(b \leqslant_a c) \wedge (c \leqslant_a b)$. We define $\vee$, $\rightarrow$, $\leftrightarrow$ as ordinary abbreviations. Our syntax is different from [8] in at least two respects. First, we do not introduce the universal quantifier for agents. This is because we considered that it is *redundant* and most of the ideas in [8] are done without quantifiers for agents when the set of agents is finite, i.e., the universal quantifier for a finite domain is just reduced to a finite conjunction. Second, we relativize the notion of reliability ordering $\leqslant$ to each agent. In order to analyze our example from a logical perspective, we need to formalize belief change of a judge of the

---

[2] An English translation of articles can be referred from http://www.thailaws.com/.

Table 1. Examples of Static Logical Formalization

$\mathsf{Bel}(a, \varphi)$	: agent a believes that φ.
$\mathsf{Sign}(a, \varphi)$	: agent a signs statement φ.
$b \leqslant_a c$	: from agent a's perspective, agent b is at least as reliable as agent c.
$\mathsf{Sign}\big(a, \mathsf{Sign}(b, \varphi)\big)$	: agent a signs statement that agent b signs statement φ.
$\mathsf{Bel}\big(a, \mathsf{Sign}(b, \varphi)\big)$	: agent a believes that agent b signs statement φ.
$\mathsf{Bel}\big(a, b \leqslant_a c\big)$	: agent a believes that from agent a's perspective, agent b is at least as reliable as agent c.

Civil Court and we regard that belief change is induced by reliability change. However, there is no need for us to change the reliability ordering of the other agents other than the judge of the Civil Court. This is why we propose the notion of reliability ordering between agents depending on a particular agent's perspective.[3]

Let us provide Kripke semantics for our syntax. A *model* $\mathfrak{M}$ is a tuple

$$\mathfrak{M} = (W, (R_a)_{a \in G}, (S_a)_{a \in G}, (\preccurlyeq_a)_{a \in G}, V),$$

where W is a non-empty set of states, called *domain*, $R_a \subseteq W \times W$ is an accessibility relation representing beliefs, $S_a \subseteq W \times W$ is an accessibility relation representing signatures, $\preccurlyeq_a$ is a function which maps from W to $\mathcal{P}(G \times G)$ representing agent a's reliability ordering between agents, and $V : \mathsf{Prop} \to \mathcal{P}(W)$ is a valuation. In what follows, we simply write $b \preccurlyeq_a^w c$ for $(b, c) \in \preccurlyeq_a (w)$. For any binary relation X on W and any state $w \in W$, we write $X(w)$ to mean $\{v \in W | (w, v) \in X\}$.

Given any model $\mathfrak{M}$, any state $w \in W$, and any formula φ, we define the *satisfaction relation* $\mathfrak{M}, w \models \varphi$ inductively as follows:

$$\begin{array}{ll}
\mathfrak{M}, w \models p & \text{iff } w \in V(p) \\
\mathfrak{M}, w \models \neg\varphi & \text{iff } \mathfrak{M}, w \not\models \varphi \\
\mathfrak{M}, w \models \varphi \wedge \psi & \text{iff } \mathfrak{M}, w \models \varphi \text{ and } \mathfrak{M}, w \models \psi \\
\mathfrak{M}, w \models b \leqslant_a c & \text{iff } b \preccurlyeq_a^w c \\
\mathfrak{M}, w \models \mathsf{Sign}(a, \varphi) & \text{iff } \mathfrak{M}, v \models \varphi \text{ for all states } v \text{ such that } wS_a v \\
\mathfrak{M}, w \models \mathsf{Bel}(a, \varphi) & \text{iff } \mathfrak{M}, v \models \varphi \text{ for all states } v \text{ such that } wR_a v
\end{array}$$

A formula φ is *valid* in a model $\mathfrak{M}$ if $\mathfrak{M}, w \models \varphi$ for all states w of $\mathfrak{M}$.

Definition 1. *A model* $\mathfrak{M} = (W, (R_a)_{a \in G}, (S_a)_{a \in G}, (\preccurlyeq_a)_{a \in G}, V)$ *is a si-model (a* model for signed information) *if the following conditions are satisfied:*

---

[3] Ghosh et al. [9] also proposed the agent-dependent notion of reliability between agents, but the agent-dependent reliability in [9] is *rigid* in the sense that the same reliability relations from agent a's perspective hold for all states, while we relativize the notion of reliability to both agents and states, and also equip it with dynamics. We note that Ghosh et al. [9] considered several modal operators for positive and negative opinions for propositions and agents.

(i) R_a is transitive (wRv and vRu jointly imply wRu for all states w, v, u)
and Euclidean (wRv and wRu jointly imply vRu for all states w, v, u).

(ii) S_a is serial (for any state w, there is some state v such that $wS_a v$), transitive and Euclidean.

(iii) $\preccurlyeq_a^w \subseteq G \times G$ is a total pre-ordering between agents, i.e., $\preccurlyeq_a^w$ is reflexive ($b \preccurlyeq_a^w b$ for all agents b), transitive, and comparable (for any agents b and c, $b \preccurlyeq_a^w c$ or $c \preccurlyeq_a^w b$).

The first and second items of this definition ensure us that we never sign a contradiction (due to seriality of S_a), and $\mathsf{Bel}(a, \cdot)$ and $\mathsf{Sign}(a, \cdot)$ are both positively and negatively introspective. Corresponding to these constraints, we can obtain the following validities.

Proposition 1. *The following are valid in all si-models: for all* a, b, $c \in G$,

(i) $\mathsf{Bel}(a, p) \to \mathsf{Bel}(a, \mathsf{Bel}(a, p))$ *and* $\neg\mathsf{Bel}(a, p) \to \mathsf{Bel}(a, \neg\mathsf{Bel}(a, p))$.

(ii) $\neg\mathsf{Sign}(a, \bot)$, $\mathsf{Sign}(a, p) \to \mathsf{Sign}(a, \mathsf{Sign}(a, p))$, *and*
$\neg\mathsf{Sign}(a, p) \to \mathsf{Sign}(a, \neg\mathsf{Sign}(a, p))$.

(iii) $b \leqslant_a b$, $(b \leqslant_a c \wedge c \leqslant_a d) \to b \leqslant_a d$, *and* $b \leqslant_a c \vee c \leqslant_a b$.

Based on Definition 1 and the idea of [8], we can rank agents by giving a partition $(\mathsf{C}_i^a)_{i \leq M}$ to G, where M is a natural number representing the maximum rank (such M always exists because G is finite) and we read $c \in \mathsf{C}_i^a$ as 'from agent a's viewpoint, the rank of agent c is i'. As a result, the agents who are equally reliable are categorized in the same group. We define $(\mathsf{C}_i^a)_{i \leq M}$ inductively as follows. C_1^a which stands for 'a group of agents which is the most reliable from a's perspective' can be defined by the following formula:

$$c \in \mathsf{C}_1^a := \bigwedge_{b \in G}(c \leqslant_a b),$$

where we recall that G is a finite set of agents and $a, b, c \in G$. Then, we can rank the group of agents C_i^a such that $i > 1$ as follows:

$$c \in \mathsf{C}_i^a := \left(\left(\bigwedge_{1 \leq j \leq i-1} \neg(c \in \mathsf{C}_j^a) \right) \wedge \left(\bigwedge_{b \in G} \left(\left(\bigwedge_{1 \leq j \leq i-1} \neg(b \in \mathsf{C}_j^a) \right) \to (c \leqslant_a b) \right) \right) \right).$$

This implies that all agents in C_i^a are equally reliable, and if $i <_\mathbb{N} j$ then $c <_a b$ for all agents $c \in \mathsf{C}_i^a$ and agent $b \in \mathsf{C}_j^a$. Note that we relativize the notion C_i^a to a specified agent a because the notion of reliability ordering $\leqslant_a$ depends on a. This is a difference from [8] because [8] did not consider C_i depending on a specified agent.

Theorem 1. *The set of all valid formulas on all si-models is axiomatized by:*

- *All instances of propositional tautologies*
- $\mathsf{Bel}(a, p \to q) \to (\mathsf{Bel}(a, p) \to \mathsf{Bel}(a, q))$ $(a \in G)$
- $\mathsf{Sign}(a, p \to q) \to (\mathsf{Sign}(a, p) \to \mathsf{Sign}(a, q))$ $(a \in G)$
- *From* φ *we may infer* $\mathsf{Bel}(a, \varphi)$ $(a \in G)$
- *From* φ *we may infer* $\mathsf{Sign}(a, \varphi)$ $(a \in G)$
- *Uniform substitution and modus ponens,*

as well as all listed formulas of Proposition 1.

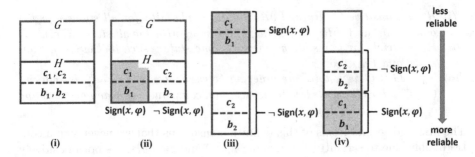

Fig. 1. Downgrading and Upgrading. (iii) is an effect of downgrading $[H \Downarrow^a_\varphi]$ to (ii), and (iv) is an effect of upgrading $[H \Uparrow^a_\varphi]$ to (ii).

3.2 Downgrade and Upgrade Operations for Agents

In order to change a reliability ordering between agents from a particular agent's perspective, we introduce two dynamic operators, i.e., the downgrade operator $[H \Downarrow^a_\varphi]$ and the upgrade operator $[H \Uparrow^a_\varphi]$, where $H \subseteq G$ is a set of agents. Our intended reading of $[H \Downarrow^a_\varphi]\psi$ is 'after the agent a downgraded such agents who sign the statement φ in H, ψ holds', and we can read $[H \Uparrow^a_\varphi]\psi$ as 'after the agent a upgraded such agents who sign the statement φ in H, ψ holds'. Semantically speaking, $[H \Downarrow^a_\varphi]$ makes such agents who sign φ in H less reliable than all the other agents, and $[H \Uparrow^a_\varphi]$ makes such agents who sign φ in H more reliable than all the other agents.

Before giving a detailed semantics, let us demonstrate the effects of $[H \Downarrow^a_\varphi]$ and $[H \Uparrow^a_\varphi]$ by figures. Firstly, we assume that a rectangle G of Fig. 1(i) represents a fixed finite set of agents. Secondly, we will select a specified set of agents in order to change their reliability ordering that can be represented by a rectangle H, and we assume that $b_1 \approx_a b_2 <_a c_1 \approx_a c_2$ holds, i.e., agents b_1 and b_2 which are equally reliable are more reliable than agents c_1 and c_2 which are equally reliable from agent a's perspective. In this sense, b_1, b_2, c_1 and c_2 are situated as in Fig. 1 (i). Then, if we focus on the agents who sign the statement φ, H is divided into two equal vertical parts by $\mathsf{Sign}(x, \varphi)$ as in Fig. 1(ii), namely by the set $\{x \in H \mid \mathfrak{M}, w \models \mathsf{Sign}(x, \varphi)\}$ and the set $\{x \in H \mid \mathfrak{M}, w \models \neg\mathsf{Sign}(x, \varphi)\}$. Next, if agent a *downgrades* all the agents signing the statement φ in H, we downgrade all of them less reliable than the other agents as in Fig. 1(iii). On the other hand, if agent a *upgrades* all the agents signing the statement φ in H, we upgrade all of them more reliable than the other agents as in Fig. 1(iv).[4]

---

[4] When $b < c$ (read: "b is more reliable than c") holds in a partial (pre-) ordering, then the first argument b comes into the lower position than the second argument c, e.g., in Hasse diagram (cf. [10]). This is the same usage as in Lorini et al. [8]. To keep our geometric intuition for '*up*-' or '*down*grading', $b < c$ may be read as "c is more reliable than b", but this would make the reader difficult to see differences and connections from the previous work.

Definition 2. *Given a Kripke model* $\mathfrak{M} = (W, (R_a)_{a \in G}, (S_a)_{a \in G}, (\preccurlyeq_d)_{d \in G}, V)$, *a semantic clause for* $[H \Downarrow_\varphi^a]$ *on* $\mathfrak{M}$ *and* $w \in W$ *is defined by:*

$$\mathfrak{M}, w \models [H \Downarrow_\varphi^a]\psi \text{ iff } \mathfrak{M}^{H \Downarrow_\varphi^a}, w \models \psi,$$

where $\mathfrak{M}^{H \Downarrow_\varphi^a} = (W, (R_a)_{a \in G}, (S_a)_{a \in G}, (\preccurlyeq'_d)_{d \in G}, V)$ *and* $\preccurlyeq'_d$ *is defined as: for all* $u \in W$:

- *if* $d \neq a$, *we put* $\preccurlyeq_d'^u = \preccurlyeq_d^u$.
- *otherwise (if* $d = a$), *we define* $b \preccurlyeq_a'^u c$ *iff*

  $\big(b, c \in H$ *and* $\mathfrak{M}, u \models \mathsf{Sign}(b, \varphi) \wedge \mathsf{Sign}(c, \varphi)$ *and* $b \preccurlyeq_a^u c\big)$ *or*

  $\big(b, c \in (G \setminus H) \cup \{x \in H \mid \mathfrak{M}, u \models \neg\mathsf{Sign}(x, \varphi)\}$ *and* $b \preccurlyeq_a^u c\big)$ *or*

  $\big(b \in (G \setminus H) \cup \{x \in H \mid \mathfrak{M}, u \models \neg\mathsf{Sign}(x, \varphi)\}$ *and* $c \in H$ *and* $\mathfrak{M}, u \models \mathsf{Sign}(c, \varphi)\big)$.[5]

Definition 3. *Given a Kripke model* $\mathfrak{M} = (W, (R_a)_{a \in G}, (S_a)_{a \in G}, (\preccurlyeq_d)_{d \in G}, V)$, *a semantic clause for* $[H \Uparrow_\varphi^a]$ *on* $\mathfrak{M}$ *and* $w \in W$ *is defined by:*

$$\mathfrak{M}, w \models [H \Uparrow_\varphi^a]\psi \text{ iff } \mathfrak{M}^{H \Uparrow_\varphi^a}, w \models \psi,$$

where $\mathfrak{M}^{H \Uparrow_\varphi^a} = (W, (R_a)_{a \in G}, (S_a)_{a \in G}, (\preccurlyeq'_d)_{d \in G}, V)$ *and* $\preccurlyeq'_d$ *is defined as: for all* $u \in W$:

- *if* $d \neq a$, *we put* $\preccurlyeq_d'^u = \preccurlyeq_d^u$.
- *otherwise (if* $d = a$), *we define* $b \preccurlyeq_a'^u c$ *iff*

  $\big(b, c \in H$ *and* $\mathfrak{M}, u \models \mathsf{Sign}(b, \varphi) \wedge \mathsf{Sign}(c, \varphi)$ *and* $b \preccurlyeq_a^u c\big)$ *or*

  $\big(b, c \in (G \setminus H) \cup \{x \in H \mid \mathfrak{M}, u \models \neg\mathsf{Sign}(x, \varphi)\}$ *and* $b \preccurlyeq_a^u c\big)$ *or*

  $\big(c \in (G \setminus H) \cup \{x \in H \mid \mathfrak{M}, u \models \neg\mathsf{Sign}(x, \varphi)\}$ *and* $b \in H$ *and* $\mathfrak{M}, u \models \mathsf{Sign}(b, \varphi)\big)$.

(see Footnote 5)

Proposition 2. *If* $\mathfrak{M}$ *is a si-model, then both* $\mathfrak{M}^{H \Uparrow_\varphi^a}$ *and* $\mathfrak{M}^{H \Downarrow_\varphi^a}$ *are si-models.*

Proposition 3 (Recursive Validities). *The following are valid on all models. Moreover, if* ψ *is valid on all models, then* $[H \Downarrow_\varphi^a]\psi$ *is also valid on all models.*

$$
\begin{array}{lll}
[H \Downarrow_\varphi^a]p & \leftrightarrow \; p & \\
[H \Downarrow_\varphi^a](b \leqslant_d c) & \leftrightarrow \; b \leqslant_d c & (d \neq a) \\
[H \Downarrow_\varphi^a](b \leqslant_a c) & \leftrightarrow \; b \leqslant_a c & (b, c \in G \setminus H) \\
[H \Downarrow_\varphi^a](b \leqslant_a c) & \leftrightarrow \; \big(\mathsf{Sign}(b, \varphi) \wedge \mathsf{Sign}(c, \varphi) \wedge (b \leqslant_a c)\big) \vee & \\
& \quad \big(\neg\mathsf{Sign}(b, \varphi) \wedge \neg\mathsf{Sign}(c, \varphi) \wedge (b \leqslant_a c)\big) \vee & \\
& \quad \big(\neg\mathsf{Sign}(b, \varphi) \wedge \mathsf{Sign}(c, \varphi)\big) & (b, c \in H) \\
[H \Downarrow_\varphi^a](b \leqslant_a c) & \leftrightarrow \; \mathsf{Sign}(c, \varphi) \vee \big(\neg\mathsf{Sign}(c, \varphi) \wedge (b \leqslant_a c)\big) & (c \in H, b \in G \setminus H) \\
[H \Downarrow_\varphi^a](b \leqslant_a c) & \leftrightarrow \; \neg\mathsf{Sign}(b, \varphi) \wedge (b \leqslant_a c) & (b \in H, c \in G \setminus H) \\
[H \Downarrow_\varphi^a]\neg\psi & \leftrightarrow \; \neg[H \Downarrow_\varphi^a]\psi & \\
[H \Downarrow_\varphi^a](\psi_1 \wedge \psi_2) & \leftrightarrow \; [H \Downarrow_\varphi^a]\psi_1 \wedge [H \Downarrow_\varphi^a]\psi_2 & \\
[H \Downarrow_\varphi^a]\mathsf{Sign}(b, \psi) & \leftrightarrow \; \mathsf{Sign}(b, [H \Downarrow_\varphi^a]\psi) & \\
[H \Downarrow_\varphi^a]\mathsf{Bel}(b, \psi) & \leftrightarrow \; \mathsf{Bel}(b, [H \Downarrow_\varphi^a]\psi) &
\end{array}
$$

---

[5] In this case, since there is no relation between agents b and c, $b \preccurlyeq_a^u c$ is omitted.

Proposition 4 (Recursive Validities). *The following are valid on all models. Moreover, if ψ is valid on all models, then $[H \Uparrow_\varphi^a]\psi$ is also valid on all models.*

$$
\begin{aligned}
&[H \Uparrow_\varphi^a]p &&\leftrightarrow\quad p \\
&[H \Uparrow_\varphi^a](b \leqslant_d c) &&\leftrightarrow\quad b \leqslant_d c &&(d \neq a) \\
&[H \Uparrow_\varphi^a](b \leqslant_a c) &&\leftrightarrow\quad b \leqslant_a c &&(b, c \in G \setminus H) \\
&[H \Uparrow_\varphi^a](b \leqslant_a c) &&\leftrightarrow\quad \big(\mathsf{Sign}(b, \varphi) \wedge \mathsf{Sign}(c, \varphi) \wedge (b \leqslant_a c)\big) \vee \\
& && \quad\quad \big(\neg\mathsf{Sign}(b, \varphi) \wedge \neg\mathsf{Sign}(c, \varphi) \wedge (b \leqslant_a c)\big) \vee \\
& && \quad\quad \big(\mathsf{Sign}(b, \varphi) \wedge \neg\mathsf{Sign}(c, \varphi)\big) &&(b, c \in H) \\
&[H \Uparrow_\varphi^a](b \leqslant_a c) &&\leftrightarrow\quad \neg\mathsf{Sign}(c, \varphi) \wedge (b \leqslant_a c) &&(c \in H, b \in G \setminus H) \\
&[H \Uparrow_\varphi^a](b \leqslant_a c) &&\leftrightarrow\quad \mathsf{Sign}(b, \varphi) \vee \big(\neg\mathsf{Sign}(b, \varphi) \wedge (b \leqslant_a c)\big) &&(b \in H, c \in G \setminus H) \\
&[H \Uparrow_\varphi^a]\neg\psi &&\leftrightarrow\quad \neg[H \Uparrow_\varphi^a]\psi \\
&[H \Uparrow_\varphi^a](\psi_1 \wedge \psi_2) &&\leftrightarrow\quad [H \Uparrow_\varphi^a]\psi_1 \wedge [H \Uparrow_\varphi^a]\psi_2 \\
&[H \Uparrow_\varphi^a]\mathsf{Sign}(b, \psi) &&\leftrightarrow\quad \mathsf{Sign}(b, [H \Uparrow_\varphi^a]\psi) \\
&[H \Uparrow_\varphi^a]\mathsf{Bel}(b, \psi) &&\leftrightarrow\quad \mathsf{Bel}(b, [H \Uparrow_\varphi^a]\psi)
\end{aligned}
$$

3.3   Private Announcements

This section introduces a new dynamic operator for private announcement $[\varphi \rightsquigarrow a]$ (whose reading is "after a private announcement of φ to agent a"), where the idea is realized by the property that the other agent than a will not notice a's belief change. One of the merits of this operator is that a sender of message φ is not specified, while a recipient is defined as agent a. This means that we may use this operator also for self-decision of agent a, i.e., the sender and the recipient are the same. This section demonstrates that $[\varphi \rightsquigarrow a]$ can capture both (i) the *tell*-action $[\mathsf{Tell}(b, a, \varphi)]$ from [8]: "agent b tells to agent a that a certain statement φ is true" and (ii) one of aggregation policies from [8] called the *careful policy*. We note that Lorini et al. [8] did not propose a logical treatment from dynamic epistemic viewpoints for any aggregation policies. Moreover, we note that the sender and the recipient are regarded as the same to capture the careful policy by our new operator $[\varphi \rightsquigarrow a]$.

Action Model for Private Announcements. In order to capture this private action, we introduce the following special structure (called *action model* in dynamic epistemic logic, the reader may find a similar structure in [4, 11]).

Definition 4. *The action model for private announcements of φ to agent a is a tuple $(E, (D_c)_{c \in G}, (U_a)_{a \in G}, \mathrm{pre})$ such that E consists of two actions: φ-announcing action $!_\varphi$ to agent a and non-announcing action $\top$, and $D_a = \{(!_\varphi, !_\varphi), (\top, \top)\}$ and $D_c = \{(!_\varphi, \top), (\top, \top)\}$ if $c \neq a$, $U_c = \{(!_\varphi, \top), (\top, \top)\}$ for all $c \in G$, and pre assigns a precondition to each action by $\mathrm{pre}(!_\varphi) = \varphi$ and $\mathrm{pre}(\top) = \top$.*

Definition 5. *Given a Kripke model $\mathfrak{M} = (W, (R_c)_{c \in G}, (S_c)_{c \in G}, (\preccurlyeq_c)_{c \in G}, V)$, a semantic clause for $[\varphi \rightsquigarrow a]\psi$ on $\mathfrak{M}$ and $w \in W$ is defined as follows:*

$$
\mathfrak{M}, w \models [\varphi \rightsquigarrow a]\psi \text{ iff } \mathfrak{M}^{\varphi \rightsquigarrow a}, (w, !_\varphi) \models \psi,
$$

where $\mathfrak{M}^{\varphi \leadsto a} = (W', (R'_c)_{c \in G}, (S'_c)_{c \in G}, (\preccurlyeq'_c)_{c \in G}, V')$ is the updated model by the action model of Definition 4, i.e.,

- $W' := W \times E = W \times \{!_\varphi, \top\}$.
- $(w, e)R'_c(v, f)$ iff $wR_c v$ and $(e, f) \in D_c$ and $\mathfrak{M}, v \models \mathrm{pre}(f)$ (for all $c \in G$).
- $(w, e)S'_c(v, f)$ iff $wS_c v$ and $(e, f) \in U_c$ (for all $c \in G$).
- $d \preccurlyeq'^{(w,e)}_c d'$ iff $d \preccurlyeq^w_c d'$.
- $(w, e) \in V'(p)$ iff $w \in V(p)$.

Proposition 5. *If $\mathfrak{M}$ is a si-model, then $\mathfrak{M}^{\varphi \leadsto a}$ is also a si-model.*

Proposition 6 (Recursive Validities). *The following are valid on all models. Moreover, if ψ is valid on all models, then $[\varphi \leadsto a]\psi$ is also valid on all models.*

$$
\begin{aligned}
[\varphi \leadsto a]p &\leftrightarrow p \\
[\varphi \leadsto a]d \leqslant_c d' &\leftrightarrow d \leqslant_c d' \\
[\varphi \leadsto a]\neg\varphi &\leftrightarrow \neg[\varphi \leadsto a]\varphi \\
[\varphi \leadsto a](\psi \wedge \theta) &\leftrightarrow [\varphi \leadsto a]\psi \wedge [\varphi \leadsto a]\theta \\
[\varphi \leadsto a]\mathsf{Bel}(a, \psi) &\leftrightarrow \mathsf{Bel}(a, \varphi \to [\varphi \leadsto a]\psi) \\
[\varphi \leadsto a]\mathsf{Bel}(c, \psi) &\leftrightarrow \mathsf{Bel}(c, \psi) \quad (a \neq c) \\
[\varphi \leadsto a]\mathsf{Sign}(c, \psi) &\leftrightarrow \mathsf{Sign}(c, \psi)
\end{aligned}
$$

Note that the axiom $[\varphi \leadsto a]\mathsf{Bel}(c, \psi) \leftrightarrow \mathsf{Bel}(c, \psi)$ captures that the action of a's privately receiving message φ will not affect of the other agents' beliefs than a.

Theorem 2. *The set of all valid formulas of the expanded syntax of $\mathcal{L}$ with $[H \Downarrow^a_\varphi]$, $[H \Uparrow^a_\varphi]$ and $[\varphi \leadsto a]$ is axiomatized by the axiomatization of Theorem 1 as well as the axioms and the rules of Propositions 3, 4, and 6.*

First Application: Tell Action. An underlying idea of tell-action is that agent b *privately* tells φ to agent a, that is, the other agents than a would not notice this action. As a result, only agent a would change her belief by φ but the other agents than a would not change their beliefs. After the action, agent a would update her belief not only by the statement φ but also by the signed statement $\mathsf{Sign}(b, \varphi)$. Now we define:

$$[\mathsf{Tell}(b, a, \varphi)]\psi := [\mathsf{Sign}(b, \varphi) \leadsto a]\psi.$$

Then, we can recover all recursion axioms in [8] by Proposition 6. Especially, we obtain the following.

Proposition 7 (Successful Telling [8]). $[\mathsf{Tell}(b, a, \varphi)]\mathsf{Bel}\big(a, \mathsf{Sign}(b, \varphi)\big)$ *is valid in all si-models.*

This proposition is the essential aspect of tell-action. That is, after agent b tells to agent a information φ, agent a believes that agent b signs φ.

Second Application: Careful Policy. Lorini et al. [8] introduced several policies, as *meta-logical* principles, in order to decide which pieces of signed information an agent should believe. A common and rational policy is called a *careful policy*. An idea of this policy is to accept, as beliefs, the statements which are universally signed by a group of agents who are equally reliable. Firstly, we define $\mathsf{Sign}(\mathsf{C}_i^a, \varphi)$ which stands for 'all agents who are in the set C_i^a sign statement φ' by:

$$\mathsf{Sign}(\mathsf{C}_i^a, \varphi) := \bigwedge_{c \in \mathsf{C}_i^a} \left(\mathsf{Sign}(c, \varphi) \right).$$

We also introduce the following abbreviation, whose reading is "a believes that φ is universally signed by a group of agents who are equally reliable":

$$\mathsf{UniSign}(\varphi, a) := \bigvee_{i \leq M} \left(\mathsf{Bel}\big(a, \mathsf{Sign}(\mathsf{C}_i^a, \varphi)\big) \wedge \mathsf{Bel}\big(a, \bigwedge_{1 \leq j \leq i-1} \neg\mathsf{Sign}(\mathsf{C}_j^a, \neg\varphi)\big) \right),$$

where M is the maximum natural number of $\{i \leq \#G \mid \mathsf{C}_i^a \neq \emptyset\}$. Then, Lorini et al. [8]'s definition of careful policy is introduced as the following implication:

$$\mathsf{UniSign}(\varphi, a) \rightarrow \mathsf{Bel}(a, \varphi).$$

However, Lorini et al. did not discuss how we can handle the idea of careful policy in terms of dynamic operators, while they used the policy as a meta-logical principle. With the help of our private announcement operator $[\varphi \rightsquigarrow a]$, we now define the careful policy as a dynamic operator as follows:

$$[\mathsf{Careful}(a, \varphi)]\psi := \mathsf{UniSign}(\varphi, a) \rightarrow [\varphi \rightsquigarrow a]\psi,$$

where we may read $[\mathsf{Careful}(a, \varphi)]\psi$ as 'after agent a aggregates signed information about φ by the careful policy, ψ holds.' By Proposition 6, we obtain the following.

Proposition 8. *The following are valid in all si-models.*

(i) $[\mathsf{Careful}(a, p)]\mathsf{Bel}(a, p)$.
(ii) $[\mathsf{Careful}(a, \mathsf{Sign}(b, \varphi))]\mathsf{Bel}(a, \mathsf{Sign}(b, \varphi))$.

The first item of this proposition says that after agent a aggregates information about p by the careful policy, agent a now believes p. However, we cannot generalize the first item to an arbitrary formula φ,[6] while the second item of this proposition still holds.

---

[6] For example, if we define a formula φ by $p \wedge \neg\mathsf{Bel}(a, p)$, then $[\mathsf{Careful}(a, \varphi)]\mathsf{Bel}(a, \varphi)$ cannot hold, since the rewritten equivalent formula (by Proposition 6) becomes $\mathsf{UniSign}(\varphi, a) \rightarrow \mathsf{Bel}(a, (p \rightarrow \mathsf{Bel}(a, p)))$, which is not valid in all *si*-models.

4   Dynamic Logical Analysis of Target Legal Case

Let us move back to our legal case. In order to analyze reliability change from the judge's perspective, we will only focus on the Civil Court. We will not consider the inquiry stage because there is no change of reliability. The Appeal Court and the Supreme Court are also excluded because they only adopted the result of the Civil Court. Furthermore, we will simplify the target legal case by removing agent f_1 in order to avoid an unnecessary complication (this is not an essential point for our analysis).

In the Civil Court, the set G of agents is $\{po, v, f_2, mo, j\}$, where we recall that po, v, f_2, mo are agents of four witnesses, and j is a judge of the Civil Court. For the statement involving the legal case, we consider only one propositional letter p whose reading is "d is the offender" that provides information who is the offender. We assume at first that all witnesses are equally reliable for j as follows:

$$\mathfrak{M}, w \models \mathsf{Bel}(j,\ v \approx_j f_2 \approx_j mo \approx_j po).$$

In the trial, the witness v told a piece of information which is different from the inquiry stage to j. The first action is $T_1 := \mathsf{Tell}(v, j, \neg p)$. Then, po was called to be a witness and told the received information in the inquiry stage to j that can be represented by the following tell-actions.

$$T_2 := \mathsf{Tell}\big(po, j, \mathsf{Sign}(v, p)\big), \ T_3 := \mathsf{Tell}\big(po, j, \mathsf{Sign}(f_2, \neg p)\big), \ T_4 := \mathsf{Tell}\big(po, j, \mathsf{Sign}(mo, p)\big)$$

After that, j will believe the following information by Proposition 7.

$$\mathfrak{M}, w \models [T_1][T_2][T_3][T_4]\mathsf{Bel}\left(\begin{array}{c} j,\ \mathsf{Sign}(v, \neg p)\ \wedge \\ \mathsf{Sign}\big(po, \mathsf{Sign}(v, p) \wedge \mathsf{Sign}(f_2, \neg p) \wedge \mathsf{Sign}(mo, p)\big)\end{array}\right)$$

Based on these pieces of information alone, j cannot decide which pieces of information should believe. This is firstly because (P1) if j considers the reliability of information sources, j cannot distinguish all witnesses in terms of the reliability ordering because they are equally reliable. Moreover, (P2) there is contradicting pieces of signed information about p from witnesses. So, j cannot decide which signed information should be in j's belief, i.e., p or $\neg p$. We use the following two ideas: (i) reliability change, and (ii) aggregation policy, to resolve the above problems (P1) and (P2).

(i) Reliability change: The downgrade and upgrade operators of Sect. 3 are applied in order to simulate the effect of reliability change of the judge in the Civil Court.[7] This allows us to solve the above problem (P1). We also note that, if we apply a framework based on [8], a reliability relation between agents is *fixed*, i.e., the reliability relation between agents cannot be changed.

---

[7] In this work, we will not analyze how an agent decides to change the reliability ordering between the other agents, as this is a psychological issue and is out of our scope.

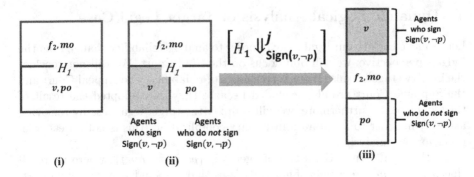

Fig. 2. Downgrading by $[H_1 \Downarrow^j_{\text{Sign}(v,\neg p)}]$

(ii) Aggregation policy: The reformulation of the careful policy (in Sect. 3) is employed in order to allow the judge of the Civil Court to decide which pieces of the received signed information should believe.

Now let us apply our two ideas to dissolve the judge's difficulty in deciding which pieces of information she should believe. In what follows, we assume that j is the judge in the Civil Court, and define $\mathfrak{M}'$ by the updated model of $\mathfrak{M}$ after the tell-actions T_1–T_4.

From the tell-actions T_1–T_4, there is conflicting information about p. That is, v told statement $\neg p$ (by T_1), while po told signed statement p by v (by T_2). So, j now believes both $\text{Sign}(v, \neg p)$ and $\text{Sign}(po, \text{Sign}(v, p))$. Since the signature operator $\text{Sign}(a, \cdot)$ is positively introspective, note that $\text{Sign}(v, \neg p)$ implies $\text{Sign}(v, \text{Sign}(v, \neg p))$. From Sect. 2, we may regard that j believes that the signed information of v in the Civil Court is less reliable than that in the inquiry stage. This means that $\text{Sign}(v, \neg p)$ is not reliable information for j, and so, we regard that j downgrades all agents between po and v who sign the statement $\text{Sign}(v, \neg p)$ by $[H_1 \Downarrow^j_{\text{Sign}(v, \neg p)}]$, where we define $H_1 = \{v, po\}$ is a set of agents of witnesses in the Civil Court (see Fig. 2(i)). Let us see a process of downgrading step by step (see Fig. 2). When we consider the agents who sign the statement $\text{Sign}(v, \neg p)$, H_1 is divided into two equal vertical parts by $\text{Sign}(x, \text{Sign}(v, \neg p))$ as in Fig. 2(ii). Next, j downgrades all agents in H_1 who sign $\text{Sign}(v, \neg p)$ (recall that $\text{Sign}(v, \text{Sign}(v, \neg p))$ holds), and the result can be shown as in Fig. 2(iii). That is, the agent v becomes less reliable than all the other agents. Note that the agents who are in the same part are equally reliable. Thus, j changes her belief about the reliability ordering as follows:

$$\mathfrak{M}', w \models [H_1 \Downarrow^j_{\text{Sign}(v, \neg p)}]\text{Bel}(j, po <_j f_2 \approx_j mo <_j v).$$

Since po now becomes the most reliable agent according to j, j can accept the signed statements by po by our careful policy as follows:

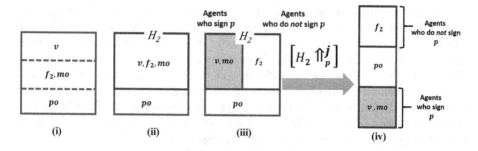

Fig. 3. Upgrading by $[H_2 \Uparrow_p^j]$

$$\mathfrak{M}', w \models [H_1 \Downarrow_{\mathsf{Sign}(v, \neg p)}^j][\mathsf{Careful}(j, \mathsf{Sign}(v, p) \wedge \mathsf{Sign}(f_2, \neg p) \wedge \mathsf{Sign}(mo, p))]$$

$$\mathsf{Bel}\left(j, \ \mathsf{Sign}(v, p) \wedge \mathsf{Sign}(f_2, \neg p) \wedge \mathsf{Sign}(mo, p)\right),$$

where we also note that the assumption of the careful policy holds, i.e., $\mathfrak{M}', w \models [H_1 \Downarrow_{\mathsf{Sign}(v, \neg p)}^j]\mathsf{UniSign}(\mathsf{Sign}(v, p) \wedge \mathsf{Sign}(f_2, \neg p) \wedge \mathsf{Sign}(mo, p), j)$ holds. Let us denote $\mathfrak{M}''$ by the updated model of $\mathfrak{M}'$ after the above downgrading and the careful policy.

Since j believes that the signed information of v in the Civil Court is less reliable than that in the inquiry stage, we can regard that j believes that the signed information p of v in the inquiry stage is more reliable. Thus, j upgrades all agents who sign the statement p by $[H_2 \Uparrow_p^j]$, where H_2 is defined by $\{v, f_2, mo\}$ as in Fig. 3(ii) (because j focuses on the inquiry stage). Figure 3(i) is the initial reliability ordering for j before upgrading. When we consider the statement p, H_2 is divided into two equal vertical parts by $\mathsf{Sign}(x, p)$ as in Fig. 3(iii). By $[H_2 \Uparrow_p^j]$, agents v and mo who sign the statement p are upgraded to be more reliable than all the other agents as in Fig. 3(iv). Consequently, j changes her reliability ordering between all witnesses as follows:

$$\mathfrak{M}'', w \models [H_2 \Uparrow_p^j]\mathsf{Bel}(\ j, \ v \approx_j mo <_j po <_j f_2).$$

Since now mo and v become most reliable agents according to j, j now successfully aggregates information p by the careful policy again and will believe that d is the offender (p) as follows:

$$\mathfrak{M}'', w \models [H_2 \Uparrow_p^j][\mathsf{Careful}(j, p)] \ \mathsf{Bel}(j, p).$$

Let us denote $\mathfrak{M}'''$ by the updated model of $\mathfrak{M}''$ after $[H_2 \Uparrow_p^j]$ and $[\mathsf{Careful}(j, p)]$. Therefore, $\mathfrak{M}''', w \models \mathsf{Bel}(j, p)$.

5   Conclusion

This work has proposed logical analysis for formalizing reliability change of an agent. We introduced two dynamic operators: downgrading $[H \Downarrow_\varphi^a]$ and upgrading $[H \Uparrow_\varphi^a]$. The first operator downgrades such agents who sign φ in H, while the second operator upgrades them. Based on these operators, we have formalized an example of a legal case. In the trials, the judge first believed that all witnesses are equally reliable. Then, the judge changed her belief about the reliability ordering between witnesses. We can successfully analyze this process by downgrading and upgrading the reliability of the witnesses. Moreover, we reformulated the careful policy [8], which allows an agent to decide which signed information should believe, in terms of dynamic operators, i.e., [Careful(a, φ)]. Our contribution is to formalize the change of the reliability ordering between the other agents depending on an agent's perspective.

In this work, we only capture an effect of reliability change on belief change, i.e., when a judge changed her reliability ordering between some witnesses, she may change her beliefs about information from those witnesses. On the other hand, belief change may affect reliability change. In this sense, our work just supposes that the judge changes her reliability based on her belief change, but does not analyze how belief change affects reliability change. Therefore, we plan to formalize an effect of belief change on reliability change by applying the notion of preference upgrade in [12]. Furthermore, this work only formalizes the reliability of agents, but does not consider the reliability of statements. That is, this work assumes that when agent a received a statement φ from agent b, agent a has already decided if the statement φ is reliable or not. If agent a considers that the statement φ is not reliable, then she believes that agent b who gives the statement φ will be unreliable. However, we may analyze such reliability change of statements by employing a preference modality based on [12] or the framework by [9].[8]

A   Complete Axiomatization of Dynamic Logic

A.1   Proof of Theorem 1

Proof. Let us write our axiomatization by $\mathbf{BS}_{\leqslant}$. We show that any unprovable formula φ in $\mathbf{BS}_{\leqslant}$ is falsified in some si-model and we basically follow the standard techniques, e.g. found in [13]. Let φ be an unprovable formula in $\mathbf{BS}_{\leqslant}$. We define the canonical model $\mathfrak{M}$ where φ is falsified at some point of $\mathfrak{M}$. We say that a set Γ of formulas is $\mathbf{BS}_{\leqslant}$-*consistent* (for short, *consistent*) if $\bigwedge \Gamma'$ is unprovable in $\mathbf{BS}_{\leqslant}$, for all finite subsets Γ' of Γ, and that Γ is *maximally consistent* if Γ is consistent and $\varphi \in \Gamma$ or $\neg\varphi \in \Gamma$ for all formulas φ. Note that ψ is unprovable in $\mathbf{BS}_{\leqslant}$ iff $\neg\psi$ is $\mathbf{BS}_{\leqslant}$-consistent, for any formula ψ. We define the canonical model $\mathfrak{M} = (W, (R_a)_{a \in G}, (S_a)_{a \in G}, (\preccurlyeq_a)_{a \in G}, V)$, for $\mathbf{BS}_{\leqslant}$ by:

- W is the set of all maximal consistent sets;
- $\Gamma R_a \Delta$ iff (Bel$(a, \psi) \in \Gamma$ implies $\psi \in \Delta$) for all ψ;
- $\Gamma S_a \Delta$ iff (Sign$(a, \psi) \in \Gamma$ implies $\psi \in \Delta$) for all ψ;

[8] We would like to give our thanks to the anonymous reviewers, who gave useful comments on this paper. We also thank the participants at JURISIN 2014 who commented on our draft. The work of the second author was partially supported by JSPS KAKENHI, Grant-in-Aid for Young Scientists (B) 24700146.

– $b \prec_a^\Gamma c$ iff $b \leqslant_a c \in \Gamma$;
– $\Gamma \in V(p)$ iff $p \in \Gamma$.

Then, we can show the following equivalence (Truth Lemma [13, Lemma 4.21]): $\mathfrak{M}, \Gamma \models \psi$ iff $\psi \in \Gamma$ for all formulas ψ and $\Gamma \in W$. Given any unprovable formula φ in $\mathbf{BS}_\leqslant$, we can find a maximal consistent set Δ such that $\neg\varphi \in \Gamma$. Then, by the equivalence above, φ is falsified at Δ of the canonical model $\mathfrak{M}$ for $\mathbf{BS}_\leqslant$, where we can assure that $\mathfrak{M}$ is our intended si-model by axioms of Proposition 1.   □

A.2   Proof of Theorem 2

Proof. By $\vdash \psi$ (or $\vdash^+ \psi$), we mean that ψ is a theorem of the axiomatization $\mathbf{BS}_\leqslant$ in the previous proof (or, the axiomatization $\mathbf{BS}_\leqslant^+$ given in the statement of Theorem 2, respectively.) As for the completeness part, we can reduce the completeness of our dynamic extension to the static counterpart (i.e., Theorem 1) as follows. With the help of the axioms of Propositions 3, 4, and 6, we can define a mapping t sending a formula ψ of the expanded syntax (we denote this by $\mathcal{L}^+$ below) possibly with three kinds of dynamic operators (i.e., $[H \Downarrow_\varphi^a]$, $[H \Uparrow_\varphi^a]$, and $[\varphi \rightsquigarrow a]$) to a formula $t(\psi)$ of the original syntax $\mathcal{L}$. For this aim, we employ *inside-out strategy*, i.e., we start rewriting the *innermost occurrences* of three kinds of dynamic operators. (So, we do not need to consider an axiom for iterated dynamic operators such as $[\varphi \rightsquigarrow a][\psi \rightsquigarrow a]$ or $[\varphi \rightsquigarrow a][H \Uparrow_\varphi^a]$.) For example, if one of the innermost dynamic operators is $[\varphi \rightsquigarrow a]$, then we cannot find any occurrences of three kinds of dynamic operators. For inside-out strategy, we need to have the following inference rules for dynamic operators:

$$\frac{\psi \leftrightarrow \psi'}{[H \Downarrow_\varphi^a]\psi \leftrightarrow [H \Downarrow_\varphi^a]\psi'} \quad \frac{\psi \leftrightarrow \psi'}{[H \Uparrow_\varphi^a]\psi \leftrightarrow [H \Uparrow_\varphi^a]\psi'} \quad \frac{\psi \leftrightarrow \psi'}{[\varphi \rightsquigarrow a]\psi \leftrightarrow [\varphi \rightsquigarrow a]\psi'},$$

to assure the replacement of equivalent formulas inside of a formula. But, these rules are derivable from the corresponding necessitation laws and the reduction axioms for the negation and the conjunction in Propositions 3, 4, and 6. Then, for this mapping t, we can show that $\psi \leftrightarrow t(\psi)$ is valid on all si-models and $\vdash^+ \psi \leftrightarrow t(\psi)$. Then, we can proceed as follows. Fix any formula ψ of $\mathcal{L}^+$ such that ψ is valid on all si-models. By the validity of $\psi \leftrightarrow t(\psi)$ on all si-models, we obtain that $t(\psi)$ is valid on all si-models. By Theorem 1, $\vdash t(\psi)$, which implies $\vdash^+ t(\psi)$. Finally, it follows from $\vdash^+ \psi \leftrightarrow t(\psi)$ that $\vdash^+ \psi$, as desired.   □

References

1. Prakken, H., Sartor, G.: The role of logic in computational models of legal argument: a critical survey. In: Kakas, A.C., Sadri, F. (eds.) Computational Logic: Logic Programming and Beyond. LNCS (LNAI), vol. 2408, pp. 342–381. Springer, Heidelberg (2002)
2. Bench-Capon, T.J.M., Prakken, H.: Introducing the logic and law corner. J. Log. Comput. **18**(1), 1–12 (2008)

3. Grossi, D., Rotolo, A.: Logic in the law: a concise overview. Log. Philos. Today. Stud. Log. **30**, 251–274 (2011)
4. van Ditmarsch, H., van der Hoek, W., Kooi, B.: Dynamic Epistemic Logic. Springer, Netherlands (2008)
5. van Benthem, J.: Dynamic logic for belief revision. J. Appl. Non-Classic Log. **14**(2), 129–155 (2004)
6. Liau, C.J.: Belief, information acquisition, and trust in multi-agent systems-a modal logic formulation. Artif. Intell. **149**(1), 31–60 (2003)
7. Perrussel, L., Lorini, E., Thévenin, J.: From signed information to belief in multi-agent systems. In: Proceedings of The Multi-Agent Logics, Languages, and Organisations Federated Workshops (MALLOW 2010), Lyon, France, August 30 - September 2, 2010 (2010)
8. Lorini, E., Perrussel, L., Thévenin, J.-M.: A modal framework for relating belief and signed information. In: Leite, J., Torroni, P., Ågotnes, T., Boella, G., van der Torre, L. (eds.) CLIMA XII 2011. LNCS, vol. 6814, pp. 58–73. Springer, Heidelberg (2011)
9. Ghosh, S., Velázquez-Quesada, F.: Merging information. In: van Benthem, J., Gupta, A., Pacuit, E. (eds.) Games, Norms and Reasons: Logic at the Crossroads, Volume 353 of Synthese Library. Springer, Netherlands (2011)
10. Brüggemann, R., Patil, G.: Ranking and Prioritization for Multi-indicator Systems: Introduction to Partial Order Applications. Environmental and Ecological Statistics. Springer (2011)
11. Baltag, A., van Ditmarsch, H.P., Moss, L.S.: Epistemic logic and information update. In: Adriaans, P., van Benthem, J. (eds.) Handbook on the Philosophy of Information, pp. 361–456. Elsevier Science Publishers, Amsterdam (2008)
12. van Benthem, J., Liu, F.: Dynamic logic of preference upgrade. J. Appl. Non-Classical Log. **17**(2), 157–182 (2007)
13. Blackburn, P., de Rijke, M., Venema, Y.: Modal Logic. Cambridge Tracts in Theoretical Computer Science. Cambridge University Press, Cambridge (2001)

GABA 2014

Workshop on Graph-Based Algorithms for Big Data and Its Applications (GABA2014)

Yoshinobu Kawahara[1], Tetsuji Kuboyama[2], and Hiroshi Sakamoto[3]([✉])

[1] The Institute of Scientific and Industrial Research, Osaka Univeristy,
8-1 Mihogaoka, Ibaraki-shi, Osaka 567-0047, Japan
`ykawahara@sanken.osaka-u.ac.jp`
[2] Computer Centre, Gakushuin University, 1-5-1 Mejiro,
Toshima-ku, Tokyo 171-8588, Japan
`ori-gaba2014@tk.cc.gakushuin.ac.jp`
[3] Graduate School of Computer Science and Systems Engineering,
Kyushu Institute of Technology, Kawazu 680-4, Iizuka-shi 820-8502, Japan
`hiroshi@ai.kyutech.ac.jp`

1 The Workshop

The Workshop on Graph-based Algorithms for Big Data and its Applications (GABA2014) was held on November 23rd at Keio University in the 6th JSAI International Symposia on AI (JSAI-isAI 2014), sponsored by the Japan Society for Artificial Intelligence (JSAI). GABA2014 is the first workshop on subjects related to developing algorithms or data structures for discovering knowledge from large-scale graphs. Intelligent pre/post-processing plays a crucial role in knowledge discovery from big data. Counting of words, compression/decompression of row data, and segmentation of time series, etc. are those concrete examples and are embedded in many important applications. However, such a task becomes a critical part when processing whole data along with an increase of the data size. For this problem, many researchers have proposed novel data structures, algorithms, and frameworks for data use. Besides, we need a new approach for handling dynamic data streams and reconstructing veracious knowledge. We welcomed interesting results and ideas based on, but not restricted to, graph structures including string, tree, bipartite- and di-graph and their applications to Machine Learning and Knowledge Discovery. We first organized the program committee consisting of 11 researchers concerning with subjects in the workshop scope, and announced a call for papers. By the PC members, 14 submitted papers were accepted. More information on GABA2014 is available at the workshop Web site[1]. The proceedings were published from JSAI[2].

---

[1] https://sites.google.com/site/graph2014workshop.
[2] ISBN 978-4-915905-65-0 C3004(JSAI).

© Springer-Verlag Berlin Heidelberg 2015
T. Murata et al. (Eds.): JSAI-isAI 2014 Workshops, LNAI 9067, pp. 293–295, 2015.
DOI: 10.1007/978-3-662-48119-6_21

2   Post-workshop Proceedings

Five papers out of 14 papers presented in the workshop were selected to be published in this post-workshop proceedings after revision. Each of them was peer reviewed by three PC members, and external reviewers, which consists of two PC members previously assigned plus another. Two of the papers are focused on the problem of tree edit distance. One is about the anchored alignment tree and the other is the mapping kernels between ordered trees. The rest of the papers are related to the dimension reduction techniques applicable to the central point selection from data objects in database, the ambiguous pattern matching in compressed data, and the anomaly detection from structured graph. The abstracts of the five papers are following.

Ishizaka *et al.* formulated the anchored alignment problem, given two rooted labeled trees and an anchoring between them, to output an anchored alignment tree if it exists, where the notion of anchoring in trees was introduced in the context of forest alignments in bioinformatics. They showed that the problem can be solved in $O(ha^2 + n + m)$ time and in $O(ha)$ space where n, m are the number of nodes in the two trees, h is the height of the trees, and a is the cardinality of an anchoring.

Jin *et al.* proposed a binary quantization to select central points in database. The Simple-Map uses the distances between central points and objects as the coordinate values, and, in the previous researches, the candidates for central points are randomly selected from data objects. They improved this selection. As they reported, the coordinate value of central points obtained after the local search tend to be the maximum or minimum ends of the space. Consequently, the computation time of the Simple-Map is reduced to one-sixth compared with the conventional method.

Hirata *et al.* investigated several mapping kernels to count all of the mappings on beyond ordered trees: the cyclically ordered trees. They designed the algorithms to compute the corresponding mapping kernels in a polynomial-time in n, m: the number of nodes in two trees, D: the maximum degree of the trees, and d: the minimum degree of the trees. They also showed the $\sharp P$-completeness of two variants of the mapping kernel.

Maeda *et al.* developed the algorithm for the ambiguous pattern matching on compressed string. Given a grammar compressed string S, a pattern P, and a threshold $d \geq 0$, the problem is to find all occurrences of P' in S with $d(P', P) \leq d$ where $d(,)$ is the Hamming distance. They proposed the algorithm for this in $O(\lg \lg n \lg^* N(m + docc_d \lg \frac{m}{d} N))$ time, where $N = |S|$, $m = |P|$, n is the size of the grammar, and occ_d is an approximated occurrences of P. They implemented this algorithm and compared with a naive filtering on grammar compression.

Sugiyama and Otaki introduced the method for detecting anomalies from structured graph. To date, there exists no efficient method that works on massive attributed graphs with millions of vertices for detecting anomalous subgraphs with an abnormal distribution of vertex attributes. Using the recent graph cut-based formulation, this problem was solved efficiently. They examined the method using various sizes of synthetic and real-world datasets and show that their method is more than five orders of magnitude faster than the state-of-the-art method.

Acknowledgments. GABA2014 was closed successfully. We are grateful for the great support received from the program committee members: Hiroki Arimura, Kouichi Hirata, Nobuhiro Kaji, Miyuki Koshimura, Yoshiaki Okubo, Takeshi Shinohara, Yasuo Tabei, and Akihiro Yamamoto.They and other anonymous reviewers belong to the Special Interest Group on Fundamental Problems in AI (SIG-FPAI).Without their cooperation, the workshop have failed. We are thankful to Prof. Tsuyoshi Murata for his organization of JSAI-isAI 2014. We also thank Prof. Daisuke Bekki and Prof. Koji Mineshima for their arrangement to publish the LNAI volume of these post-workshop proceedings. Finally, we thank all speakers and all audiences who attended the workshop.

Anchored Alignment Problem for Rooted Labeled Trees

Yuma Ishizaka[1], Takuya Yoshino[1], and Kouichi Hirata[2]([✉])

[1] Graduate School of Computer Science and Systems Engineering,
Kyushu Institute of Technology, Kawazu 680-4, Iizuka 820-8502, Japan
{y_ishizaka,yoshino}@dumbo.ai.kyutech.ac.jp
[2] Department of Artificial Intelligence,
Kyushu Institute of Technology, Kawazu 680-4, Iizuka 820-8502, Japan
hirata@dumbo.ai.kyutech.ac.jp

Abstract. An *anchored alignment tree* between two rooted labeled trees
with respect to a mapping that is a correspondence between nodes in two
trees, called an *anchoring*, is an alignment tree which contains a node
labeled by a pair of labels for every pair of nodes in the anchoring. In
this paper, we formulate an *anchored alignment problem* as the problem,
when two rooted labeled trees and an anchoring between them are given
as input, to output an anchored alignment tree if there exists; to return
"no" otherwise. Then, we show that the anchored alignment problem
can be solved in $O(h\alpha^2 + n + m)$ time and in $O(h\alpha)$ space, where n
is the number of nodes in a tree, m is the number of nodes in another
tree, h is the maximum height of two trees and α is the cardinality of an
anchoring.

1 Introduction

An *anchored alignment tree* between two rooted labeled trees (trees, for short)
with respect to a mapping that is a correspondence between nodes in two trees,
called an *anchoring*, has been introduced by Schiermer and Giegerich [5] in the
context of forest alignments in bioinformatics. Then, the anchored alignment
tree is an alignment tree which contains a node labeled by a pair of labels for
every pair of nodes in the anchoring. By using an anchoring whose number is
α, we can obtain the anchored alignment tree in α times faster than the case
without using an anchoring [5].

Note first that an arbitrary anchoring between two trees does not always
provide an anchored alignment tree; If an anchoring is not less-constrained [4],
then there exists no alignment tree between them, because the less-constrained
mapping coincides with an alignable mapping [3], and then we can construct an
alignment tree from no less-constrained anchoring. Then, in this paper, we deal
with the following *anchored alignment problem*.

This work is partially supported by Grant-in-Aid for Scientific Research 24240021,
24300060, 25540137, 26280085 and 26370281 from the Ministry of Education, Culture, Sports, Science and Technology, Japan.

T. Murata et al. (Eds.): JSAI-isAI 2014 Workshops, LNAI 9067, pp. 296–309, 2015.
DOI: 10.1007/978-3-662-48119-6_22

ANCHOREDALIGNMENT
INSTANCE: Two trees T_1 and T_2, and a mapping $M \subseteq V(T_1) \times V(T_2)$, called an *anchoring*.
SOLUTION: Find an *anchored alignment tree* $\mathcal{T}$ of T_1 and T_2 such that $\mathcal{T}$ contains a node labeled by $(l(v), l(w))$ for every $(v, w) \in M$ if $\mathcal{T}$ exists; return "no" otherwise.

Note that the anchored alignment tree as output is not necessary to be optimum in the sense of the alignment distance or the minimum cost alignment [2]; it is just an alignment tree between two trees containing nodes labeled by every pair of labels in an anchoring.

In order to solve the anchored alignment problem, in this paper, we provide an alternative proof that a less-constrained mapping coincides with an alignable mapping [3]. In this proof, first we introduce the cover sequences consisting of nodes of complete subtrees from a node in a mapping to the root. Then, we show that a mapping is less-constrained if and only if, for every pair of nodes in the mapping, the cover sequence of a tree and one in another tree are comparable. By using this property, we can prove the above theorem, according to following algorithm to solve the problem of ANCHOREDALIGNMENT. Here, n is the number of nodes in T_1, m is the number of nodes in T_2, h is the maximum height of T_1 and T_2 and α is the cardinality of an anchoring M.

First, we compute cover sequences of an anchoring and determine whether or not they are comparable in $O(h\alpha)$ time and space. If so, then next we construct an alignment subtree by aligning these cover sequences and by merging them in $O(h^2\alpha)$ time. Finally, we complete an anchored alignment tree by adding appropriate alignment subtrees to the merged alignment subtree in $O(n + m)$ time. Hence, we can solve the problem of ANCHOREDALIGNMENT in $O(h\alpha^2 + n + m)$ time and in $O(h\alpha)$ space.

Schiermer and Giegerich [5] have introduced the anchoring to divide the dynamic programming to compute the alignment distance [2] into α parts and claimed to reduce the time complexity from $O(nmD^2)$ time [2] to $O(nmD^2/\alpha)$ time, where D is the maximum degree of two trees. However, since the anchoring is not always less-constrained, we cannot guarantee that the division is correct. On the other hand, this paper determines whether or not the anchoring is less-constrained and, if so, then uses it to find the anchored alignment tree directly and correctly in $O(h\alpha^2 + n + m)$ time. When $n \geq m$, we can roughly estimate that $O(nmD^2/\alpha) = O(h\alpha^2 + n + m) = O(n^3)$. Hence, we can find the anchored alignment tree as fast as [5] even if an anchoring is less-constrained.

2   Preliminaries

A *tree* is a connected graph without cycles. For a tree $T = (V, E)$, we denote V and E by $V(T)$ and $E(T)$, respectively. We sometimes denote $v \in V(T)$ by $v \in T$. We denote an empty tree by $\emptyset$.

A *rooted tree* is a tree with one node r chosen as its *root*. We denote the root of a rooted tree T by $r(T)$. For each node v in a rooted tree with the root r, let

$UP_r(v)$ be the unique path from v to r. If $UP_r(v)$ has exactly k edges, then we say that the *depth* of v is k and denote it by $d(v) = k$. The *height* of T, denoted by $h(T)$, is defined as $\max\{dep(v) \mid v \in T\}$. The *parent* of $v(\neq r)$, which we denote by $par(v)$, is its adjacent node on $UP_r(v)$ and the *ancestors* of $v(\neq r)$ are the nodes on $UP_r(v) - \{v\}$. We say that u is a *child* of v if v is the parent of u. In this paper, we use the ancestor orders $<$ and $\leq$, that is, $u < v$ if v is an ancestor of u and $u \leq v$ if $u < v$ or $u = v$. In particular, we denote neither $u \leq v$ nor $v \leq u$ by $u \# v$. We say that w is the *least common ancestor* of u and v, denoted by $u \sqcup v$, if $u \leq w$, $v \leq w$ and there exists no w' such that $w' < w$, $u \leq w'$ and $v \leq w'$. A *(complete) subtree* of $T = (V, E)$ rooted *at* v, denoted by $T[v]$, is a tree $T' = (V', E')$ such that $r(T') = v$, $V' = \{u \in V \mid u \leq v\}$ and $E' = \{(u, w) \in E \mid u, w \in V'\}$.

A rooted tree is *labeled* if every node is labeled by some alphabet. A rooted tree is *ordered* if a left-to-right order among siblings is fixed; *unordered* otherwise. In particular, for nodes u and v in an ordered tree, u is *to the left of* v, denoted by $u \preceq v$, if $pre(u) \leq pre(v)$ and $post(u) \leq post(v)$ for the preorder number pre and the postorder number $post$. In this paper, we call a rooted labeled tree a *tree* simply. If it is necessary to distinguish, we call either ordered trees or unordered trees.

We say that two sets A and B are *incomparable* if none of $A \subset B$, $A = B$ and $B \subset A$ holds, that is, there exist both $a \in A \setminus B$ and $b \in B \setminus A$; *comparable* otherwise. Also we say that two sequences $A_1, \ldots, A_n$ and $B_1, \ldots, B_m$ of sets are *incomparable* if there exist i and j ($1 \leq i \leq n$, $1 \leq j \leq m$) such that A_i and B_j are incomparable; *comparable* otherwise. Furthermore, we call a sequence $A_1, \ldots, A_n$ of sets such that $A_i \subseteq A_{i+1}$ ($1 \leq i \leq n - 1$) *increasing*.

3   Less-Constrained Mapping

In this section, we introduce a less-constrained mapping and characterize it as cover sequences.

Definition 1 (Mapping [6]). Let T_1 and T_2 be trees and $M \subseteq V(T_1) \times V(T_2)$. We say that a triple (M, T_1, T_2) is a *Tai mapping* between T_1 and T_2 if every pair (v_1, w_1) and (v_2, w_2) in M satisfies the following conditions.

1. $v_1 = v_2$ iff $w_1 = w_2$ (one-to-one condition).
2. $v_1 \leq v_2$ iff $w_1 \leq w_2$ (ancestor condition).
3. $v_1 \preceq v_2$ iff $w_1 \preceq w_2$ (sibling condition).

For unordered trees, the condition 3 is omitted. We will use M instead of (M, T_1, T_2) when there is no confusion. Furthermore, we denote the set $\{v \in T_1 \mid (v, w) \in M\}$ by $M|_1$ and the set $\{w \in T_2 \mid (v, w) \in M\}$ by $M|_2$.

Definition 2 (Less-Constrained Mapping [3,4]). Let T_1 and T_2 be trees. We say that a mapping M between T_1 and T_2 is a *less-constrained mapping* if M satisfies the following condition.

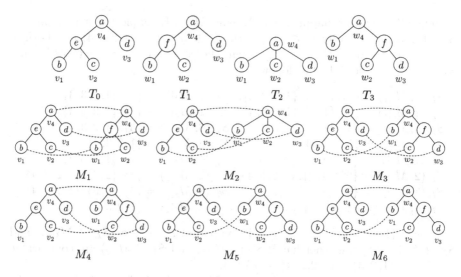

Fig. 1. Trees T_0, T_1, T_2 and T_3 (upper), mappings M_1, M_2 and M_3 (center) and mappings M_4, M_5 and M_6 (lower) in Example 1.

$$\forall (v_1, w_1), (v_2, w_2), (v_3, w_3) \in M \Big(v_1 \sqcup v_2 < v_1 \sqcup v_3 \implies w_2 \sqcup w_3 = w_1 \sqcup w_3 \Big).$$

Or equivalently [3]:

$$\forall (v_1, w_1), (v_2, w_2), (v_3, w_3) \in M \Big(w_1 \sqcup w_2 < w_1 \sqcup w_3 \implies v_2 \sqcup v_3 = v_1 \sqcup v_3 \Big).$$

Example 1. Consider trees T_0, T_1, T_2 and T_3 in Fig. 1 (upper). Also suppose that M_i is a mapping $\{(v_1, w_1), (v_2, w_2), (v_3, w_3), (v_4, w_4)\}$ between T_0 and T_i ($i = 1, 2, 3$) in Fig. 1 (center). Then, M_1 and M_2 are less-constrained mapping, while M_3 is not, because $v_1 \sqcup v_2 < v_1 \sqcup v_3$ but $w_2 \sqcup w_3 < w_1 \sqcup w_3$.

Furthermore, let $M_4 = M_3 - \{(v_1, w_1)\}$, $M_5 = M_3 - \{(v_2, w_2)\}$ and $M_6 = M_3 - \{(v_3, w_3)\}$ in Fig. 1 (lower). Then, we can show that M_4, M_5 and M_6 are less-constrained.

Definition 3 (Cover Set and Cover Sequence). Let T be a tree with the root r, v a node in T and U a set of nodes in T. Also suppose that $UP_r(v)$ is $v = v_1, \ldots, v_n = r$.

Then, we call a set $\{w \in T[v] \mid w \in U\}$ (or equivalently, $T[v] \cap U$) the *cover set* of v in T w.r.t. U and denote it by $C_T(v, U)$. Also we call a sequence $C_1, \ldots, C_n$ such that $C_i = C_T(v_i, U)$ for every i ($1 \le i \le n$) the *cover sequence* of v in T w.r.t. U and denote it by $S_T(v, U)$.

In particular, we use the cover sequences concerned with a mapping M between T_1 and T_2, that is, $S_{T_1}(v, M|_1)$ and $S_{T_2}(w, M|_2)$ for $(v, w) \in M$. For $r_1 = r(T_1)$ and $r_2 = r(T_2)$, we call $UP_{r_1}(v)$ and $UP_{r_2}(w)$ *paths* of $S_{T_1}(v, M|_1)$ and $S_{T_2}(w, M|_2)$, respectively, and denote them by $P_{T_1}(v)$ and $P_{T_2}(w)$, respectively.

Example 2. Consider mapping M_1, M_2 and M_3 in Example 1. For mapping M_i ($i = 1, 2, 3$), we identify $v_j \in M_i|_1$ with $w_j \in M_i|_2$ ($j = 1, 2, 3, 4$) and both of them are denoted by the index j. Then, the cover sequences $S_{T_0}(j, M_i|_1)$ and $S_{T_i}(j, M_i|_2)$ are described as follows.

$S_{T_0}(1, M_i|_1) = \{1\}, \{1, 2\}, \{1, 2, 3, 4\}.$ $S_{T_2}(1, M_2|_2) = \{1\}, \{1, 2, 3, 4\}.$
$S_{T_0}(2, M_i|_1) = \{2\}, \{1, 2\}, \{1, 2, 3, 4\}.$ $S_{T_2}(2, M_2|_2) = \{2\}, \{1, 2, 3, 4\}.$
$S_{T_0}(3, M_i|_1) = \{3\}, \{1, 2, 3, 4\}.$ $S_{T_2}(3, M_2|_2) = \{3\}, \{1, 2, 3, 4\}.$
$S_{T_0}(4, M_i|_1) = \{1, 2, 3, 4\}.$ $S_{T_2}(4, M_2|_2) = \{1, 2, 3, 4\}.$
$S_{T_1}(1, M_1|_2) = \{1\}, \{1, 2\}, \{1, 2, 3, 4\}.$ $S_{T_3}(1, M_3|_2) = \{1\}, \{1, 2, 3, 4\}.$
$S_{T_1}(2, M_1|_2) = \{2\}, \{1, 2\}, \{1, 2, 3, 4\}.$ $S_{T_3}(2, M_3|_2) = \{2\}, \{2, 3\}, \{1, 2, 3, 4\}.$
$S_{T_1}(3, M_1|_2) = \{3\}, \{1, 2, 3, 4\}.$ $S_{T_3}(3, M_3|_2) = \{3\}, \{2, 3\}, \{1, 2, 3, 4\}.$
$S_{T_1}(4, M_1|_2) = \{1, 2, 3, 4\}.$ $S_{T_3}(4, M_3|_2) = \{1, 2, 3, 4\}.$

Since $\{1, 2\}$ and $\{2, 3\}$ are incomparable, so are $S_{T_0}(2, M_3|_1)$ and $S_{T_3}(2, M_3|_2)$. On the other hand, $S_{T_0}(j, M_i|_1)$ and $S_{T_i}(j, M_i|_2)$ are comparable for $(i, j) \in \{1, 2, 3\} \times \{1, 2, 3, 4\} - \{(3, 2)\}$.

Furthermore, consider mappings M_4, M_5 and M_6 in Example 1. Then, the cover sequences $S_{T_0}(j, M_i|_1)$ and $S_{T_2}(j, M_i|_2)$ are described as follows, where $j \in I_i$ and $I_4 = \{2, 3, 4\}$, $I_5 = \{1, 3, 4\}$ and $I_6 = \{1, 2, 4\}$. All of them are comparable.

$S_{T_0}(2, M_4|_1) = \{2\}, \{2\}, \{2, 3, 4\}.$ $S_{T_3}(2, M_4|_2) = \{2\}, \{2, 3\}, \{2, 3, 4\}.$
$S_{T_0}(3, M_4|_1) = \{3\}, \{2, 3, 4\}.$ $S_{T_3}(3, M_4|_2) = \{3\}, \{2, 3\}, \{2, 3, 4\}.$
$S_{T_0}(4, M_4|_1) = \{2, 3, 4\}.$ $S_{T_3}(4, M_4|_2) = \{2, 3, 4\}.$
$S_{T_0}(1, M_5|_1) = \{1\}, \{1\}, \{1, 3, 4\}.$ $S_{T_3}(1, M_5|_2) = \{1\}, \{1, 3, 4\}.$
$S_{T_0}(3, M_5|_1) = \{3\}, \{1, 3, 4\}.$ $S_{T_3}(3, M_5|_2) = \{3\}, \{3\}, \{1, 3, 4\}.$
$S_{T_0}(4, M_5|_1) = \{1, 3, 4\}.$ $S_{T_3}(4, M_5|_2) = \{1, 3, 4\}.$
$S_{T_0}(1, M_6|_1) = \{1\}, \{1, 2\}, \{1, 2, 4\}.$ $S_{T_3}(1, M_6|_2) = \{1\}, \{1, 2, 4\}.$
$S_{T_0}(2, M_6|_1) = \{2\}, \{1, 2\}, \{1, 2, 4\}.$ $S_{T_3}(2, M_6|_2) = \{2\}, \{2\}, \{1, 2, 4\}.$
$S_{T_0}(4, M_6|_1) = \{1, 2, 4\}.$ $S_{T_3}(4, M_6|_2) = \{1, 2, 4\}.$

Theorem 1. *Let T_1 and T_2 be trees. Also let M be a mapping between T_1 and T_2. Then, M is not a less-constrained mapping between T_1 and T_2 if and only if there exists a pair $(v, w) \in M$ such that $S_{T_1}(v, M|_1)$ and $S_{T_2}(w, M|_2)$ are incomparable.*

Proof. Suppose that there exists a pair $(v_1, w_1) \in M$ such that cover sets $C_1 \in S_{T_1}(v_1, M|_1)$ and $C_2 \in S_{T_2}(w_1, M|_2)$ are incomparable. Then, there exist $v_2 \in M|_1$ and $w_3 \in M|_2$ such that $v_2 \in C_1 - C_2$ and $w_3 \in C_2 - C_1$. Let v^* and w^* be nodes $v_1 \sqcup v_2$ and $w_1 \sqcup w_3$, respectively. Then, we can assume that $C_1 = C_{T_1}(v^*, M|_1)$ and $C_2 = C_{T_2}(w^*, M|_2)$. Also consider v_3 and w_2.

Since $v_3 \notin C_1$, it holds that $v^* < v_1 \sqcup v_3$. Since $w_2 \notin C_2$, it holds that $w^* < w_2 \sqcup w_3$. Hence, even if $v^* = v_1 \sqcup v_2 < v_2 \sqcup v_3$, it holds that $w^* = w_1 \sqcup w_3 < w_2 \sqcup w_3$, which implies that M is not a less-constrained mapping.

Conversely, suppose that M is not a less-constrained mapping. Then, there exist $v_1, v_2, v_3 \in M|_1$ and $w_1, w_2, w_3 \in M|_2$ such that (1) $v_1 \sqcup v_2 < v_1 \sqcup v_3$ holds and either (2) $w_2 \sqcup w_3 < w_1 \sqcup w_3$ or (3) $w_2 \sqcup w_3 > w_1 \sqcup w_3$ holds.

By the condition (1), the cover sequences $S_{T_1}(v_1, M|_1)$ and $S_{T_1}(v_2, M|_1)$ contain a cover set C_1 such that $\{v_1, v_2\} \subseteq C_1$ and $v_3 \notin C_1$. On the other hand, by the condition (2), the cover sequence $S_{T_2}(w_2, M|_2)$ contains a cover set C_2 such that $\{w_2, w_3\} \subseteq C_2$ and $w_1 \notin C_2$, which implies that C_1 and C_2 are incomparable. Also, by the condition (3), the cover sequence $S_{T_2}(w_1, M|_2)$ contains a cover set C_3 such that $\{w_1, w_3\} \subseteq C_3$ and $w_2 \notin C_3$, which implies that C_1 and C_3 are incomparable. □

Corollary 1. *Let T_1 and T_2 be trees. Also let M be a mapping between T_1 and T_2. Then, M is a less-constrained mapping between T_1 and T_2 if and only if, for every pair $(v, w) \in M$, $S_{T_1}(v, M|_1)$ and $S_{T_2}(w, M|_2)$ are comparable.*

4   Alignable Mapping and Alignment Tree

Let T_1 and T_2 be trees. We say that I is a *root-preserving mapping* from T_1 to T_2 if I is a mapping between T_1 and T_2 and $r(T_1) \in I|_1$ always holds. In particular, for $v \in T_1$, we denote the node $w \in T_2$ such that $(v, w) \in I$ by $I(v)$. Note that T_2 is not necessary to be labeled.

Definition 4 (Alignable Mapping [3]). Let T_1 and T_2 be trees. We say that M is an *alignable* mapping between T_1 and T_2 if there exist a tree $\mathcal{T}$ (not necessary to be labeled) and root-preserving mappings I_1 from T_1 to $\mathcal{T}$ and I_2 from T_2 to $\mathcal{T}$ satisfying that $I_1(v) = I_2(w)$ for every $(v, w) \in M$. In particular, we call the tree $\mathcal{T}$ an *aligned tree* between T_1 and T_2 and the root-preserving mappings I_1 and I_2 *side mappings* of M from T_1 and T_2, respectively.

Let M be an alignable mapping between T_1 and T_2 and I_i a side mapping of M from T_i ($i = 1, 2$). Then, it holds that $M|_1 = \{v \in T_1 \mid I_1(v) = I_2(w)\}$ and $M|_2 = \{w \in T_2 \mid I_1(v) = I_2(w)\}$. For an aligned tree $\mathcal{T}$, we denote the inverse image of I_i from $V(\mathcal{T})$ to $V(T_i)$ by I_i^{-1}. In particular, when no $v \in T_i$ such that $I_i(v) = u$ exists for a node $u \in \mathcal{T}$, we denote $I_i^{-1}(u)$ by $\emptyset$ and $l(I_i^{-1}(u))$ by ε.

Definition 5 (Alignment Tree [2]). Let M be an alignable mapping between T_1 and T_2, I_i a side mapping of M from T_i ($i = 1, 2$) and $\mathcal{T}$ an aligned tree between T_1 and T_2. Then, we call the tree obtained by replacing every label of $u \in \mathcal{T}$ with $(l(I_1^{-1}(u)), l(I_2^{-1}(u)))$ an *alignment tree* between T_1 and T_2. For an alignment tree $\mathcal{T}$, we denote a mapping M between T_1 and T_2 constructed from $\mathcal{T}$ such that $(v, w) \in M$ iff $(l(v), l(w)) \in \mathcal{T}$ by $M_{\mathcal{T}}$.

Example 3. Consider mappings M_i ($i = 4, 5, 6$) in Example 1 (Fig. 1). Then, every mapping M_i is an alignable mapping. Also, the tree $\mathcal{T}_i$ in Fig. 2 is an alignment tree between T_0 and T_3 in Example 1 corresponding to M_i.

Let $\varepsilon \notin \Sigma$ denote a special *blank* symbol and define $\Sigma_\varepsilon = \Sigma \cup \{\varepsilon\}$. Then, we define a *cost function* $\gamma : (\Sigma_\varepsilon \times \Sigma_\varepsilon - \{(\varepsilon, \varepsilon)\}) \mapsto \mathbf{R}^+$ on pairs of labels. We constrain γ to be a *metric*, that is, $\gamma(l_1, l_2) \geq 0$, $\gamma(l_1, l_1) = 0$, $\gamma(l_1, l_2) = \gamma(l_2, l_1)$ and $\gamma(l_1, l_3) \leq \gamma(l_1, l_2) + \gamma(l_2, l_3)$. In particular, the *unit cost function* μ such

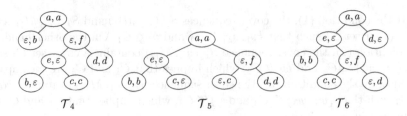

Fig. 2. The alignment trees $\mathcal{T}_i$ between T_0 and T_3 in Example 1.

that $\mu(a,b) = 0$ if $a = b$ and $\mu(a,b) = 1$ if $a \neq b$ is the most famous cost function. The *cost* of an alignment tree $\mathcal{T}$ under γ, denoted by $\gamma(\mathcal{T})$, is the sum of the costs of all labels in $\mathcal{T}$. The minimum cost of all the possible alignment trees is known to an alignment distance [2].

5   An Alternative Proof of Theorem 2

In this section, by using the cover sequence, Theorem 1 and Corollary 1, we give an alternative proof of the following Theorem 2.

Theorem 2 (Kuboyama [3]). *Let T_1 and T_2 be trees and M a mapping between T_1 and T_2. Then, M is less-constrained if and only if M is alignable.*

First, we show the if-direction of Theorem 2.

Lemma 1. *Let T_1 and T_2 be trees and M an alignable mapping between T_1 and T_2. Then, M is also a less-constrained mapping.*

Proof. For an alignable mapping M, there exists an alignment tree $\mathcal{T}$ such that $M = M_{\mathcal{T}}$. Also suppose that M is not a less-constrained mapping. By Theorem 1, there exists a pair $(v,w) \in M$ such that cover sets $C_1 \in S_{T_1}(v, M|_1)$ and $C_2 \in S_{T_2}(w, M|_2)$ are incomparable. Then, there exist $v_1 \in M|_1$ and $w_2 \in M|_2$ such that $v_1 \in C_1 - C_2$ and $w_2 \in C_2 - C_1$. Let v' and w' denote $v \sqcup v_1$ and $w \sqcup w_2$. Then, we can assume that $C_1 = C_{T_1}(v', M|_1)$ and $C_2 = C_{T_2}(w', M|_2)$. Also consider w_1 and v_2 such that $(v_1, w_1) \in M$ and $(v_2, w_2) \in M$.

Since $M = M_{\mathcal{T}}$, both $(l(v_1), l(w_1))$ and $(l(v_2), l(w_2))$ occur in $\mathcal{T}$. Since $v_1 \in C_1$, it holds that $(l(v), l(w)) < (l(v_1), l(w_1))$ in $\mathcal{T}$. Since $w_2 \in C_2$, it holds that $(l(v), l(w)) < (l(v_2), l(w_2))$ in $\mathcal{T}$. Furthermore, since $v_1 \in C_1 - C_2$ and $w_2 \in C_2 - C_1$, we can show that $(l(v_1), l(w_1)) \# (l(v_2), l(w_2))$ in $\mathcal{T}$ as follows. Note here that we identify v_i with w_i $(i = 1, 2)$.

If $(l(v_1), l(w_1)) < (l(v_2), l(w_2))$ in $\mathcal{T}$, then it holds that $v < v_1 < v_2$ in T_1 and $w < w_1 < w_2$ in T_2. Then, since $v' = v_1$ and $w' = w_2$, it holds that $C_1 = C_{T_1}(v', M|_1) \subseteq C_{T_1}(v_2, M|_1) = C_{T_2}(w', M|_2) = C_2$. If $(l(v_2), l(w_2)) < (l(v_1), l(w_1))$ in $\mathcal{T}$, then it holds that $v < v_2 < v_1$ in T_1 and $w < w_2 < w_1$ in T_2. Then, since $v' = v_1$ and $w' = w_2$, it holds that $C_2 = C_{T_2}(w', M|_2) \subseteq C_{T_2}(w_1, M|_2) = C_{T_1}(v', M|_1) = C_1$. These imply a contradiction that C_1 and C_2 are incomparable.

Hence, it holds that $(l(v), l(w)) < (l(v_1), l(w_1))$, $(l(v), l(w)) < (l(v_2), l(w_2))$ and $(l(v_1), l(w_1)) \# (l(v_2), l(w_2))$ in T for $(l(v), l(w)), (l(v_1), l(w_1)), (l(v_2), l(w_2)) \in T$, which is a contradiction that T is a tree. □

In order to show the only-if-direction of Theorem 2, we start the following lemma.

Lemma 2. *Let T_1 and T_2 be trees and M a less-constrained mapping between T_1 and T_2. Then, for $(v, w) \in M$, both $S_{T_1}(v, M|_1)$ and $S_{T_2}(w, M|_2)$ are comparable increasing such that the last element of $S_{T_1}(v, M|_1)$ (resp., $S_{T_2}(w, M|_2)$) is $M|_1$ (resp., $M|_2$).*

Let M be a mapping between T_1 and T_2. Then, for every $(v_j, w_j) \in M$, we sometimes identify $v_j \in M|_1$ with $w_j \in M|_2$ and both of them are denoted by the index j. Under such an identification, we can regard that $M|_1 = M|_2$. Then, we introduce the following *aligned sequence* and *aligned path* for comparable increasing sequences of sets.

Definition 6 (Aligned Sequence, Aligned Path). Let $S_1 = A_1, \ldots, A_n$ and $S_2 = B_1, \ldots, B_m$ be comparable increasing sequences of sets such that $A_n = B_m$. Then, we call the sequences $S_1' = A_1', \ldots, A_k'$ and $S_2' = B_1', \ldots, B_k'$ obtained from S_1 and S_2 by the procedure ALNSQ in Algorithm 1 *aligned sequences* of S_1 and S_2. Furthermore, for the aligned sequences $S_1' = A_1', \ldots, A_k'$ and $S_2' = B_1', \ldots, B_k'$ of S_1 and S_2, we define the *aligned path of S_1 and S_2* as a rooted labeled path $P = (V, E)$ such that $V = \{p_1, \ldots, p_k\}$, $E = \{(p_i, p_{i+1}) \mid 1 \le i \le k-1\}$, the root of P is p_1 and the label of p_i is (A_{k-i+1}', B_{k-i+1}') for $1 \le i \le k$. We sometimes denote such a path by $[p_1, \ldots, p_k]$.

Example 4. Consider a mapping M_2 in Example 1. By Lemma 2, $S_{T_0}(j, M_2|_1)$ and $S_{T_2}(j, M_2|_2)$ in Example 2 are comparable increasing. Then, we can obtain the aligned sequences $S_{T_0}'(j, M_2|_1)$ and $S_{T_2}'(j, M_2|_2)$ illustrated in Fig. 3 (upper). Also, for an aligned path $P_{M_2}(j) = [p_1, p_2, p_3]$ of $S_{T_0}(j, M_2|_1)$ and $S_{T_2}(j, M_2|_2)$,

```
procedure ALNSQ(S_1, S_2)
      /* S_1 = A_1,...,A_n, S_2 = B_1,...,B_m */
1     i ← 1; j ← 1; k ← 1;
2     while i ≤ n + 1 and j ≤ m + 1 do
3         if i = n + 1 then A'_k ← λ; B'_k ← B_j; j++;
4         else if j = m + 1 then A'_k ← A_i; B'_k ← λ; i++;
5         else if A_i = B_j then A'_k ← A_i; B'_k ← B_j; i++; j++;
6         else if A_i ⊂ B_j then A'_k ← A_i; B'_k ← λ; i++;
7         else if A_i ⊃ B_j then A'_k ← λ; B'_k ← B_j; j++;
8         k++;
9     return S'_1 = A'_1,...,A'_k and S'_2 = B'_1,...B'_k;
```

Algorithm 1. ALNSQ.

$S'_{T_0}(1, M_2|_1) = \{1\}, \{1, 2\}, \{1, 2, 3, 4\}.$ $S'_{T_2}(1, M_2|_2) = \{1\}, \lambda, \{1, 2, 3, 4\}.$
$S'_{T_0}(2, M_2|_1) = \{2\}, \{1, 2\}, \{1, 2, 3, 4\}.$ $S'_{T_2}(2, M_2|_2) = \{2\}, \lambda, \{1, 2, 3, 4\}.$
$S'_{T_0}(3, M_2|_1) = \{3\}, \{1, 2, 3, 4\}.$ $S'_{T_2}(3, M_2|_2) = \{3\}, \{1, 2, 3, 4\}.$
$S'_{T_0}(4, M_2|_1) = \{1, 2, 3, 4\}.$ $S'_{T_2}(4, M_2|_2) = \{1, 2, 3, 4\}.$

$P_{M_2}(j)$	$l(p_1)$	$l(p_2)$	$l(p_3)$
$P_{M_2}(1)$	$(\{1, 2, 3, 4\}, \{1, 2, 3, 4\})$	$(\{1, 2\}, \lambda)$	$(\{1\}, \{1\})$
$P_{M_2}(2)$	$(\{1, 2, 3, 4\}, \{1, 2, 3, 4\})$	$(\{1, 2\}, \lambda)$	$(\{2\}, \{2\})$
$P_{M_2}(3)$	$(\{1, 2, 3, 4\}, \{1, 2, 3, 4\})$	$(\{3\}, \{3\})$	
$P_{M_2}(4)$	$(\{1, 2, 3, 4\}, \{1, 2, 3, 4\})$		

Fig. 3. The aligned sequences $S'_{T_0}(j, M_2|_1)$ and $S'_{T_2}(j, M_2|_2)$ of $S_{T_0}(j, M_2|_1)$ and $S_{T_2}(j, M_2|_2)$ (upper) and the labels in aligned path $P_{M_2}(j)$ of $S_{T_0}(j, M_2|_1)$ and $S_{T_i}(j, M_2|_2)$ (lower) in Example 4.

every label $l(p_i)$ of a vertex p_i $(i = 1, 2, 3)$ in $P_{M_2}(j)$ is illustrated in Fig. 3 (lower).

Consider mappings M_4, M_5 and M_6 in Example 1. By Lemma 2, $S_{T_0}(j, M_i|_1)$ and $S_{T_2}(j, M_i|_2)$ $(i = 4, 5, 6, j \in I_i)$ in Example 2 are comparable increasing. Then, we can obtain the aligned sequences $S'_{T_0}(j, M_i|_1)$ and $S'_{T_2}(j, M_i|_2)$ illustrated in Fig. 4 (upper). Also, for an aligned path $P_{M_i}(j) = [p_1, p_2, p_3, p_4]$ of $S_{T_0}(j, M_i|_1)$ and $S_{T_2}(j, M_i|_2)$, every label $l(p_i)$ of a vertex p_i in $P_{M_i}(j)$ are illustrated in Fig. 4 (lower).

Let M be a less-constrained mapping between T_1 and T_2, where $r_1 = r(T_1)$ and $r_2 = r(T_2)$. By Lemma 2, for $S_{T_1}(v, M|_1) = A_1, \ldots, A_n$ and $S_{T_2}(w, M|_2) = B_1, \ldots, B_m$ for every $(v, w) \in M$, there exist paths $P_{T_1}(v) = v_1, \ldots, v_n$ and $P_{T_2}(w) = w_1, \ldots, w_m$ such that $v_1 = v$, $w_1 = w$, $v_n = r_1$ and $w_m = r_2$. Also, by identifying $v' \in M|_1$ with $w' \in M|_2$ for $(v', w') \in M$, it holds that $A_1 = B_1 = \{v\} = \{w\}$ and $A_n = B_m = M|_1 = M|_2$.

Furthermore, suppose that the aligned sequences $S'_{T_1}(v, M|_1)$ of $S_{T_1}(v, M|_1)$ and $S'_{T_2}(w, M|_2)$ of $S_{T_2}(w, M|_2)$ are of the forms $A'_1, \ldots, A'_k$ and $B'_1, \ldots, B'_k$, respectively. Then, we denote the corresponding path of $S'_{T_1}(v, M|_1)$ in T_1 including λ by $P'_{T_1}(v) = v'_1, \ldots, v'_k$ such that $v'_i = \lambda$ if $A'_i = \lambda$ and $v'_i = v_{i'}$ otherwise, where $i' = |\{l \mid 1 \leq l \leq i, v'_l \neq \lambda\}|$. Also we denote the corresponding path of $S_{T_2}(v, M|_1)$ in T_2 including λ by $P'_{T_2}(w) = w'_1, \ldots, w'_k$ such that $w'_j = \lambda$ if $B'_j = \lambda$ and $w'_j = w_{j'}$ otherwise, where $j' = |\{l \mid 1 \leq l \leq j, w'_l \neq \lambda\}|$.

Lemma 3. Let M be a less-constrained mapping between T_1 and T_2. Also, for $(v_1, w_1), (v_2, w_2) \in M$, let:
$S'_{T_1}(v_1, M|_1) = A'_1, \ldots, A'_k, S'_{T_2}(w_1, M|_2) = B'_1, \ldots, B'_k,$
$S'_{T_1}(v_2, M|_1) = C'_1, \ldots, C'_h, S'_{T_2}(w_2, M|_2) = D'_1, \ldots, D'_h.$

Then, for the maximum indices i and j $(2 \leq i \leq k, 2 \leq j \leq h)$ such that $(A'_i, B'_i) = (C'_j, B'_j)$ and $(A'_{i-1}, B'_{i-1}) \neq (C'_{j-1}, B'_{j-1})$, it holds that $(A'_a, B'_a) \neq (C'_b, B'_b)$ for every a $(1 \leq a \leq i-1)$ and b $(1 \leq b \leq j-1)$.

$S'_{T_0}(2, M_4|_1) = \{2\}, \{2\}, \lambda, \{2, 3, 4\}.$ $S'_{T_3}(2, M_4|_2) = \{2\}, \lambda, \{2, 3\}, \{2, 3, 4\}.$
$S'_{T_0}(3, M_4|_1) = \{3\}, \lambda, \{2, 3, 4\}.$ $S'_{T_3}(3, M_4|_2) = \{3\}, \{2, 3\}, \{2, 3, 4\}.$
$S'_{T_0}(4, M_4|_1) = \{2, 3, 4\}.$ $S'_{T_3}(4, M_4|_2) = \{2, 3, 4\}.$
$S'_{T_0}(1, M_5|_1) = \{1\}, \{1\}, \{1, 3, 4\}.$ $S'_{T_3}(1, M_5|_2) = \{1\}, \lambda, \{1, 3, 4\}.$
$S'_{T_0}(3, M_5|_1) = \{3\}, \lambda, \{1, 3, 4\}.$ $S'_{T_3}(3, M_5|_2) = \{3\}, \{3\}, \{1, 3, 4\}.$
$S'_{T_0}(4, M_5|_1) = \{1, 3, 4\}.$ $S'_{T_3}(4, M_5|_2) = \{1, 3, 4\}.$
$S'_{T_0}(1, M_6|_1) = \{1\}, \{1, 2\}, \{1, 2, 4\}.$ $S'_{T_3}(1, M_6|_2) = \{1\}, \lambda, \{1, 2, 4\}.$
$S'_{T_0}(2, M_6|_1) = \{2\}, \lambda, \{1, 2\}, \{1, 2, 4\}.$ $S'_{T_3}(2, M_6|_2) = \{2\}, \{2\}, \lambda, \{1, 2, 4\}.$
$S'_{T_0}(4, M_6|_1) = \{1, 2, 4\}.$ $S'_{T_3}(4, M_6|_2) = \{1, 2, 4\}.$

$P_{M_i}(j)$	$l(p_1)$	$l(p_2)$	$l(p_3)$	$l(p_4)$
$P_{M_4}(2)$	$(\{2,3,4\}, \{2,3,4\})$	$(\lambda, \{2,3\})$	$(\{2\}, \lambda)$	$(\{2\}, \{2\})$
$P_{M_4}(3)$	$(\{2,3,4\}, \{2,3,4\})$	$(\lambda, \{2,3\})$	$(\{3\}, \{3\})$	
$P_{M_4}(4)$	$(\{2,3,4\}, \{2,3,4\})$			
$P_{M_5}(1)$	$(\{1,3,4\}, \{1,3,4\})$	$(\{1\}, \lambda)$	$(\{1\}, \{1\})$	
$P_{M_5}(3)$	$(\{1,3,4\}, \{1,3,4\})$	$(\lambda, \{3\})$	$(\{3\}, \{3\})$	
$P_{M_5}(4)$	$(\{1,3,4\}, \{1,3,4\})$			
$P_{M_6}(1)$	$(\{1,2,4\}, \{1,2,4\})$	$(\{1,2\}, \lambda)$	$(\{1\}, \{1\})$	
$P_{M_6}(2)$	$(\{1,2,4\}, \{1,2,4\})$	$(\{1,2\}, \lambda)$	$(\{1,2\}, \{2\})$	$(\{2\}, \{2\})$
$P_{M_6}(4)$	$(\{1,2,4\}, \{1,2,4\})$			

Fig. 4. The aligned sequences $S'_{T_0}(j, M_i|_1)$ and $S'_{T_2}(j, M_i|_2)$ of $S_{T_0}(j, M_i|_1)$ and $S_{T_2}(j, M_i|_2)$ (upper) and the labels in aligned path $P_{M_i}(j)$ of $S_{T_0}(j, M_i|_1)$ and $S_{T_2}(j, M_i|_2)$ (lower) in Example 4.

Proof. Let $A = A'_1 \cup \cdots \cup A'_{i-1} - \{\lambda\}$, $B = B'_1 \cup \cdots \cup B'_{i-1} - \{\lambda\}$, $C = C'_1 \cup \cdots \cup C'_{j-1} - \{\lambda\}$ and $D = D'_1 \cup \cdots \cup D'_{j-1} - \{\lambda\}$. Then, we show that $A \cap C = \emptyset$ and $B \cap D = \emptyset$.

Suppose that $A \cap C \neq \emptyset$. Then, there exist a vertex $v \in T_1$ such that $v \in A \cap C$. Then, it holds that $v \in P'_{T_1}(v_1)$ and $v \in P'_{T_1}(v_2)$. Since T_1 is a rooted tree, $v_1 \sqcup v$ and $v_2 \sqcup v$ satisfy one of the statements of $v_1 \sqcup v < v_2 \sqcup v$, $v_2 \sqcup v < v_1 \sqcup v$ and $v_1 \sqcup v = v_2 \sqcup v$. For $P'_{T_1}(v_1) = p'_1, \ldots, p'_k$ and $P'_{T_1}(v_2) = q'_1, \ldots, q'_h$ such that $p'_1 = v_1$, $q'_1 = v_2$, $p'_k = q'_h = r(T_1)$, let $v^* = p'_{i'+1} = q'_{j'+1} \in T_1$. Then, it holds that $v_1 \sqcup v_2 = v^*$.

If $v_1 \sqcup v < v_2 \sqcup v$ holds, then it holds that $v_1 \sqcup v < v^*$, which means that $v \notin C$. If $v_2 \sqcup v < v_1 \sqcup v$ holds, then it holds that $v_2 \sqcup v < v^*$, which means that $v \notin A$. If $v_1 \sqcup v = v_2 \sqcup v$, then it holds that $v^* \leq v_1 \sqcup v = v_2 \sqcup v$, which means that $v \notin A \cap C$. Hence, it holds that $A \cap C = \emptyset$.

By the same way, we can show that $B \cap D = \emptyset$. Hence, the statement holds. $\square$

Definition 7 (Merged Graph). For a less-constrained mapping M between T_1 and T_2, let $\mathcal{P}_M$ be the set of all aligned paths concerned with M. Then, we define the *merged graph* $\mathcal{G}_M$ of M as a rooted graph obtained by identifying vertices with the same labels in $\mathcal{P}_M$, where the label of the root is $(M|_1, M|_2)$.

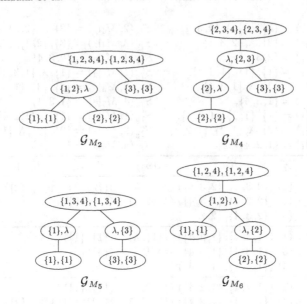

Fig. 5. The merged graphs $\mathcal{G}_{M_i}$ ($i = 2, 4, 5, 6$) in Example 5.

Lemma 4. *Let T_1 and T_2 be trees and M a less-constrained mapping between T_1 and T_2. Then, the merged graph $\mathcal{G}_M$ of M is a rooted labeled tree.*

Example 5. Consider the mappings M_2, M_4, M_5 and M_6 in Example 1. By Example 4, it holds that $\mathcal{P}_{M_2} = \{P_{M_2}(1), P_{M_2}(2), P_{M_2}(3), P_{M_2}(4)\}$ and $\mathcal{P}_{M_i} = \{P_{M_i}(j) \mid j \in I_i\}$ ($i = 4, 5, 6$). Then, the merged graphs $\mathcal{G}_{M_i}$ of $\mathcal{P}_{M_i}$ are illustrated in Fig. 5.

Let T_1 and T_2 be trees, M a less-constrained mapping between T_1 and T_2 and $\mathcal{G}_M$ the merged graph of M. Then, we denote the tree obtained by removing all of the labels in $\mathcal{G}_M$ by $\mathcal{G}_M^-$. Also, for a vertex $u \in \mathcal{G}_M$, the label of u is of the form (A, B), where $A \subseteq M_1$ or $A = \lambda$ and $B \subseteq M_2$ or $B = \lambda$. When $A \neq \lambda$ (*resp.*, $B \neq \lambda$), there exists a unique vertex in T_1 corresponding to A (*resp.*, in T_2 corresponding to B), which we denote such a vertex by $v_{T_1}(A)$ (*resp.*, $v_{T_2}(B)$).

For every vertex $u \in \mathcal{G}_M$, consider to replace the label (A, B) of u with $(l(v_{T_1}(A)), l(v_{T_2}(B)))$ if $A \neq \lambda$ and $B \neq \lambda$; $(\varepsilon, l(v_{T_2}(B)))$ if $A = \lambda$ and $B \neq \lambda$; $(l(v_{T_1}(A)), \varepsilon)$ if $A \neq \lambda$ and $B = \lambda$. We denote the tree obtained by this replacement of labels in every $u \in \mathcal{G}_M$ from $\mathcal{G}_M$ by $\mathcal{G}_M^*$.

Lemma 5. *Let T_1 and T_2 be trees, M a less-constrained mapping between T_1 and T_2 and $\mathcal{G}_M$ the merged graph of M. Then, $\mathcal{G}_M^-$ is a subtree of the aligned tree between T_1 and T_2, and $\mathcal{G}_M^*$ is a subtree of the alignment tree between T_1 and T_2.*

Proof. Note that the labels of vertices in $\mathcal{G}_M$ are of the form (A, B). Then, by the definition of $\mathcal{G}_M$ and Lemma 4, we obtain a subtree of T_1 (*resp.*, T_2) by first connecting $v_{T_1}(A)$ (*resp.*, $v_{T_2}(B)$) for every A (*resp.*, B) and then by deleting λ and, for a vertex v whose label is λ, connecting the children of v to the parent of v. Hence, the statement holds. □

Let $\mathcal{P}_1$ be the set of all rooted maximal paths in $T_1 - \{P_{T_1}(v) \mid v \in M|_1\}$ and $\mathcal{P}_2$ the set of all rooted maximal paths in $T_2 - \{P_{T_2}(w) \mid w \in M|_2\}$. For every $P = [p_1, \ldots, p_k] \in \mathcal{P}_1$ such that $r(P) = p_1$, there exists a vertex $v \in T_1$ such that v is a parent of p_1 in T_1, which we denote by $par_{T_1}(P)$. Similarly, for every $Q = [q_1, \ldots, q_k] \in \mathcal{P}_2$ such that $r(Q) = q_1$, there exists a vertex $v \in T_2$ such that v is a parent of q_1 in T_2, which we denote by $par_{T_2}(Q)$.

Furthermore, for every $P = [p_1, \ldots, p_k] \in \mathcal{P}_1$, we denote a labeled path obtained by replacing $l(p_i)$ with $(l(p_i), \varepsilon)$ by $\langle P, \varepsilon \rangle$, and, for every $Q = [q_1, \ldots, q_k] \in \mathcal{P}_2$, we denote a labeled path obtained by replacing $l(q_i)$ with $(\varepsilon, l(q_i))$ by $\langle \varepsilon, Q \rangle$.

Lemma 6. *Let T_1 and T_2 be trees and M a less-constrained mapping between T_1 and T_2. Then, M is also an alignable mapping.*

Proof. It is sufficient to construct an alignment tree between T_1 and T_2 from M. By Lemma 5, $\mathcal{G}_M^*$ is a subtree of the alignment tree between T_1 and T_2. In order to complete the alignment tree, it is necessary to insert the paths not "covered by" M, which are denoted by the above $\mathcal{P}_1$ and $\mathcal{P}_2$. Hence, by inserting paths $\langle P, \varepsilon \rangle$ to the appropriate child of $par_{T_1}(P)$ in $\mathcal{G}_M^*$ for every $P \in \mathcal{P}_1$ and $\langle \varepsilon, Q \rangle$ to the appropriate child of $par_{T_1}(P)$ in $\mathcal{G}_M^*$ for every $Q \in \mathcal{P}_2$, we can obtain the alignment tree between T_1 and T_2. □

It is not necessary for Theorem 2 to distinguish that trees are ordered or unordered.

6   Anchored Alignment Problem

Finally, we discuss the *anchored alignment problem* introduced in Sect. 1.

Theorem 3. *Let $n = |T_1|$, $m = |T_2|$, $h = \max\{h(T_1), h(T_2)\}$ and $\alpha = |M|$. Then, the problem of ANCHOREDALIGNMENT can be solved in $O(h\alpha^2 + n + m)$ time and in $O(h\alpha)$ space for both ordered and unordered trees.*

Proof. Since the correctness is shown in Sect. 5, it is sufficient to show the time complexity. We use α-bits $\{0, 1\}$ vectors for set operations in totally $O(h\alpha)$ space, which we can prepare in $O(h\alpha)$ time.

For an anchoring M, first check whether or not M is less-constrained by using Corollary 1. If M is not less-constrained, then return "no," which runs in $O(h\alpha)$ time. Otherwise, construct a partial alignment tree T between T_1 and T_2 by using the algorithm ALNSQ in Algorithm 1. We can check whether $A_i = B_j$, $A_i \subset B_j$ or $A_i \supset B_j$ in Algorithm 1 in $O(\alpha)$ time, so the running time of Algorithm 1 is $O(h\alpha)$ and the total running time in this process is $O(h\alpha^2)$. Next, construct the merged graph $\mathcal{G}_M$ and the replacement $\mathcal{G}_M^*$ of $\mathcal{G}_M$, which runs in $O(h\alpha)$ time for ordered trees (just checking adjacent nodes in postorder) and in $O(h\alpha^2)$ time for unordered trees. Finally, add $\langle P, \varepsilon \rangle$ and $\langle \varepsilon, Q \rangle$ to $\mathcal{T}$ according to Lemma 6, which runs in $O(n + m)$ time.

Hence, the time complexity is $O(h\alpha) + O(h\alpha^2) + O(h\alpha) + O(n+m) = O(h\alpha^2 + n + m)$ for ordered trees and $O(h\alpha) + O(h\alpha^2) + O(h\alpha^2) + O(n+m) = O(h\alpha^2 + n + m)$ for unordered trees. □

7   Conclusion

In this paper, first we have provided an alternative proof that a mapping is less-constrained iff it is alignable, by using cover sequences and merged graphs. Then, we have formulated the problem of ANCHOREDALIGNMENT and then shown that we can solve it in $O(h\alpha^2 + n + m)$ time and in $O(h\alpha)$ space for both ordered and unordered trees. Note that, if a given anchoring is optimum, that is, the cost of a given anchoring is minimum [2], then the problem of ANCHOREDALIGNMENT is corresponding to the traceback of the alignment.

As stated in Sect. 1, an anchored alignment tree as output is not necessary to be optimum; it is just an alignment tree between two trees containing nodes labeled by pairs of labels in an anchoring. For example, consider the trees T_0 and T_3 in Fig. 1 and let M_7 in Fig. 6 (left) be an anchor between T_0 and T_3. Then, the anchored alignment tree of M_7 is $\mathcal{T}_7$ in Fig. 6 (right) such that $\mu(\mathcal{T}_7) = 8$ under a unit cost function μ. On the other hand, $\mathcal{T}_5$ in Fig. 2 is the optimum alignment tree between T_0 and T_3 such that $\mu(\mathcal{T}_5) = 4$.

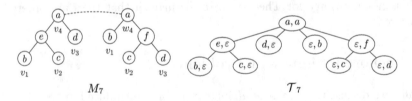

Fig. 6. A mapping M_7 and the anchored alignment tree of M_7.

Then, it is a future work to discuss the problem of ANCHOREDALIGNMENT such that the anchored alignment tree is optimum. Also, it is a future work to investigate whether or not we can improve the time complexity to find the alignment distance for ordered trees, by using the algorithm to solve the problem of ANCHOREDALIGNMENT. Furthermore, it is a future work to discuss the relationship between the results of this paper and the maximum agreement supertrees [1].

References

1. Berry, V., Nicolas, F.: Maximum agreement and comaptible supertrees. J. Discrete Algorothms **5**, 564–591 (2007)
2. Jiang, T., Wang, L., Zhang, K.: Alignment of trees - an alternative to tree edit. Theoret. Comput. Sci. **143**, 137–148 (1995)
3. Kuboyama, T.: Matching and learning in trees. Ph.D. thesis, University of Tokyo (2007)
4. Lu, C.L., Su, Z.-Y., Tang, C.Y.: A new measure of edit distance between labeled trees. In: Wang, J. (ed.) COCOON 2001. LNCS, vol. 2108, pp. 338–348. Springer, Heidelberg (2001)
5. Schiermer, S., Giegerich, R.: Forest alignment with affine gaps and anchors, applied in RNA structure comparision. Theoret. Comput. Sci. **483**, 51–67 (2013)
6. Tai, K.-C.: The tree-to-tree correction problem. J. ACM **26**, 422–433 (1979)

Central Point Selection in Dimension Reduction Projection Simple-Map with Binary Quantization

Quming Jin[1], Masaya Nakashima[1,3], Takeshi Shinohara[1](✉), Kouichi Hirata[1], and Tetsuji Kuboyama[2]

[1] Department of Artificial Intelligence, Kyushu Institute of Technology,
Kawazu 680-4, Iizuka 820-8502, Japan
{n673006k,shino,hirata}@ai.kyutech.ac.jp
[2] Computer Center, Gakushuin University, Mejiro 1-5-1,
Toshima, Tokyo 171-8588, Japan
kuboyama@gakushuin.ac.jp
[3] Icom Systech Co., Ltd., Tokyo, Japan

Abstract. A Simple-Map (S-Map, for short), which is one of dimension reduction techniques applicable to any metric space, uses the distances between central points and objects as the coordinate values. S-Map with multiple central points is a projection to multidimensional L_∞ space. In the previous researches for S-Map, the candidates for central points are randomly selected from data objects in database, and the summation of projective distances between sampled pairs of points is used as the scoring function to be maximized. We can improve the above method to select central points by using local search. The coordinate values of central points obtained after local search tend to be the maximum or the minimum ends of the space. By focusing on this tendency, in this paper, we propose a binary quantization to select central points divided into the maximum values and the minimum values based on whether the coordinate value of an object in database is greater than the threshold or not.

1 Introduction

At present, it is possible for computers to process large numbers of multi-dimensional media data such as sound and image because of an enormous increase of the computing power and storage capacity. Under such circumstances that a great deal of data mixed in the world, it is necessary to master information retrieval techniques that just search large numbers of data for what you need.

To search the multi-dimensional media data for query points by computer, it is more important to search for approximated data points than exactly identical

This work is partially supported by Grant-in-Aid for Scientific Research 24240021, 24300060, 25540137, 26280085, 26280090 and 26370281 from the Ministry of Education, Culture, Sports, Science and Technology, Japan.

T. Murata et al. (Eds.): JSAI-isAI 2014 Workshops, LNAI 9067, pp. 310–316, 2015.
DOI: 10.1007/978-3-662-48119-6_23

ones only because there were many instances of such data deterioration and processing by compression techniques and others. We can determine by precise indications such as the distance if we search approximately by computer. Index structures such as R-trees [3] which are extended to multi-dimension from B-trees [1] are typically used if we search through large numbers of multi-dimensional media data approximately at high speed. However, the efficiency of search would be worse if an R-tree processed high-dimensional feature data. Therefore, we make use of dimension reduction which projects onto low-dimensional space to prevent index structure from worsening of the efficiency of search.

In this paper we make use of a method of dimension reduction named *Simple-Map* (*S-Map*, for short) [4]. It is necessary to consider how the distances in the projective space are kept from the corresponding distances in the original space. S-Map, which is applicable to any metric space, uses the distances between central points and objects as the coordinate values. S-Map with multiple central points is a projection to multi-dimensional L_∞ space. We use a heuristic method on S-Map because there is no analytical technique which determines the optimum projection such as principle component analysis [2]. In this paper we investigate methods to select central points for increasing the performance of S-Map.

In the previous researches [5,6] for the S-Map, the candidates for central points are randomly selected from data objects in database, and the summation of projective distances between sampled pairs of points is used as the scoring function to be maximized. We can improve the above method to select central points by using local search. The coordinate values of central points obtained after local search tend to be the maximum or the minimum ends of the space. By focusing on this tendency, in this paper, we propose *binary quantization* to select central points divided into the maximum values and the minimum values based on whether the coordinate value of an object in database is greater than the threshold or not. As experimental results, by using image data extracted from video, we observe that binary quantization approach reduces the search time about 30 % than using objects in database as are. Unfortunately, it has almost no difference between conventional approach and binary quantization by using local search.

2 Preliminaries

Approximate search is to draw out data which is approximated with query points using dissimilarity (namely, distance) of data. Make all of feature space which is based on database of approximate search to be $\mathcal{U} = \mathbb{R}^n$. $\mathbb{R}$ is for all real numbers and n is the number of dimensions of feature data here. In this paper, to search approximately, the following conditions, distance axiom about distance function d must be present when d indicated by index of the dissimilarity between any two objects is $d : \mathcal{U} \times \mathcal{U} \to \mathbb{R}^+$ and metric space is $\mathcal{D} = (\mathcal{S}, d)$.

- $d(X, Y) \geq 0$ (Non-negativity)
- $d(X, Y) = d(Y, X)$ (Symmetry)
- $d(X, Y) \leq d(X, Z) + d(Z, Y)$ (Triangle inequality)

- $d(X,Y) = 0 \Leftrightarrow X = Y$ (Identity)

$X, Y, Z \in \mathcal{D}$ here. Triangle inequality is the most important one in the above conditions.

And then, let us introduce distance measurement in this experiment. If any object in feature space is x, the feature of x is expressed by $(x_{(1)}, x_{(2)}, \ldots, x_{(n)})$, a set of n real numbers. The following three functions of distance measurement satisfy the distance axiom.

- L_1 distance : $D(x,y) = \sum_{i=1}^{n} |x_{(i)} - y_{(i)}|$

- L_2 distance : $D(x,y) = \sqrt{\sum_{i=1}^{n} (x_{(i)} - y_{(i)})^2}$

- L_∞ distance : $D(x,y) = \max_{i=1}^{n} |x_{(i)} - y_{(i)}|$

There are two approaches, range queries and neighborhood queries, principally used in approximate search. In this paper, we make use of *nearest neighbor queries* $NN(\mathcal{D}, Q)$ in the neighborhood queries, obtained from the object which has the least distance from query point Q to experiment.

- $NN(\mathcal{D}, Q) = \{O_i \in \mathcal{S} \mid O_i \text{ is the smallest one in } d(O_i, Q)\}$

It is set that search range r is infinity for nearest neighbor queries in the beginning period. And then it starts searching thus, shrinks search range to the distance between query point and interim solution that the object is logged on if there is one found in search range. Repeat the above steps until the new one could not be found in search range.

R-tree is a type of balanced tree with multi-branches that it is possible to insert and delete objects dynamically. As a target, space is cut up in n-dimensional ultra-rectangular named MBR (Minimum Bounding Rectangle). Here n is the number of dimensions of feature. Every node contains MBR which includes all of the child-nodes, and child-pointer, except that leaf nodes contain tuple-identifiers, pointers to objects in database. It is necessary to pay attention that it is possible for MBRs which include different nodes each other to superpose. In higher dimensional space, the performance of search would be worse as a result of the phenomenon named curse of dimensionality. We consider that it is because the form of MBR is no longer a normalized form and overlap redundantly in higher dimensional space. In this paper, we make use of dimension reduction technique to project from higher dimensional feature space to lower dimensional space in order to alleviate the worsening of retrieval performance in higher dimensional space.

In the search of R-tree, the nodes that queries range intersects with MBR are only visited in accordance with the order of distance between query point and MBR. If dimension reduction technique is used in constructing R-tree, to investigate the distance in projective space before calculating the distance between query point and object in leaf nodes, the efficient search will be achieved because the number of times to calculate actual range is decreased to the minimum.

3   Simple Map: Dimension Reduction Projection

It is well known that the performance of search would be worse as a result of curse of dimensionality if spatial index structure such as R-tree is used in higher dimensional feature space. Dimension reduction technique which projects from feature space to lower dimensional space is a method to alleviate the curse of dimensionality. In this paper, dimension reduction we used is S-Map. We call feature space before projecting the *original space* and space after projecting the *projective space*.

An actual distance between a central point and an object is used for projection function in S-Map. S-Map also applies to any metric space which satisfies the distance axiom. The distance between any two points of object images in projective space may be shrunk much more than the one in the original space. It is called *shrinkage of distance*.

If central point used to project is p, the coordinate value of any object O in projective space is defined as $\varphi_p(O) = d(p, O)$. By triangle inequality, the distance $d(X, Y)$ between any two points X and Y of objects may be shrunk in projective space. That is, if the distance in projective space is $d'(X, Y)$, then

$$- \ d'(X, Y) = |\varphi_p(X) - \varphi_p(Y)| \le d(X, Y).$$

When n' central points are used, n' is the dimensionality of projective space by S-Map. If the set of central points is $P = \{p_1, p_2, \ldots, p_{n'}\}$, then the projection function is

$$- \ f_P(X) = (\varphi_{p_1}(X), \varphi_{p_2}(X), \ldots, \varphi_{p_{n'}}(X)).$$

The distance between two points in projective space is

$$- \ d'(X, Y) = \max_{p \in P} |\varphi_p(X) - \varphi_p(Y)| \le d(X, Y).$$

Thus, the distance in projective space is measured by L_∞ distance. The larger number of central points, the more information of distances in projective space is kept. Namely the more central points, the less shrinkage of the distance it has. However, too high dimensionality of projective space makes the costs of computing projected distances large.

Search efficiency widens the gap according to the selected central point when high dimensional data projects to low dimensional space. Therefore, it is important for fast search to take hold of a method for selecting the central point. *Score* is used in the indication of performance of central points. The method of scoring is used with projection distance between any pairs of objects in database. The higher score, the better central points they are.

The above method of central point selection is to choose candidates randomly from the objects in database. However, it is greatly possible that more suitable central points were missed because points existed in data is too few in the whole space.

Thus, in *local search*, we vary the value of each dimension little by little, around the objects selected as the candidates for central points, in order to find the more suitable central points. If it is found that the point which has the higher score than the candidate of central points, we make that point to be the new one of candidate and repeat the process.

After applying local search to S-Map, the coordinate values of central points tend to be the maximum or the minimum ends of the space. Based on this fact, we propose a method of *binary quantization* to select central points divided into the maximum values and the minimum values based on whether the coordinate value of an object in database is greater than the threshold or not. In fact, the limitation of feature data could be relaxed easily, which is to be calculated in any L_1 distance with coordinate display. Furthermore, in term of feature subjects, the metric space used to calculate could be L_1 distance, L_2 distance and any other data without coordinate display besides, such as edit distance.

4   Experimental Results

We use, as objects in database, about 7 million image feature data extracted from about 2800 videos in experiments. Image feature is a 2 dimensional frequency spectrum in 64 dimensions, each of which is represented by an 8-bit integer from 0 to 255. We first reduce dimensions of objects by S-Map with the conventional and the proposed methods. For the image features, we adopt 8 as the most effective dimension of projective space by S-Map. Then we construct R-trees for projective space to search through nearest neighbor queries. Finally we compare two methods of selecting central points for S-Map, the conventional and the proposed binary quantization. Three types of queries are prepared. Near queries for which there exist very close image flames in database, quasi near queries for which there exist relatively close image flames, and far queries for which there exist only far away image flames. The number of each query types is about 30 thousand. Our computer environment is that CPU is Intel Core i7-975 3.33 GHz and RAM is 9 GB.

First, we compare efficiencies of S-Maps by binary quantization with several thresholds. The experimental result is indicated in Table 1, where shrinkage of distance shows the percentage of projective distance to original and the ratio of speeding up indicates the percentage decreased from search time in the conventional approach. The bold one shows the peak performance. As a consequence, we recognize that the best performance could improve about 32 % of search time if the quantization threshold value is 88.

We run another experiment to effectiveness of local search. We compare three methods (1), (2) and (3) for selecting central points.

(1) The binary quantization with threshold value 88.
(2) The binary quantization improved by local search.
(3) The conventional method improved by local search.

The results are indicated in Table 2, where computation time is time to select central points. As a consequence, we recognize that the computation time of (1) is shorter but the search time is longer than (3). Unfortunately, there are almost no differences of computation time and search time between (2) and (3).

Table 1. Shrinkage of distance and ratio of speeding up search in each threshold value

Threshold value	Shrinkage of distance (%)	Ratio of speeding up (%)
68	49.84	20.34
78	53.57	30.52
88	54.11	32.20
92[a]	53.72	31.66
98[b]	53.27	30.84
108	52.53	27.79
118	50.03	18.42
128[c]	47.18	1.39

[a]median, [b]average, [c]center.

Table 2. The comparison of the methods of selecting central points

Method	Computation time (min)	Shrinkage of distance (%)	Ratio of speeding up (%)
(1)	2.5	54.11	32.20
(2)	16.2	56.82	40.73
(3)	16.2	56.25	39.68

5   Conclusion and Future Works

The computation time for S-Map by binary quantization without local search is one-sixth compared with the conventional approach using local search. The ratio of speeding up is about 8 % inferior to the conventional approach. Thus, proposed method can select relatively efficient S-Maps in short computation time. Furthermore, if the proposed method and the local search are used simultaneously, both the search time and the computation time are almost equal with the ones by the conventional approach with local search, even though the better one of central points has been selected before doing local search.

As the challenge for future, we will make use of sound data instead of image data for verification, and establish the method to select the appropriate threshold value from the database of objects searched. And also, it may not be restricted to binary discretization but others such as ternary-valued or quaternary-valued.

References

1. Bayer, R., McCreight, E.: Organization and maintenance of large ordered indexes. Acta Inform. **1**, 173–189 (1972)
2. Fukunaga, K.: Statistical Pattern Recognition, 2nd edn. Academic Press, San Diego (1990)
3. Guttman, A.: R-trees: a dynamic index structure for spatial searching. In: Proceedings of SIGMOD 1984, pp. 47–57 (1984)

4. Shinohara, T., Ishizaka, H.: On dimension reduction mappings for approximate retrieval of multi-dimensional data. In: Arikawa, S., Shinohara, A. (eds.) Progress in Discovery science. LNCS (LNAI), vol. 2281, pp. 224–231. Springer, Heidelberg (2002)
5. Nakanishi, Y.: A study on central point selection of dimension reduction projection simple-map for fast similarity search in high-dimensional data, Master Thesis, Kyushu Institute of Technology (2006)
6. Ogawa, F.: A study on simulated annealing and local search for central point selection of dimension reduction projection Simple-Map, Graduation Thesis, Kyushu Institute of Technology (2014)

Mapping Kernels Between Rooted Labeled Trees Beyond Ordered Trees

Kouichi Hirata[1]($\boxtimes$), Tetsuji Kuboyama[2], and Takuya Yoshino[1]

[1] Kyushu Institute of Technology, Kawazu 680-4, Iizuka 820-8502, Japan
{hirata,yoshino}@dumbo.ai.kyutech.ac.jp
[2] Gakushuin University, Mejiro 1-5-1, Toshima, Tokyo 171-8588, Japan
ori-gaba2014@tk.cc.gakushuin.ac.jp

Abstract. In this paper, we investigate several mapping kernels to count all of the mappings between two rooted labeled trees beyond ordered trees, that is, *cyclically ordered trees* such as *biordered trees*, *cyclic-ordered trees* and *cyclic-biordered trees*, and *degree-bounded unordered trees*. Then, we design the algorithms to compute a *top-down mapping kernel*, an *LCA-preserving segmental mapping kernel*, an *LCA-preserving mapping kernel*, an *accordant mapping kernel* and an *isolated-subtree mapping kernel* for biordered trees in $O(nm)$ time and ones for cyclic-ordered and cyclic-biordered trees in $O(nmdD)$ time, where n is the number of nodes in a tree, m is the number of nodes in another tree, D is the maximum value of the degrees in two trees and d is the minimum value of the degrees in two trees. Also we design the algorithms to compute the above kernels for degree-bounded unordered trees in $O(nm)$ time. On the other hand, we show that the problem of computing *label-preserving leaf-extended* top-down mapping kernel and *label-preserving* bottom-up mapping kernel is #P-complete.

1 Introduction

A *tree kernel* is one of the fundamental method to classify *rooted labeled trees* (*trees*, for short) through support vector machines (SVMs). Many researches to design tree kernels for *ordered* trees, in which an order among siblings is fixed, have been developed (*cf.*, [2,6,14–17]). We call them *ordered tree kernels*.

A *mapping kernel* [15–17] is a powerful and general framework for tree kernels based on counting all of the *mappings* (and their variations) as the set of one-to-one node correspondences [18]. It is known that the minimum cost of (Tai) mappings coincides with an edit distance between trees. Also, as the properties of mapping kernels, almost ordered tree kernels are classified into the framework of mapping kernels [15], and a mapping kernel is *positive definite* if and only if the mapping is *transitive*, that is, *closed under the composition* [16,17].

This work is partially supported by Grant-in-Aid for Scientific Research 24240021, 24300060, 25540137, 26280085 and 26370281 from the Ministry of Education, Culture, Sports, Science and Technology, Japan.

T. Murata et al. (Eds.): JSAI-isAI 2014 Workshops, LNAI 9067, pp. 317–330, 2015.
DOI: 10.1007/978-3-662-48119-6_24

On the other hand, few researches to design tree kernels for *unordered* trees, in which an order among siblings is arbitrary, have been developed. We call them *unordered tree kernels*. One of the reasons is that the problem of counting all of the subtrees for unordered trees is #P-complete [6].

In order to avoid such difficulty, the unordered tree kernel have been developed as counting all of the specific substructures. For example, Kuboyama *et al.* [9] and Kimura *et al.* [7] have designed the unordered tree kernel counting all of the *bifoliate q-grams* and all of the *subpaths*, respectively.

As a tractable mapping kernel for unordered trees, Hamada *et al.* [3] have introduced an *agreement-subtree mapping kernel* for *phylogenetic trees* (leaf-labeled binary unordered trees). Also they have given a new proof of intractability of computing a mapping kernel for unordered trees, simpler than Kashima *et al.* [6], such that the problem of counting the number of leaves with the same labels in leaf-labeled tree is #P-complete, which is based on the problem of counting all of the matchings in a bipartite graph.

It is known that, by introducing several conditions to mappings, we deal with several variations of mappings and they form the hierarchy of mappings [5, 8, 21, 23]. Every variation of mappings provides not only a variation of the edit distance as the minimum cost of all the mappings [5, 8, 22, 23] but also a tree kernel as the number of all the mappings [8, 10, 15].

Note that the problem of computing the tractable variations of the edit distance between unordered trees such as a top-down distance [1, 13], an LCA-preserving segmental distance [23], an LCA-preserving distance [27], an accordant distance [8, 10, 22] and an isolated-subtree distance [25, 26] is essential to solve the minimum weighted maximum matching in a bipartite graph [22, 26, 27]. On the other hand, it is essential for the above #P-completeness [3, 6] to reduce from the problem of counting all of the matchings in a bipartite graph.

Recently, as trees extended from ordered trees and restricted to unordered trees, Yoshino and Hirata [24] have introduced the following three kinds of a *cyclically ordered tree* that is an unordered tree preserving the adjacency among siblings in a tree as possible. Let $v_1, \ldots, v_n$ be siblings from left to right. We say that a tree is *biordered* if it allows two orders $v_1, \ldots, v_n$ and $v_n, \ldots, v_1$. Also we say that a tree is *cyclic-ordered* if it allows a cyclic order $v_i, \ldots, v_n, v_1, \ldots, v_{i-1}$ for every i $(1 \leq i \leq n)$. Furthermore, we say that a tree is *cyclic-biordered* if it allows cyclic orders $v_i, \ldots, v_n, v_1, \ldots, v_{i-1}$ and $v_{i-1}, \ldots, v_1, v_n, \ldots, v_i$ for every i $(1 \leq i \leq n)$. Then, they have designed the algorithm to compute an *alignment distance* [4] between cyclically ordered trees in polynomial time. Note that the algorithm does not use the maximum matching for a bipartite graph. It is a simple extension of the algorithm (or recurrences) of computing the alignment distance between ordered trees [4].

Hence, in this paper, we first investigate several mapping kernels such as a *top-down mapping kernel*, an *LCA-preserving segmental mapping kernel*, an *LCA-preserving mapping kernel*, an *accordant mapping kernel* and an *isolated-subtree mapping kernel* for cyclically ordered trees. Then, we design the algorithms to compute all of the above mapping kernels for biordered trees in $O(nm)$

time and ones for cyclic-ordered and cyclic-biordered trees in $O(nmdD)$ time, where n is the number of nodes in a tree, m is the number of nodes in another tree, D is the maximum value of the degrees in two trees and d is the minimum value of the degrees in two trees.

Next, by focusing that the agreement subtree mapping kernel is applied to full binary trees, we investigate the above kernels for bounded-degree unordered trees. Then, we design the algorithms to compute all of the above mapping kernels in $O(nm)$ time, which follows from the algorithms to compute ones for unordered trees in $O(nmD^D)$ time, which is exponential to D.

On the other hand, for unordered trees, we show that the problem of computing the *label-preserving leaf-extended* top-down mapping kernel and the *label-preserving* bottom-up mapping kernel is #P-complete. Note here that the proof of the above #P-completeness [3,6] cannot apply to top-down and bottom-up mapping kernels for unordered tree directly. Also, the degrees of unordered trees in this proof are not bounded.

2   Preliminaries

A *tree* is a connected graph without cycles. For a tree $T = (V, E)$, we denote V and E by $V(T)$ and $E(T)$, respectively. Also the *size* of T is $|V|$ and denoted by $|T|$. We sometime denote $v \in V(T)$ by $v \in T$. We denote an empty tree by $\emptyset$.

A *rooted tree* is a tree with one node r chosen as its *root*. We denote the root of a rooted tree T by $r(T)$. A(n ordered) *forest* is a sequence $[T_1, \dots, T_n]$ of trees which we denote by $T_1 \bullet \dots \bullet T_n$ or $\bullet_{i=1}^n T_i$. In particular, for two forests $F_1 = T_1 \bullet \dots \bullet T_n$ and $F_2 = S_1 \bullet \dots \bullet S_m$, we denote the forest $T_1 \bullet \dots \bullet T_n \bullet S_1 \bullet \dots \bullet S_m$ by $F_1 \bullet F_2$. For a forest F, we denote the tree rooted by v whose children are trees in F by $v(F)$.

For each node v in a rooted tree with the root r, let $UP_r(v)$ be the unique path (as trees) from v to r. The *parent* of $v(\neq r)$, which we denote by $par(v)$, is its adjacent node on $UP_r(v)$ and the *ancestors* of $v(\neq r)$ are the nodes on $UP_r(v) - \{v\}$. We denote the set of all ancestors of v by $anc(v)$. We say that u is a *child* of v if v is the parent of u. The set of children of v is denoted by $ch(v)$. A *leaf* is a node having no children. We denote the set of all leaves in T by $lv(T)$. A node that is neither a leaf nor a root is called an *internal node*. We call the number of children of v the *degree* of v and denote it by $d(v)$, that is, $d(v) = |ch(v)|$. Also we define $d(T) = \max\{d(v) \mid v \in T\}$ and call it the *degree* of T.

In this paper, we use the ancestor orders $<$ and $\leq$, that is, $u < v$ if v is an ancestor of u and $u \leq v$ if $u < v$ or $u = v$. We say that w is the *least common ancestor* (LCA for short) of u and v, denoted by $u \sqcup v$, if $u \leq w$, $v \leq w$ and there exists no w' such that A *(complete)* $w' < w$, $u \leq w'$ and $v \leq w'$. A *(complete) subtree of* $T = (V, E)$ *rooted by* v, denoted by $T[v]$, is a tree $T' = (V', E')$ such that $r(T') = v$, $V' = \{u \in V \mid u \leq v\}$ and $E' = \{(u, w) \in E \mid u, w \in V'\}$.

We say that a rooted tree is *labeled* if each node is assigned a symbol from a fixed finite alphabet Σ. For a node v, we denote the label of v by $l(v)$, and sometimes identify v with $l(v)$. Also let $\varepsilon \notin \Sigma$ denote a special *blank* symbol and define $\Sigma_\varepsilon = \Sigma \cup \{\varepsilon\}$.

Let $v \in T$ and $v_i, v_j \in ch(v)$ such that v_i the i-th child of v and v_j the j-th child of v. Then, we say that v_i is *to the left of* v_j if $i \leq j$. Then, for every $u, v \in T$, $u \preceq v$ if either u is to the left of v (when both u and v are the children of the same node in T) or there exist $u', v' \in ch(u \sqcup v)$ such that $u \leq u'$, $v \leq v'$ and u' is to the left of v'. Hence, we say that a rooted tree is *ordered* if a left-to-right order among siblings is fixed; *unordered* otherwise. Furthermore, in this paper, we introduce *cyclically ordered trees* by using the following functions $\sigma^+_{p,n}(i)$ and $\sigma^-_{p,n}(i)$ for $1 \leq i, p \leq n$.

$$\sigma^+_{p,n}(i) = ((i + p - 1) \bmod n) + 1, \quad \sigma^-_{p,n}(i) = ((n - i - p + 1) \bmod n) + 1.$$

Definition 1 (Cyclically Ordered Trees). Let T be a tree and suppose that $v_1, \ldots, v_n$ are the children of $v \in T$ from left to right.

1. We say that T is *biordered* if T allows the orders of both $v_1, \ldots, v_n$ and $v_n, \ldots, v_1$.
2. We say that T is *cyclic-ordered* if T allows the orders $v_{\sigma^+_{p,n}(1)}, \ldots, v_{\sigma^+_{p,n}(n)}$ for every $1 \leq p \leq n$.
3. We say that T is *cyclic-biordered* if T allows the orders $v_{\sigma^+_{p,n}(1)}, \ldots, v_{\sigma^+_{p,n}(n)}$ and $v_{\sigma^-_{p,n}(1)}, \ldots, v_{\sigma^-_{p,n}(n)}$ for every $1 \leq p \leq n$.

Sometimes we use the scripts o, b, c, cb, u, and the notation of $\pi \in \{o, b, c, cb, u\}$.

It is obvious that the cyclically ordered trees are an extension of ordered trees and a restriction of unordered trees. The number of orders among siblings of a node v in ordered trees, biordered trees, cyclic-ordered trees, cyclic-biordered trees and unordered trees is $1, 2, d(v), 2d(v)$ and $d(v)!$, respectively. Also it holds that, when $d(T) = 2$, T is unordered iff it is biordered, cyclic-ordered or cyclic-biordered, and when $d(T) = 3$, T is unordered iff it is cyclic-biordered.

3 Mapping

In this section, we introduce a *Tai mapping* and its variations, and then the distance as the minimum cost of all the mappings.

Definition 2 (Tai Mapping [18]). Let T_1 and T_2 be trees and $M \subseteq V(T_1) \times V(T_2)$.

1. We say that a triple (M, T_1, T_2) is an *ordered Tai mapping* from T_1 to T_2, denoted by $M \in \mathcal{M}^o_{\mathrm{TAI}}(T_1, T_2)$, if every pair (u_1, v_1) and (u_2, v_2) in M satisfies the following conditions.
    (i) $u_1 = u_2$ iff $v_1 = v_2$ (one-to-one condition).
    (ii) $u_1 \leq u_2$ iff $v_1 \leq v_2$ (ancestor condition).
    (iii) $u_1 \preceq u_2$ iff $v_1 \preceq v_2$ (sibling condition).

2. We say that a triple (M, T_1, T_2) is an *unordered Tai mapping* from T_1 to T_2, denoted by $M \in \mathcal{M}^u_{\text{TAI}}(T_1, T_2)$, if M satisfies the conditions (i) and (ii).

In the following, let $u_1, u_2, u_3, u_4 \in ch(u)$ and $v_1, v_2, v_3, v_4 \in ch(v)$.

3. We say that a triple (M, T_1, T_2) is a *biordered Tai mapping* from T_1 to T_2, denoted by $M \in \mathcal{M}^b_{\text{TAI}}(T_1, T_2)$, if M satisfies the above conditions (i) and (ii) and the following condition (iv).

(iv) For every $u \in T_1$ and $v \in T_2$ such that $(u_1, v_1), (u_2, v_2), (u_3, v_3) \in M$, one of the following statements holds.
   1. $u_1 \preceq u_2 \preceq u_3$ iff $v_1 \preceq v_2 \preceq v_3$.
   2. $u_1 \preceq u_2 \preceq u_3$ iff $v_3 \preceq v_2 \preceq v_1$.

4. We say that a triple (M, T_1, T_2) is a *cyclic-ordered Tai mapping* from T_1 to T_2, denoted by $M \in \mathcal{M}^c_{\text{TAI}}(T_1, T_2)$, if M satisfies the above conditions (i) and (ii) and the following condition (v).

(v) For every $u \in T_1$ and $v \in T_2$ such that $(u_1, v_1), (u_2, v_2), (u_3, v_3) \in M$, one of the following statements holds.
   1. $u_1 \preceq u_2 \preceq u_3$ iff $v_1 \preceq v_2 \preceq v_3$.
   2. $u_1 \preceq u_2 \preceq u_3$ iff $v_2 \preceq v_3 \preceq v_1$.
   3. $u_1 \preceq u_2 \preceq u_3$ iff $v_3 \preceq v_1 \preceq v_2$.

5. We say that a triple (M, T_1, T_2) is a *cyclic-biordered Tai mapping* from T_1 to T_2, denoted by $M \in \mathcal{M}^{cb}_{\text{TAI}}(T_1, T_2)$, if M satisfies the above conditions (i) and (ii) and the following condition (vi).

(vi) For every $u \in T_1$ and $v \in T_2$ such that $(u_1, v_1), (u_2, v_2), (u_3, v_3), (u_4, v_4) \in M$, one of the following statements holds.
   1. $u_1 \preceq u_2 \preceq u_3 \preceq u_4$ iff $v_1 \preceq v_2 \preceq v_3 \preceq v_4$.
   2. $u_1 \preceq u_2 \preceq u_3 \preceq u_4$ iff $v_2 \preceq v_3 \preceq v_4 \preceq v_1$.
   3. $u_1 \preceq u_2 \preceq u_3 \preceq u_4$ iff $v_3 \preceq v_4 \preceq v_1 \preceq v_2$.
   4. $u_1 \preceq u_2 \preceq u_3 \preceq u_4$ iff $v_4 \preceq v_1 \preceq v_2 \preceq v_3$.
   5. $u_1 \preceq u_2 \preceq u_3 \preceq u_4$ iff $v_4 \preceq v_3 \preceq v_2 \preceq v_1$.
   6. $u_1 \preceq u_2 \preceq u_3 \preceq u_4$ iff $v_3 \preceq v_2 \preceq v_1 \preceq v_4$.
   7. $u_1 \preceq u_2 \preceq u_3 \preceq u_4$ iff $v_2 \preceq v_1 \preceq v_4 \preceq v_3$.
   8. $u_1 \preceq u_2 \preceq u_3 \preceq u_4$ iff $v_1 \preceq v_4 \preceq v_3 \preceq v_2$.

We will use M instead of (M, T_1, T_2) simply and call a Tai mapping a *mapping* simply.

Definition 3 (Variations of Tai Mapping). Let T_1 and T_2 be trees, $\pi \in \{o, b, c, cb, u\}$ and $M \in \mathcal{M}^\pi_{\text{TAI}}(T_1, T_2)$. Here, we denote $M - \{(r(T_1), r(T_2))\}$ by M^-.

1. We say that M is a *top-down mapping* [1,13] (or a *degree-1 mapping*), denoted by $M \in \mathcal{M}^\pi_{\text{TOP}}(T_1, T_2)$, if M satisfies the following condition.

$$\forall (u, v) \in M^- \Big((par(u), par(v)) \in M \Big).$$

2. We say that M is an *LCA-preserving segmental mapping* [23], denoted by $M \in \mathcal{M}^\pi_{\text{LCASG}}(T_1, T_2)$, if there exists a pair $(u, v) \in T_1 \times T_2$ such that $M \in \mathcal{M}^\pi_{\text{TOP}}(T_1[u], T_2[v])$.

3. We say that M is an *LCA-preserving mapping* (or a *degree-2 mapping* [27]), denoted by $M \in \mathcal{M}_{\text{LCA}}^{\pi}(T_1, T_2)$, if M satisfies the following condition.

$$\forall (u_1, v_1), (u_2, v_2) \in M\Big((u_1 \sqcup u_2, v_1 \sqcup v_2) \in M\Big).$$

4. We say that M is an *accordant mapping* [8] (or a *Lu's mapping* [12]), denoted by $M \in \mathcal{M}_{\text{ACC}}^{\pi}(T_1, T_2)$, if M satisfies the following condition.

$$\forall (u_1, v_1), (u_2, v_2), (u_3, v_3) \in M\Big(u_1 \sqcup u_2 = u_1 \sqcup u_3 \iff v_1 \sqcup v_2 = v_1 \sqcup v_3\Big).$$

5. We say that M is an *isolated-subtree mapping* [21] (or a *constrained mapping* [25,26]), denoted by $M \in \mathcal{M}_{\text{ILST}}^{\pi}(T_1, T_2)$, if M satisfies the following condition.

$$\forall (u_1, v_1), (u_2, v_2), (u_3, v_3) \in M\Big(u_3 < u_1 \sqcup u_2 \iff v_3 < v_1 \sqcup v_2\Big).$$

6. We say that M is a *bottom-up mapping* [8,20,22], denoted by $M \in \mathcal{M}_{\text{BOT}}^{\pi}(T_1, T_2)$, if M satisfies the following condition.

$$\forall (u, v) \in M \begin{pmatrix} \forall u' \in T_1[u] \exists v' \in T_2[v]\big((u', v') \in M\big) \\ \wedge \forall v' \in T_2[v] \exists u' \in T_1[u]\big((u', v') \in M\big) \end{pmatrix}.$$

Proposition 1 (*cf.* [8,23]). *For* $\pi \in \{o, b, c, cb, u\}$ *and trees* T_1 *and* T_2, *the following statement holds:*

$$\mathcal{M}_{\text{TOP}}^{\pi}(T_1, T_2) \subset \mathcal{M}_{\text{LCASG}}^{\pi}(T_1, T_2) \subset \mathcal{M}_{\text{LCA}}^{\pi}(T_1, T_2)$$
$$\subset \mathcal{M}_{\text{ACC}}^{\pi}(T_1, T_2) \subset \mathcal{M}_{\text{ILST}}^{\pi}(T_1, T_2).$$

Furthermore, for $A \in \{\text{TOP}, \text{LCASG}, \text{LCA}, \text{ACC}, \text{ILST}\}$, $\mathcal{M}_{\text{BOT}}^{\pi}(T_1, T_2)$ *is incomparable with* $\mathcal{M}_{A}^{\pi}(T_1, T_2)$

4   Mapping Kernels

Let $\pi \in \{o, b, c, cb, u\}$ and $A \in \{\text{TOP}, \text{LCASG}, \text{LCA}, \text{ACC}, \text{ILST}\}$ unless otherwise noted. A *mapping* between forests F_1 and F_2 is defined as a mapping M between trees $v(F_1)$ and $v(F_2)$ such that $(v, v) \notin M$. We define $\mathcal{M}_{A}^{\pi}(F_1, F_2)$ as similar as $\mathcal{M}_{A}^{\pi}(T_1, T_2)$. Let $\sigma : \Sigma \times \Sigma \to \mathbf{R}^+$ be a similarity function. The similarity $\sigma(M)$ of a mapping $M \in \mathcal{M}_{A}^{\pi}(T_1, T_2)$ between two trees T_1 and T_2 is defined as $\sigma(M) = \prod_{(u,v) \in M} \sigma(l(u), l(v))$. The *similarity* between two forests F_1 and F_2 is defined as follows:

$$\mathcal{K}_{A}^{\pi}(F_1, F_2) = \sum_{M \in \mathcal{M}_{A}^{\pi}(F_1, F_2)} \sigma(M).$$

Corollary 1. *For* $\pi \in \{o, b, c, cb, u\}$ *and* $\mathtt{A} \in \{\textsc{Top}, \textsc{LcaSg}, \textsc{Lca}, \textsc{Acc}, \textsc{Ilst}\}$, $\mathcal{K}_{\mathtt{A}}^{\pi}$ *is positive definite.*

Proof. Since $\mathcal{M}_{\mathtt{A}}^{\pi}$ is closed under the composition [8,23,27] and by [16], the statement holds. $\square$

Kuboyama [8] has introduced the recurrences to compute $\mathcal{K}_{\mathtt{A}}^{o}(T_1, T_2)$ for $\mathtt{A} \in \{\textsc{Top}, \textsc{LcaSg}, \textsc{Lca}\}$ implicitly and $\mathtt{A} \in \{\textsc{Acc}, \textsc{Ilst}\}$ explicitly illustrated in Fig. 1. Note the underlined formulas that denote the difference between similar formulas.

$$
\begin{aligned}
\mathcal{K}_{\textsc{Top}}^{o}(u(F_1), v(F_2)) &= \sigma(l(u), l(v)) \cdot (1 + \mathcal{F}_{\textsc{Top}}^{o}(F_1, F_2)), \\
\mathcal{F}_{\textsc{Top}}^{o}(\emptyset, F) &= \mathcal{F}_{\textsc{Top}}^{o}(F, \emptyset) = 0, \\
\mathcal{F}_{\textsc{Top}}^{o}(T_1 \bullet F_1, T_2 \bullet F_2) &= \mathcal{K}_{\textsc{Top}}^{o}(T_1, T_2) \cdot (1 + \mathcal{F}_{\textsc{Top}}^{o}(F_1, F_2)) \\
&\quad + \mathcal{F}_{\textsc{Top}}^{o}(F_1, T_2 \bullet F_2) + \mathcal{F}_{\textsc{Top}}^{o}(T_1 \bullet F_1, F_2) - \underline{\mathcal{F}_{\textsc{Top}}^{o}(F_1, F_2)}.
\end{aligned}
$$

$$
\mathcal{K}_{\textsc{LcaSg}}^{o}(T_1, T_2) = \sum_{u \in T_1} \sum_{v \in T_2} \mathcal{K}_{\textsc{Top}}^{o}(T_1[u], T_2[v]).
$$

$$
\begin{aligned}
\mathcal{K}_{\textsc{Lca}}^{o}(T_1, T_2) &= \sum_{u \in T_1} \sum_{v \in T_2} \mathcal{T}_{\textsc{Lca}}^{o}(T_1[u], T_2[v]), \\
\mathcal{T}_{\textsc{Lca}}^{o}(u(F_1), v(F_2)) &= \sigma(l(u), l(v)) \cdot (1 + \mathcal{F}_{\textsc{Lca}}^{o}(F_1, F_2)), \\
\mathcal{F}_{\textsc{Lca}}^{o}(\emptyset, F) &= \mathcal{F}_{\textsc{Lca}}^{o}(F, \emptyset) = 0, \\
\mathcal{F}_{\textsc{Lca}}^{o}(T_1 \bullet F_1, T_2 \bullet F_2) &= \mathcal{K}_{\textsc{Lca}}^{o}(T_1, T_2) \cdot (1 + \mathcal{F}_{\textsc{Lca}}^{o}(F_1, F_2)) \\
&\quad + \mathcal{F}_{\textsc{Lca}}^{o}(F_1, T_2 \bullet F_2) + \mathcal{F}_{\textsc{Lca}}^{o}(T_1 \bullet F_1, F_2) - \underline{\mathcal{F}_{\textsc{Lca}}^{o}(F_1, F_2)}.
\end{aligned}
$$

$$
\begin{aligned}
\mathcal{K}_{\textsc{Acc}}^{o}(\emptyset, F) &= \mathcal{K}_{\textsc{Acc}}^{o}(F, \emptyset) = \mathcal{TF}_{\textsc{Acc}}^{o}(T, \emptyset) = 0, \\
\mathcal{K}_{\textsc{Acc}}^{o}(T_1 \bullet F_1, T_2 \bullet F_2) &= \mathcal{T}_{\textsc{Acc}}^{o}(T_1, T_2) \cdot (\mathcal{F}_{\textsc{Acc}}^{o}(F_1, F_2) - 1) \\
&\quad + \mathcal{TF}_{\textsc{Acc}}^{o}(T_1, T_2 \bullet F_2) - \mathcal{TF}_{\textsc{Acc}}^{o}(T_1, F_2) \\
&\quad + \mathcal{TF}_{\textsc{Acc}}^{o}(T_2, T_1 \bullet F_1) - \mathcal{TF}_{\textsc{Acc}}^{o}(T_2, F_1) \\
&\quad + \mathcal{K}_{\textsc{Acc}}^{o}(T_1 \bullet F_1, F_2) - \mathcal{K}_{\textsc{Acc}}^{o}(F_1, F_2), \\
\mathcal{TF}_{\textsc{Acc}}^{o}(v(F_1), T_2 \bullet F_2) &= \mathcal{T}_{\textsc{Acc}}^{o}(v(F_1), T_2) - \mathcal{TF}_{\textsc{Acc}}^{o}(T_2, F_1) + \mathcal{TF}_{\textsc{Acc}}^{o}(v(F_1), F_2) \\
&\quad - \mathcal{K}_{\textsc{Acc}}^{o}(F_1, F_2) + \mathcal{K}_{\textsc{Acc}}^{o}(F_1, T_1 \bullet F_2), \\
\mathcal{T}_{\textsc{Acc}}^{o}(u(F_1), v(F_2)) &= \sigma(l(u), l(v)) \cdot (1 + \mathcal{F}_{\textsc{Acc}}^{o}(F_1, F_2)) + \mathcal{TF}_{\textsc{Acc}}^{o}(u(F_1), F_2) \\
&\quad + \mathcal{TF}_{\textsc{Acc}}^{o}(v(F_2), F_1) - \mathcal{K}_{\textsc{Acc}}^{o}(F_1, F_2), \\
\mathcal{F}_{\textsc{Acc}}^{o}(\emptyset, F) &= \mathcal{F}_{\textsc{Acc}}^{o}(F, \emptyset) = 0, \\
\mathcal{F}_{\textsc{Acc}}^{o}(T_1 \bullet F_1, T_2 \bullet F_2) &= \mathcal{T}_{\textsc{Acc}}^{o}(T_1, T_2) \cdot (1 + \mathcal{F}_{\textsc{Acc}}^{o}(F_1, F_2)) \\
&\quad + \mathcal{F}_{\textsc{Acc}}^{o}(F_1, T_2 \bullet F_2) + \mathcal{F}_{\textsc{Acc}}^{o}(T_1 \bullet F_1, F_2) - \mathcal{F}_{\textsc{Acc}}^{o}(F_1, F_2).
\end{aligned}
$$

$$
\begin{aligned}
\mathcal{K}_{\textsc{Ilst}}^{o}(\emptyset, F) &= \mathcal{K}_{\textsc{Ilst}}^{o}(F, \emptyset) = \mathcal{TF}_{\textsc{Ilst}}^{o}(T, \emptyset) = 0, \\
\mathcal{K}_{\textsc{Ilst}}^{o}(T_1 \bullet F_1, T_2 \bullet F_2) &= \mathcal{T}_{\textsc{Ilst}}^{o}(T_1, T_2) \cdot (\mathcal{F}_{\textsc{Ilst}}^{o}(F_1, F_2) - 1) \\
&\quad + \mathcal{TF}_{\textsc{Ilst}}^{o}(T_1, T_2 \bullet F_2) - \mathcal{TF}_{\textsc{Ilst}}^{o}(T_1, F_2) \\
&\quad + \mathcal{TF}_{\textsc{Ilst}}^{o}(T_2, T_1 \bullet F_1) - \mathcal{TF}_{\textsc{Ilst}}^{o}(T_2, F_1) \\
&\quad + \mathcal{K}_{\textsc{Ilst}}^{o}(T_1 \bullet F_1, F_2) - \mathcal{K}_{\textsc{Ilst}}^{o}(F_1, F_2), \\
\mathcal{TF}_{\textsc{Ilst}}^{o}(v(F_1), T_2 \bullet F_2) &= \mathcal{T}_{\textsc{Ilst}}^{o}(v(F_1), T_2) - \mathcal{TF}_{\textsc{Ilst}}^{o}(T_2, F_1) + \mathcal{TF}_{\textsc{Ilst}}^{o}(v(F_1), F_2) \\
&\quad - \mathcal{K}_{\textsc{Ilst}}^{o}(F_1, F_2) + \mathcal{K}_{\textsc{Ilst}}^{o}(F_1, T_1 \bullet F_2), \\
\mathcal{T}_{\textsc{Ilst}}^{o}(u(F_1), v(F_2)) &= \sigma(l(u), l(v)) \cdot (1 + \underline{\mathcal{K}_{\textsc{Ilst}}^{o}(F_1, F_2)}) + \mathcal{TF}_{\textsc{Ilst}}^{o}(u(F_1), F_2) \\
&\quad + \mathcal{TF}_{\textsc{Ilst}}^{o}(v(F_2), F_1) - \underline{\mathcal{K}_{\textsc{Ilst}}^{o}(F_1, F_2)}, \\
\mathcal{F}_{\textsc{Ilst}}^{o}(\emptyset, F) &= \mathcal{F}_{\textsc{Ilst}}^{o}(F, \emptyset) = 0, \\
\mathcal{F}_{\textsc{Ilst}}^{o}(T_1 \bullet F_1, T_2 \bullet F_2) &= \mathcal{T}_{\textsc{Ilst}}^{o}(T_1, T_2) \cdot (1 + \mathcal{F}_{\textsc{Ilst}}^{o}(F_1, F_2)) \\
&\quad + \mathcal{F}_{\textsc{Ilst}}^{o}(F_1, T_2 \bullet F_2) + \mathcal{F}_{\textsc{Ilst}}^{o}(T_1 \bullet F_1, F_2) - \mathcal{F}_{\textsc{Ilst}}^{o}(F_1, F_2).
\end{aligned}
$$

Fig. 1. The recurrences of computing $\mathcal{K}_{\mathtt{A}}^{o}(T_1, T_2)$ for $\mathtt{A} \in \{\textsc{Top}, \textsc{LcaSg}, \textsc{Lca}, \textsc{Acc}, \textsc{Ilst}\}$ [8].

Theorem 1 (cf., Kuboyama [8]). *For* $\mathtt{A} \in \{\text{TOP}, \text{LCASG}, \text{LCA}, \text{ACC}, \text{ILST}\}$, *the recurrences in Fig. 1 correctly compute* $\mathcal{K}_{\mathtt{A}}^o(T_1, T_2)$ *in* $O(nm)$ *time, where* $n = |T_1|$ *and* $m = |T_2|$.

4.1 Mapping Kernels for Cyclically Ordered Trees

In this section, we extend the recurrences in Fig. 1 to the recurrences to compute $\mathcal{K}_{\mathtt{A}}^\pi(T_1, T_2)$ for $\pi \in \{o, b, c, cb\}$ and $\mathtt{A} \in \{\text{TOP}, \text{LCASG}, \text{LCA}, \text{ACC}, \text{ILST}\}$.

For $u(F_1)$ and $v(F_2)$, let $F_1 = [T_1[u_1], \ldots, T_1[u_s]]$ and $F_2 = [T_2[v_1], \ldots, T_2[v_t]]$, that is, $ch(u) = \{u_1, \ldots, u_s\}$, $ch(v) = \{v_1, \ldots, v_t\}$, $d(u) = s$ and $d(v) = t$. Also let $1 \leq p \leq s$ and $1 \leq q \leq t$. We denote the forests $[T_1[u_{\sigma_{p,s}^+(1)}], \ldots, T_1[u_{\sigma_{p,s}^+(s)}]]$ and $[T_2[v_{\sigma_{q,t}^+(1)}], \ldots, T_2[v_{\sigma_{q,t}^+(t)}]]$ by F_1^p and F_2^q. Furthermore, we denote the forests $[T_1[u_{\sigma_{p,s}^-(1)}], \ldots, T_1[u_{\sigma_{p,s}^-(s)}]]$ and $[T_2[v_{\sigma_{q,t}^-(1)}], \ldots, T_2[v_{\sigma_{q,t}^-(t)}]]$ by F_1^{-p} and F_2^{-q}. It is obvious that $F_1 = F_1^1$ and $F_2 = F_2^1$.

Furthermore, the values of p and q in F_1^p, and F_2^q are (1) $p = q = 1$ if $\pi = o$, (2) $p = \pm 1$ and $q = \pm 1$ if $\pi = b$, (3) $1 \leq p \leq s$ and $1 \leq q \leq t$ if $\pi = c$ and (4) $1 \leq p \leq s$, $-s \leq p \leq -1$, $1 \leq q \leq t$ and $-t \leq q \leq -1$ if $\pi = cb$. Hence, we prepare the following sets: (1) $o(s) = o(t) = \{1\}$, (2) $b(s) = b(t) = \{-1, 1\}$, (3) $c(s) = \{1, \ldots, s\}$, $c(t) = \{1, \ldots, t\}$, and (4) $cb(s) = \{-s, \ldots, -1, 1, \ldots, s\}$, $cb(t) = \{-t, \ldots, -1, 1, \ldots, t\}$. We refer these sets to $\pi(s)$ and $\pi(t)$ for $\pi \in \{o, b, c, cb\}$.

Then, we design the recurrences to compute $\mathcal{K}_{\mathtt{A}}^\pi(T_1, T_2)$ illustrated in Fig. 2.

Theorem 2. *For* $\mathtt{A} \in \{\text{TOP}, \text{LCASG}, \text{LCA}, \text{ACC}, \text{ILST}\}$, *the recurrences in Fig. 2 correctly compute* $\mathcal{K}_{\mathtt{A}}^b(T_1, T_2)$ *in* $O(nm)$ *time and* $\mathcal{K}_{\mathtt{A}}^c(T_1, T_2)$ *and* $\mathcal{K}_{\mathtt{A}}^{cb}(T_1, T_2)$ *in* $O(nmdD)$ *time, where* $n = |T_1|$, $m = |T_2|$, $d = \min\{d(T_1), d(T_2)\}$ *and* $D = \max\{d(T_1), d(T_2)\}$.

Proof. In the formulas of $\mathcal{K}_{\text{TOP}}^\pi$ and $\mathcal{T}_{\text{LCA}}^\pi$, the number of $\mathcal{F}_{\text{TOP}}^\pi(F_1^p, F_2^q)$ and $\mathcal{F}_{\text{LCA}}^\pi(F_1^p, F_2^q)$ is 1 if $\pi = o$, 4 if $\pi = b$, $d(u) \cdot d(v)$ if $\pi = c$ and $2d(u) \cdot 2d(v)$ if $\pi = bc$. Also in the formulas of $\mathcal{T}_{\text{ACC}}^\pi$ and $\mathcal{T}_{\text{ILST}}^\pi$, the number of $\mathcal{F}_{\text{ACC}}^\pi(F_1^p, F_2^q)$ and $\mathcal{F}_{\text{ILST}}^\pi(F_1^p, F_2^q)$ is 1 if $\pi = o$, $4 + 2 + 2 + 4 = 12$ if $\pi = b$, $d(u) \cdot d(v) + d(u) + d(v) + d(u) \cdot d(v) = 2d(u) \cdot d(v) + d(u) + d(v)$ if $\pi = c$ and $2d(u) \cdot 2d(v) + 2d(u) + 2d(v) + 2d(u) \cdot 2d(v) = 8d(u) \cdot d(v) + 2d(u) + 2d(v)$ if $\pi = bc$. Then, we can compute these recurrences in $O(1)$ time if $\pi \in \{o, b\}$, whereas in $O(d(u) \cdot d(v)) = O(dD)$ time if $\pi \in \{c, cb\}$. Hence, the time complexity in the statement holds. Also we can show the correctness by extending Theorem 1. □

4.2 Mapping Kernels for Bounded-Degree Unordered Trees

In this section, we extend the recurrences in Fig. 1 to the recurrences to compute $\mathcal{K}_{\mathtt{A}}^u(T_1, T_2)$ for $\mathtt{A} \in \{\text{TOP}, \text{LCASG}, \text{LCA}, \text{ACC}, \text{ILST}\}$.

For nonnegative integers s and t, let $B_{s,t}$ be a complete bipartite graph $(X \cup Y, E)$ such that $X = \{1, \ldots, s\}$ and $Y = \{1, \ldots, t\}$, and $BM(s, t)$ the set of

$$\mathcal{K}^{\pi}_{\mathrm{Top}}(u(F_1), v(F_2)) = \sigma(l(u), l(v)) \cdot \left(1 + \sum_{p \in \pi(d(u))} \sum_{q \in \pi(d(v))} \mathcal{F}^{\pi}_{\mathrm{Top}}(F_1^p, F_2^q)\right),$$

$$\mathcal{F}^{\pi}_{\mathrm{Top}}(\emptyset, F) = \mathcal{F}^{\pi}_{\mathrm{Top}}(F, \emptyset) = 0,$$

$$\mathcal{F}^{\pi}_{\mathrm{Top}}(T_1 \bullet F_1, T_2 \bullet F_2) = \mathcal{K}^{\pi}_{\mathrm{Top}}(T_1, T_2) \cdot (1 + \mathcal{F}^{\pi}_{\mathrm{Top}}(F_1, F_2))$$
$$+ \mathcal{F}^{\pi}_{\mathrm{Top}}(F_1, T_2 \bullet F_2) + \mathcal{F}^{\pi}_{\mathrm{Top}}(T_1 \bullet F_1, F_2) - \mathcal{F}^{\pi}_{\mathrm{Top}}(F_1, F_2).$$

$$\mathcal{K}^{\pi}_{\mathrm{LcaSg}}(T_1, T_2) = \sum_{u \in T_1} \sum_{v \in T_2} \mathcal{K}^{\pi}_{\mathrm{Top}}(T_1[u], T_2[v]).$$

$$\mathcal{K}^{\pi}_{\mathrm{Lca}}(T_1, T_2) = \sum_{u \in T_1} \sum_{v \in T_2} \mathcal{T}^{\pi}_{\mathrm{Lca}}(T_1[u], T_2[v]),$$

$$\mathcal{T}^{\pi}_{\mathrm{Lca}}(u(F_1), v(F_2)) = \sigma(l(u), l(v)) \cdot \left(1 + \sum_{p \in \pi(d(u))} \sum_{q \in \pi(d(v))} \mathcal{F}^{\pi}_{\mathrm{Lca}}(F_1^p, F_2^q)\right),$$

$$\mathcal{F}^{\pi}_{\mathrm{Lca}}(\emptyset, F) = \mathcal{F}^{\pi}_{\mathrm{Lca}}(F, \emptyset) = 0,$$

$$\mathcal{F}^{\pi}_{\mathrm{Lca}}(T_1 \bullet F_1, T_2 \bullet F_2) = \mathcal{K}^{\pi}_{\mathrm{Lca}}(T_1, T_2) \cdot (1 + \mathcal{F}^{\pi}_{\mathrm{Lca}}(F_1, F_2))$$
$$+ \mathcal{F}^{\pi}_{\mathrm{Lca}}(F_1, T_2 \bullet F_2) + \mathcal{F}^{\pi}_{\mathrm{Lca}}(T_1 \bullet F_1, F_2) - \mathcal{F}^{\pi}_{\mathrm{Lca}}(F_1, F_2).$$

$$\mathcal{K}^{\pi}_{\mathrm{Acc}}(\emptyset, F) = \mathcal{K}^{\pi}_{\mathrm{Acc}}(F, \emptyset) = \mathcal{TF}^{\pi}_{\mathrm{Acc}}(T, \emptyset) = 0,$$

$$\mathcal{K}^{\pi}_{\mathrm{Acc}}(T_1 \bullet F_1, T_2 \bullet F_2) = \mathcal{T}^{\pi}_{\mathrm{Acc}}(T_1, T_2) \cdot (\mathcal{F}^{\pi}_{\mathrm{Acc}}(F_1, F_2) - 1)$$
$$+ \mathcal{TF}^{\pi}_{\mathrm{Acc}}(T_1, T_2 \bullet F_2) - \mathcal{TF}^{\pi}_{\mathrm{Acc}}(T_1, F_2)$$
$$+ \mathcal{TF}^{\pi}_{\mathrm{Acc}}(T_2, T_1 \bullet F_1) - \mathcal{TF}^{\pi}_{\mathrm{Acc}}(T_2, F_1)$$
$$+ \mathcal{K}^{\pi}_{\mathrm{Acc}}(T_1 \bullet F_1, F_2) - \mathcal{K}^{\pi}_{\mathrm{Acc}}(F_1, F_2),$$

$$\mathcal{TF}^{\pi}_{\mathrm{Acc}}(v(F_1), T_2 \bullet F_2) = \mathcal{T}^{\pi}_{\mathrm{Acc}}(v(F_1), T_2) - \mathcal{TF}^{\pi}_{\mathrm{Acc}}(T_2, F_1) + \mathcal{TF}^{\pi}_{\mathrm{Acc}}(v(F_1), F_2)$$
$$- \mathcal{K}^{\pi}_{\mathrm{Acc}}(F_1, F_2) + \mathcal{K}^{\pi}_{\mathrm{Acc}}(F_1, T_1 \bullet F_2),$$

$$\mathcal{T}^{\pi}_{\mathrm{Acc}}(u(F_1), v(F_2)) = \sigma(l(u), l(v)) \cdot \left(1 + \sum_{p \in \pi(d(u))} \sum_{q \in \pi(d(v))} \mathcal{F}^{\pi}_{\mathrm{Acc}}(F_1^p, F_2^q)\right)$$
$$+ \sum_{q \in \pi(d(v))} \mathcal{TF}^{\pi}_{\mathrm{Acc}}(u(F_1), F_2^q) + \sum_{p \in \pi(d(u))} \mathcal{TF}^{\pi}_{\mathrm{Acc}}(v(F_2), F_1^p)$$
$$- \sum_{p \in \pi(d(u))} \sum_{q \in \pi(d(v))} \mathcal{K}^{\pi}_{\mathrm{Acc}}(F_1^p, F_2^q),$$

$$\mathcal{F}^{\pi}_{\mathrm{Acc}}(\emptyset, F) = \mathcal{F}^{\pi}_{\mathrm{Acc}}(F, \emptyset) = 0,$$

$$\mathcal{F}^{\pi}_{\mathrm{Acc}}(T_1 \bullet F_1, T_2 \bullet F_2) = \mathcal{T}^{\pi}_{\mathrm{Acc}}(T_1, T_2) \cdot (1 + \mathcal{F}^{\pi}_{\mathrm{Acc}}(F_1, F_2))$$
$$+ \mathcal{F}^{\pi}_{\mathrm{Acc}}(F_1, T_2 \bullet F_2) + \mathcal{F}^{\pi}_{\mathrm{Acc}}(T_1 \bullet F_1, F_2) - \mathcal{F}^{\pi}_{\mathrm{Acc}}(F_1, F_2).$$

$$\mathcal{K}^{\pi}_{\mathrm{Ilst}}(\emptyset, F) = \mathcal{K}^{\pi}_{\mathrm{Ilst}}(F, \emptyset) = \mathcal{TF}^{o}_{\mathrm{Ilst}}(T, \emptyset) = 0,$$

$$\mathcal{K}^{\pi}_{\mathrm{Ilst}}(T_1 \bullet F_1, T_2 \bullet F_2) = \mathcal{T}^{\pi}_{\mathrm{Ilst}}(T_1, T_2) \cdot (\mathcal{F}^{\pi}_{\mathrm{Ilst}}(F_1, F_2) - 1)$$
$$+ \mathcal{TF}^{\pi}_{\mathrm{Ilst}}(T_1, T_2 \bullet F_2) - \mathcal{TF}^{\pi}_{\mathrm{Ilst}}(T_1, F_2)$$
$$+ \mathcal{TF}^{\pi}_{\mathrm{Ilst}}(T_2, T_1 \bullet F_1) - \mathcal{TF}^{\pi}_{\mathrm{Ilst}}(T_2, F_1)$$
$$+ \mathcal{K}^{\pi}_{\mathrm{Ilst}}(T_1 \bullet F_1, F_2) - \mathcal{K}^{\pi}_{\mathrm{Ilst}}(F_1, F_2),$$

$$\mathcal{TF}^{\pi}_{\mathrm{Ilst}}(v(F_1), T_2 \bullet F_2) = \mathcal{T}^{\pi}_{\mathrm{Ilst}}(v(F_1), T_2) - \mathcal{TF}^{\pi}_{\mathrm{Ilst}}(T_2, F_1) + \mathcal{TF}^{\pi}_{\mathrm{Ilst}}(v(F_1), F_2)$$
$$- \mathcal{K}^{\pi}_{\mathrm{Ilst}}(F_1, F_2) + \mathcal{K}^{\pi}_{\mathrm{Ilst}}(F_1, T_1 \bullet F_2),$$

$$\mathcal{T}^{\pi}_{\mathrm{Ilst}}(u(F_1), v(F_2)) = \sigma(l(u), l(v)) \cdot \left(1 + \sum_{p \in \pi(d(u))} \sum_{q \in \pi(d(v))} \mathcal{K}^{\pi}_{\mathrm{Ilst}}(F_1^p, F_2^q)\right)$$
$$+ \sum_{q \in \pi(d(v))} \mathcal{TF}^{\pi}_{\mathrm{Ilst}}(u(F_1), F_2^q) + \sum_{p \in \pi(d(u))} \mathcal{TF}^{\pi}_{\mathrm{Ilst}}(v(F_2), F_1^p)$$
$$- \sum_{p \in \pi(d(u))} \sum_{q \in \pi(d(v))} \mathcal{K}^{\pi}_{\mathrm{Ilst}}(F_1^p, F_2^q),$$

$$\mathcal{F}^{\pi}_{\mathrm{Ilst}}(\emptyset, F) = \mathcal{F}^{\pi}_{\mathrm{Ilst}}(F, \emptyset) = 0,$$

$$\mathcal{F}^{\pi}_{\mathrm{Ilst}}(T_1 \bullet F_1, T_2 \bullet F_2) = \mathcal{T}^{\pi}_{\mathrm{Ilst}}(T_1, T_2) \cdot (1 + \mathcal{F}^{\pi}_{\mathrm{Ilst}}(F_1, F_2))$$
$$+ \mathcal{F}^{\pi}_{\mathrm{Ilst}}(F_1, T_2 \bullet F_2) + \mathcal{F}^{\pi}_{\mathrm{Ilst}}(T_1 \bullet F_1, F_2) - \mathcal{F}^{\pi}_{\mathrm{Ilst}}(F_1, F_2).$$

Fig. 2. The recurrences of computing $\mathcal{K}^{\pi}_{\mathtt{A}}(T_1, T_2)$ for $\pi \in \{o, b, c, cb\}$ and $\mathtt{A} \in \{\mathrm{Top}, \mathrm{LcaSg}, \mathrm{Lca}, \mathrm{Acc}, \mathrm{Ilst}\}$.

all maximum matchings in $B_{s,t}$. For every $M \in BM(s, t)$, it holds that $M \subset E$ and $|M| = \min\{s, t\}$.

For $u(F_1)$ and $v(F_2)$, let $F_1 = [T_1[u_1], \ldots, T_1[u_s]]$ and $F_2 = [T_2[v_1], \ldots, T_2[v_t]]$, that is, $ch(u) = \{u_1, \ldots, u_s\}$, $ch(v) = \{v_1, \ldots, v_t\}$, $d(u) = s$ and $d(v) = t$. Then, for $M \in BM(s, t)$, we denote the ordered forests $\bullet_{(i,j) \in M} T_1[u_i]$ and $\bullet_{(i,j) \in M} T_2[v_j]$ by F_1^M and F_2^M, where we assume that trees in a forest are ordered along the order of M. Furthermore, for an ordered forest F, let $pm(F)$ be the set of all

$$\mathcal{K}^u_{\mathrm{Top}}(u(F_1), v(F_2)) = \sigma(l(u), l(v)) \cdot \left(1 + \sum_{M \in BM(d(u), d(v))} \mathcal{F}^u_{\mathrm{Top}}(F_1^M, F_2^M)\right),$$
$$\mathcal{F}^u_{\mathrm{Top}}(\emptyset, F) = \mathcal{F}^u_{\mathrm{Top}}(F, \emptyset) = 0,$$
$$\mathcal{F}^u_{\mathrm{Top}}(T_1 \bullet F_1, T_2 \bullet F_2) = \mathcal{K}^u_{\mathrm{Top}}(T_1, T_2) \cdot (1 + \mathcal{F}^u_{\mathrm{Top}}(F_1, F_2))$$
$$+ \mathcal{F}^u_{\mathrm{Top}}(F_1, T_2 \bullet F_2) + \mathcal{F}^u_{\mathrm{Top}}(T_1 \bullet F_1, F_2) - \mathcal{F}^u_{\mathrm{Top}}(F_1, F_2).$$

$$\mathcal{K}^u_{\mathrm{LcaSg}}(T_1, T_2) = \sum_{u \in T_1} \sum_{v \in T_2} \mathcal{K}^u_{\mathrm{Top}}(T_1[u], T_2[v]).$$

$$\mathcal{K}^u_{\mathrm{Lca}}(T_1, T_2) = \sum_{u \in T_1} \sum_{v \in T_2} \mathcal{T}^u_{\mathrm{Lca}}(T_1[u], T_2[v]),$$
$$\mathcal{T}^u_{\mathrm{Lca}}(u(F_1), v(F_2)) = \sigma(l(u), l(v)) \cdot \left(1 + \sum_{M \in BM(d(s), d(t))} \mathcal{F}^u_{\mathrm{Lca}}(F_1^M, F_2^M)\right),$$
$$\mathcal{F}^u_{\mathrm{Lca}}(\emptyset, F) = \mathcal{F}^u_{\mathrm{Lca}}(F, \emptyset) = 0,$$
$$\mathcal{F}^u_{\mathrm{Lca}}(T_1 \bullet F_1, T_2 \bullet F_2) = \mathcal{K}^u_{\mathrm{Lca}}(T_1, T_2) \cdot (1 + \mathcal{F}^u_{\mathrm{Lca}}(F_1, F_2))$$
$$+ \mathcal{F}^u_{\mathrm{Lca}}(F_1, T_2 \bullet F_2) + \mathcal{F}^u_{\mathrm{Lca}}(T_1 \bullet F_1, F_2) - \mathcal{F}^u_{\mathrm{Lca}}(F_1, F_2).$$

$$\mathcal{K}^u_{\mathrm{Acc}}(\emptyset, F) = \mathcal{K}^u_{\mathrm{Acc}}(F, \emptyset) = \mathcal{TF}^u_{\mathrm{Acc}}(T, \emptyset) = 0,$$
$$\mathcal{K}^u_{\mathrm{Acc}}(T_1 \bullet F_1, T_2 \bullet F_2) = \mathcal{T}^u_{\mathrm{Acc}}(T_1, T_2) \cdot (\mathcal{F}^u_{\mathrm{Acc}}(F_1, F_2) - 1)$$
$$+ \mathcal{TF}^u_{\mathrm{Acc}}(T_1, T_2 \bullet F_2) - \mathcal{TF}^u_{\mathrm{Acc}}(T_1, T_2)$$
$$+ \mathcal{TF}^u_{\mathrm{Acc}}(T_2, T_1 \bullet F_1) - \mathcal{TF}^u_{\mathrm{Acc}}(T_2, T_1)$$
$$+ \mathcal{K}^u_{\mathrm{Acc}}(T_1 \bullet F_1, F_2) - \mathcal{K}^u_{\mathrm{Acc}}(F_1, F_2),$$
$$\mathcal{TF}^u_{\mathrm{Acc}}(v(F_1), T_2 \bullet F_2) = \mathcal{T}^u_{\mathrm{Acc}}(v(F_1), T_2) - \mathcal{TF}^u_{\mathrm{Acc}}(T_2, F_1) + \mathcal{TF}^u_{\mathrm{Acc}}(v(F_1), F_2)$$
$$- \mathcal{K}^u_{\mathrm{Acc}}(F_1, F_2) + \mathcal{K}^u_{\mathrm{Acc}}(F_1, T_1 \bullet F_2),$$
$$\mathcal{T}^u_{\mathrm{Acc}}(u(F_1), v(F_2)) = \sigma(l(u), l(v)) \cdot \left(1 + \sum_{M \in BM(d(s), d(t))} \mathcal{F}^u_{\mathrm{Acc}}(F_1', F_2')\right)$$
$$+ \sum_{F_2' \in pm(F_2)} \mathcal{TF}^u_{\mathrm{Acc}}(u(F_1), F_2') + \sum_{F_1' \in pm(F_1)} \mathcal{TF}^u_{\mathrm{Acc}}(v(F_2), F_1')$$
$$- \sum_{M \in BM(d(s), d(t))} \mathcal{K}^u_{\mathrm{Acc}}(F_1^M, F_2^M),$$
$$\mathcal{F}^u_{\mathrm{Acc}}(\emptyset, F) = \mathcal{F}^u_{\mathrm{Acc}}(F, \emptyset) = 0,$$
$$\mathcal{F}^u_{\mathrm{Acc}}(T_1 \bullet F_1, T_2 \bullet F_2) = \mathcal{T}^u_{\mathrm{Acc}}(T_1, T_2) \cdot (1 + \mathcal{F}^u_{\mathrm{Acc}}(F_1, F_2))$$
$$+ \mathcal{F}^u_{\mathrm{Acc}}(F_1, T_2 \bullet F_2) + \mathcal{F}^u_{\mathrm{Acc}}(T_1 \bullet F_1, F_2) - \mathcal{F}^u_{\mathrm{Acc}}(F_1, F_2).$$

$$\mathcal{K}^u_{\mathrm{Ilst}}(\emptyset, F) = \mathcal{K}^u_{\mathrm{Ilst}}(F, \emptyset) = \mathcal{TF}^u_{\mathrm{Ilst}}(T, \emptyset) = 0,$$
$$\mathcal{K}^u_{\mathrm{Ilst}}(T_1 \bullet F_1, T_2 \bullet F_2) = \mathcal{T}^u_{\mathrm{Ilst}}(T_1, T_2) \cdot (\mathcal{F}^u_{\mathrm{Ilst}}(F_1, F_2) - 1)$$
$$+ \mathcal{TF}^u_{\mathrm{Ilst}}(T_1, T_2 \bullet F_2) - \mathcal{TF}^u_{\mathrm{Ilst}}(T_1, T_2)$$
$$+ \mathcal{TF}^u_{\mathrm{Ilst}}(T_2, T_1 \bullet F_1) - \mathcal{TF}^u_{\mathrm{Ilst}}(T_2, T_1)$$
$$+ \mathcal{K}^u_{\mathrm{Ilst}}(T_1 \bullet F_1, F_2) - \mathcal{K}^u_{\mathrm{Ilst}}(F_1, F_2),$$
$$\mathcal{TF}^u_{\mathrm{Ilst}}(v(F_1), T_2 \bullet F_2) = \mathcal{T}^u_{\mathrm{Ilst}}(v(F_1), T_2) - \mathcal{TF}^u_{\mathrm{Ilst}}(T_2, F_1) + \mathcal{TF}^u_{\mathrm{Ilst}}(v(F_1), F_2)$$
$$- \mathcal{K}^u_{\mathrm{Ilst}}(F_1, F_2) + \mathcal{K}^u_{\mathrm{Ilst}}(F_1, T_1 \bullet F_2),$$
$$\mathcal{T}^u_{\mathrm{Ilst}}(u(F_1), v(F_2)) = \sigma(l(u), l(v)) \cdot \left(1 + \sum_{M \in BM(d(s), d(t))} \mathcal{K}^u_{\mathrm{Ilst}}(F_1^M, F_2^M)\right)$$
$$+ \sum_{F_2' \in pm(F_2)} \mathcal{TF}^u_{\mathrm{Ilst}}(u(F_1), F_2') + \sum_{F_1' \in pm(F_1)} \mathcal{TF}^u_{\mathrm{Ilst}}(v(F_2), F_1')$$
$$- \sum_{M \in BM(d(s), d(t))} \mathcal{K}^u_{\mathrm{Ilst}}(F_1^M, F_2^M),$$
$$\mathcal{F}^u_{\mathrm{Ilst}}(\emptyset, F) = \mathcal{F}^u_{\mathrm{Ilst}}(F, \emptyset) = 0,$$
$$\mathcal{F}^u_{\mathrm{Ilst}}(T_1 \bullet F_1, T_2 \bullet F_2) = \mathcal{T}^u_{\mathrm{Ilst}}(T_1, T_2) \cdot (1 + \mathcal{F}^u_{\mathrm{Ilst}}(F_1, F_2))$$
$$+ \mathcal{F}^u_{\mathrm{Ilst}}(F_1, T_2 \bullet F_2) + \mathcal{F}^u_{\mathrm{Ilst}}(T_1 \bullet F_1, F_2) - \mathcal{F}^u_{\mathrm{Ilst}}(F_1, F_2).$$

Fig. 3. The recurrences of computing $\mathcal{K}^u_{\mathtt{A}}(T_1, T_2)$ for $\mathtt{A} \in \{\mathrm{Top}, \mathrm{LcaSg}, \mathrm{Lca}, \mathrm{Acc}, \mathrm{Ilst}\}$.

permuted forests of F. Then, Fig. 3 illustrates the recurrences of computing $\mathcal{K}^u_{\mathtt{A}}(T_1, T_2)$.

Theorem 3. *For* $\mathtt{A} \in \{\mathrm{Top}, \mathrm{LcaSg}, \mathrm{Lca}, \mathrm{Acc}, \mathrm{Ilst}\}$, *the recurrences in Fig. 3 correctly compute* $\mathcal{K}^u_{\mathtt{A}}(T_1, T_2)$ *in* $O(nmD^D)$ *time, where* $n = |T_1|$, $m = |T_2|$ *and* $D = \max\{d(T_1), d(T_2)\}$. *Hence, if the degrees of unordered trees are bounded by some constant, then we can compute* $\mathcal{K}^u_{\mathtt{A}}(T_1, T_2)$ *in* $O(nm)$ *time.*

Proof. Since $|BM(s,t)| = {}_sP_t$ and the number of all permuted forests of F_1 (*resp.*, F_2) is ${}_sP_1$ (*resp.*, ${}_tP_1$), the number of occurrences of the formula $\mathcal{F}_{\mathtt{A}}^u(F_1^M, F_2^M)$ is bounded by D^D and the number of occurrences of the formulas $\mathcal{F}_{\mathtt{A}}^u(u(F_1), F_2')$ and $\mathcal{F}_{\mathtt{A}}^u(F_1', v(F_2))$ for $\mathtt{A} \in \{\text{ACC}, \text{ILST}\}$ is bounded by D^D. In both cases, the number of occurrences of the formulas is $O(D^D)$. Since every pair $(u, v) \in T_1 \times T_2$ is called just once, we can compute $K_{\mathtt{A}}^u(T_1, T_2)$ in $O(nmD^D)$ time by using dynamic programming. Hence, the time complexity in the statement holds. Also we can show the correctness by extending Theorem 1. $\qquad\square$

5   #P-Completeness for Unordered Trees

Since we cannot apply the #P-completeness of [3,6] to the top-down mapping kernel for unordered trees directly, in this section, we show that the problem of counting all the specific top-down mappings (or bottom-up mappings) is #P-complete.

Let M be a mapping between T_1 and T_2. We say that M is *label-preserving* (or an *indel mapping*) if it always holds that $l(u) = l(v)$ for every $(u, v) \in M$. Also we say that M is *leaf-extended* if, for every $(u, v) \in M$, there exists $(u', v') \in M$ such that $u \in anc(u')$, $v \in anc(v')$, $u' \in lv(T_1)$ and $v' \in lv(T_2)$. Then, we deal with a label-preserving leaf-extended top-down mapping M between unordered trees T_1 and T_2, which we denote $M \in \mathcal{M}_{\text{LLTop}}^u(T_1, T_2)$.

Theorem 4 (*cf.*, [3]). *The problem of counting all the mappings in $\mathcal{M}_{\text{LLTop}}^u(T_1, T_2)$ is #P-complete.*

Proof. Valiant [19] has shown that the problem of counting all the matchings in a bipartite graph, which we denote #BIPARTITEMATCHING, is #P-complete. Then, we give two trees such that the number of all the label-preserving leaf-extended top-down mapping between them is equal to the output of #BIPARTITE MATCHING. Here, for a forest F and a node v such that $l(v) = a$, we denote $v(F)$ by $a(F)$.

Let $G = (X \cup Y, E)$ be a bipartite graph. For $v \in X \cup Y$, we denote a neighbor of v by $N(v)$. It is obvious that $N(v) \subseteq Y$ if $v \in X$ and $N(v) \subseteq X$ if $v \in Y$. Then, we construct $T_x = a(\{xy \mid y \in N(x)\})$ for every $x \in X$ and $T_1 = a(\{T_x \mid x \in X\})$. Similarly, we construct $T_y = a(\{xy \mid x \in N(y)\})$ for every $y \in Y$ and $T_2 = a(\{T_y \mid y \in Y\})$. Here, we regard an edge xy in G as the label of a leaf in T_x and T_y. Figure 4 illustrates an example of the above construction of T_1 and T_2 from a bipartite graph G.

For a matching $B \subseteq E$ in G we construct the label-preserving leaf-extended top-down mapping M between T_1 and T_2 such that:

$$M = \begin{cases} \emptyset & \text{if } B = \emptyset, \\ \{(r(T_1), r(T_2))\} \cup \displaystyle\bigcup_{xy \in B} M_{xy} & \text{if } B \neq \emptyset, \end{cases}$$

$$M_{xy} = \left\{ \begin{array}{l} (u_1, v_1), (u_2, v_2) \\ \in V(T_x) \times V(T_y) \end{array} \middle| \begin{array}{l} u_1 = par(u_2), v_1 = par(v_2), \\ u_2 \in lv(T_x), v_2 \in lv(T_y) \\ l(u_1) = l(v_1) = a, l(u_2) = l(v_2) = xy \end{array} \right\}.$$

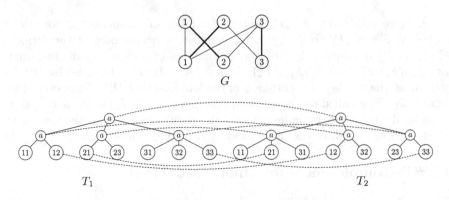

Fig. 4. A bipartite graph G and the trees T_1 and T_2.

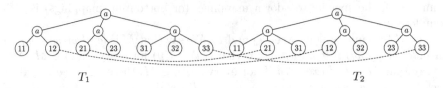

Fig. 5. Trees T_1 and T_2 in Corollary 2.

For example, let B be a matching $\{12, 21, 33\}$ in G illustrated in Fig. 4 as think lines. Then, the label-preserving leaf-extended top-down mapping M between T_1 and T_2 is illustrated by dashed lines.

Note that, by the definition of T_x and T_y, M_{xy} is a label-preserving leaf-extended top-down mapping between T_x and T_y. Also M_{xy} is corresponding to an element xy in a matching of G. Furthermore, no label-preserving leaf-extended top-down mapping M_{xy} between T_1 and T_2 contains more than one path from the root to leaves in T_x or T_y, that is, M_{xy} contains zero or one path in T_x and T_y.

Hence, a matching B in G determines the label-preserving leaf-extended top-down mapping M between T_1 and T_2 uniquely and vice versa. Then, the number of all the matchings in G which is the output of #BIPARTITEMATCHING is equal to the number of all the label-preserving leaf-extended top-down mappings between T_1 and T_2. Hence, the statement holds.   □

Finally, we denote all the *label-preserving bottom-up mappings* between unordered trees T_1 and T_2 by $\mathcal{M}^u_{\text{LBOT}}(T_1, T_2)$. Then, the proofs of [3,6] or the above proof imply the following corollary. Here, it is sufficient to construct a matching B in Fig. 4 to a mapping $\bigcup_{xy \in B} \{(u, v) \in lv(T_x) \times lv(T_y) \mid l(u) = l(v) = xy\}$ as Fig. 5, for example.

Corollary 2 (*cf.*, [3,6]). *The problem of counting all the mappings in* $\mathcal{M}^u_{\text{LBOT}}(T_1, T_2)$ *is #P-complete.*

6   Conclusion

In this paper, for mapping $A \in \{\text{TOP}, \text{LCASG}, \text{LCA}\}$, we have designed the recurrences to compute $\mathcal{K}_A^o(T_1, T_2)$ and $\mathcal{K}_A^b(T_1, T_2)$ in $O(nm)$ time and to compute $\mathcal{K}_A^c(T_1, T_2)$ and $\mathcal{K}_A^{cb}(T_1, T_2)$ in $O(nmdD)$ time. Also, we have designed the recurrences to compute $\mathcal{K}_A^u(T_1, T_2)$ in $O(nmD^D)$ time, which implies that we can compute $\mathcal{K}_A^u(T_1, T_2)$ in $O(nm)$ time if the degrees of T_1 and T_2 are bounded by some constant. On the other hand, we show that the problem of computing $\mathcal{K}_{\text{LLTOP}}^u(T_1, T_2)$ and $\mathcal{K}_{\text{LBOT}}^u(T_1, T_2)$ are #P-complete.

For $\mathcal{M}_{\text{ALN}}$ (alignable mapping [8], less-constrained mapping [11]), from [4,24], we conjecture that we can compute $\mathcal{K}_{\text{ALN}}^b(T_1, T_2)$ in $O(nmD^2)$ time, $\mathcal{K}_{\text{ALN}}^\pi(T_1, T_2)$ in $O(nmD^3)$ time ($\pi \in \{c, cb\}$) and $\mathcal{K}_{\text{ALN}}^u(T_1, T_2)$ in polynomial time if the degrees of T_1 and T_2 are bounded by some constant. Hence, it is a future work to investigate whether or not the above conjecture is correct.

In the proof of Theorem 4 and Corollary 2, the condition of label-preserving and leaf-extended are essential. If these conditions are not met, we must count all the other (standard) top-down or bottom-up mappings that are not label-preserving or leaf-extended. In order to show that the problem of counting all the mappings in $\mathcal{M}_{\text{TOP}}^u(T_1, T_2)$, $\mathcal{M}_{\text{BOT}}^u(T_1, T_2)$ and then $\mathcal{K}_A^u(T_1, T_2)$ for $A \in \{\text{LCASG}, \text{LCA}, \text{ACC}, \text{ILST}, \text{ALN}\}$ are all #P-complete, we must use the Cook-reduction [6,19] from #BIPARTITEMATCHING, which is more complex than the proof of Theorem 4. On the other hand, this paper has shown that we can compute $\mathcal{K}_A^u(T_1, T_2)$ for bounded-degree unordered trees. Hence, it is an important future work to investigate whether or not the problem of computing $\mathcal{K}_A^u(T_1, T_2)$ is #P-complete when degrees are unbounded.

References

1. Chawathe, S.S.: Comparing hierarchical data in external memory. In: Proceedings of the VLDB 1999, pp. 90–101 (1999)
2. Gärtner, T.: Kernels for Structured Data. World Scientific Publishing, Singapore (2008)
3. Hamada, I., Shimada, T., Nakata, D., Hirata, K., Kuboyama, T.: Agreement subtree mapping kernel for phylogenetic trees. In: Nakano, Y., Satoh, K., Bekki, D. (eds.) JSAI-isAI 2013. LNCS, vol. 8417, pp. 321–336. Springer, Heidelberg (2014)
4. Jiang, T., Wang, L., Zhang, K.: Alignment of trees - an alternative to tree edit. Theoret. Comput. Sci. **143**, 137–148 (1995)
5. Kan, T., Higuchi, S., Hirata, K.: Segmental mapping and distance for rooted ordered labeled trees. Fundamenta Informaticae **132**, 1–23 (2014)
6. Kashima, H., Sakamoto, H., Koyanagi, T.: Tree kernels. J. JSAI **21**, 1–9 (2006). (in Japanese)
7. Kimura, D., Kuboyama, T., Shibuya, T., Kashima, H.: A subpath kernel for rooted unordered trees. J. JSAI **26**, 473–482 (2011). (in Japanese)
8. Kuboyama, T.: Matching and learning in trees. Ph.D. thesis, University of Tokyo (2007)

9. Kuboyama, T., Hirata, K., Aoki-Kinoshita, K.F.: An efficient unordered tree kernel and its application to glycan classification. In: Washio, T., Suzuki, E., Ting, K.M., Inokuchi, A. (eds.) PAKDD 2008. LNCS (LNAI), vol. 5012, pp. 184–195. Springer, Heidelberg (2008)

10. Kuboyama, T., Shin, K., Kashima, H.: Flexible tree kernels based on counting the number of tree mappings. In: Proceedings of the MLG 2006, pp. 61–72 (2006)

11. Lu, C.L., Su, Z.-Y., Tang, C.Y.: A new measure of edit distance between labeled trees. In: Wang, J. (ed.) COCOON 2001. LNCS, vol. 2108, pp. 338–348. Springer, Heidelberg (2001)

12. Lu, S.-Y.: A tree-to-tree distance and its application to cluster analysis. IEEE Trans. Pattern Anal. Mach. Intell. 1, 219–224 (1979)

13. Selkow, S.M.: The tree-to-tree editing problem. Inform. Process. Lett. 6, 184–186 (1977)

14. Shawe-Taylor, J., Cristianini, N.: Kernel methods for pattern analysis. Cambridge University Press, Cambridge (2004)

15. Shin, K.: Engineering positive semedefinite kernels for trees - a framework and a survey. J. JSAI 24, 459–468 (2009). (in Japanese)

16. Shin, K., Cuturi, M., Kuboyama, T.: Mapping kernels for trees. In: Proceedings of ICML 2011 (2011)

17. Shin, K., Kuboyama, T.: A generalization of Haussler's convolutioin kernel - Mapping kernel and its application to tree kernels. J. Comput. Sci. Tech. 25, 1040–1054 (2010)

18. Tai, K.-C.: The tree-to-tree correction problem. J. ACM 26, 422–433 (1979)

19. Valiant, L.G.: The complexity of enumeration and reliablity problems. SIAM J. Comput. 8, 410–421 (1979)

20. Valiente, G.: An efficient bottom-up distance between trees. In: Proceedings of SPIRE 2001, pp. 212–219 (2001)

21. Wang, J.T.L., Zhang, K.: Finding similar consensus between trees: an algorithm and a distance hierarchy. Pattern Recogn. 34, 127–137 (2001)

22. Yamamoto, Y., Hirata, K., Kuboyama, T.: Tractable and intractable variations of unordered tree edit distance. Int. J. Found. Comput. Sci. 25, 307–329 (2014)

23. Yoshino, T., Hirata, K.: Hierarchy of segmental and alignable mapping for rooted labeled trees. In: Procedings of DDS 2013, pp. 62–69 (2013)

24. Yoshino, T., Hirata, K.: Alignment of cyclically ordered trees. In: Proceedings of ICPRAM 2015 (2015, to appear)

25. Zhang, K.: Algorithms for the constrained editing distance between ordered labeled trees and related problems. Pattern Recogn. 28, 463–474 (1995)

26. Zhang, K.: A constrained edit distance between unordered labeled trees. Algorithmica 15, 205–222 (1996)

27. Zhang, K., Wang, J., Shasha, D.: On the editing distance between undirected acyclic graphs. Int. J. Found. Comput. Sci. 7, 43–58 (1996)

Finding Ambiguous Patterns on Grammar Compressed String

Koji Maeda[1], Yoshimasa Takabatake[1], Yasuo Tabei[2],
and Hiroshi Sakamoto[1(✉)]

[1] Department of Artificial Intelligence, Kyushu Institute of Technology,
Kawazu 680-4, Iizuka 820, Japan
{k_maeda,takabatake}@donald.ai.kyutech.ac.jp, hiroshi@ai.kyutech.ac.jp
[2] PRESTO, JST, Tokyo, Japan
yasuo.tabei@gmail.com

Abstract. Given a grammar compressed string S, a pattern P, and $d \geq 0$, we consider the problem of finding all occurrences of P' in S such that $d(P, P') \leq d$ with respect to Hamming distance. We propose an algorithm for this problem in $O(\lg \lg n \lg^* N(m + d \, occ_d \lg \frac{m}{d} \lg N))$ time, where $N = |S|$, $m = |P|$, n is the number of variables in the grammar compression, and occ_d is the frequency of an *evidence* of a substring of P. We implement this algorithm and compare with a naive filtering on the grammar compression.

Keywords: Grammar compression · Hamming distance · Pigeonhole principle

1 Introduction

In this paper, we consider an approximate pattern matching on grammar compressed data. A sting S is represented by a CFG G beforehand, where S is derived from G deterministically. Given G, a pattern $P \in \Sigma^*$, and $d \geq 0$, the algorithm is required to output all occurrences of P' in S such that $d(P, P') \leq d$ where $d(P, P')$ means the Hamming distance between patterns P and P'.

In the last decade, many algorithms for grammar compression have been proposed [1,5,8,10,11,13,14], and based on them, grammar based self-index have been also proposed [2,4,7,9,12]. A self-index is a data structure that efficiently supports the following operations without explicitly holding the original text S: counting/locationg of a pattern P, and extractiong a segment $S[i, j]$.

ESP-index [9] is one of such grammar compressed self-indexes using a special parsing technique called *edit-sensitive parsing (ESP)* [3]. When the ESP tree T_S for S is constructed, for any substring P of S, we can obtain the *evidence* as a necessary and sufficient condition for the occurrence of $P = S[i, j]$. An evidence is a sequence of substrees in T_S with length k, say, it is represented as $E = (e_1, \ldots, e_k)$ by the roots e_i of subtrees. Then, it holds that $P = S[i, j]$ iff E is adjacently embedded into $T_S[i, j]$ where $T[i, j]$ denotes the maximal subtree

© Springer-Verlag Berlin Heidelberg 2015
T. Murata et al. (Eds.): JSAI-isAI 2014 Workshops, LNAI 9067, pp. 331–339, 2015.
DOI: 10.1007/978-3-662-48119-6_25

decomposition for the range $[i, j]$ of leaves. Moreover, it also holds that the length of E is significantly smaller than $m = |P|$, i.e., $k = O(\lg^* N \lg m)(N = |S|)$ [9], where $\lg^*$ is the number of iteration of $\lg$ untill being smaller than one[1].

Using this mechanism, we can achieve a faster locating/counting for the compressed S than naive pattern search. The searching algorithm is, however, not directly applicable to the ambiguity matching since the ESP cannot obtain evidences from patterns containing mismatch symbols. Thus, we improve the algorithm that allows the ambiguity search for Hamming distance.

Given $P \in \Sigma^*$, T_S, and d, for a candidate occurrence $S[i, j]$, a naive algorithm will try to embed the evidence E of P into $T_S[i, j]$. In this problem, however, mismatch between P and $S[i, j]$ might occur, and then, the algorithm expands the whole T_S to compare P and $S[i, j]$. To reduce this time consumption, we propose a pruning algorithm to avoid expanding whole subtrees in $T_S[i, j]$. In our method, P is decomposed into $d + 1$ blocks $P_0, \ldots, P_d$, where by the pigeonhole principle, there is at least one P_k with no mismatch with $S[i, j]$. Then, the algorithm finds all occurrences $P_k = S[i, j]$, and then, it tries to search the embedding $P_0, \ldots, P_{k-1}$ to the left and $P_k, \ldots, P_d$ to the right allowing at most distance d. Here, the embedding of the prefix and suffix are executed by expanding the corresponding subtrees where a subtree rooted by e_t can be embedded, it is not expanded since there is no mismatch.

We estimate the time complexity of the proposed algorithm. We implement proposed algorithm using ESP-index and additional hash table for maintaining all position for the decomposed P_k. We compare the improvement with the naive algorithm and empirical results are shown.

2 Preliminary

2.1 Grammar Compression

Σ is a set of alphabet symbols. Let $\mathcal{X}$ be a set of variables with $\Sigma \cap \mathcal{X} = \emptyset$. A sequence of symbols from $\Sigma \cup \mathcal{X}$ is called a string. The set of all possible strings from Σ is denoted by Σ^*. For a string S, the expressions $|S|$, $S[i]$, and $S[i, j]$ denote the length of S, the i-th symbol of S, and the substring of S from $S[i]$ to $S[j]$, respectively. Let $[S]$ be the set of symbols composing S. A string of length two is called a *digram*.

A CFG (context-free grammar) is represented by $G = (\Sigma, V, P, X_s)$, where V is a finite subset of $\mathcal{X}$, P is a finite subset of $V \times (V \cup \mathcal{X})^*$, and $X_s \in V$. A member of P is called a production rule. The set of strings in Σ^* derived from X_s by G is denoted by $L(G)$. A CFG G is called *admissible* if exactly one $X \to \alpha \in P$ exists for any $X \in V$ and $|L(G)| = 1$. An admissible G deriving S is called a *grammar compression* of S for any $X \in V$. The size of G, $|G|$, is the sum of $|\alpha|$ for all $X \to \alpha \in P$.

We consider only the case where $|\alpha| = 2$ for any production rule $X \to \alpha$ because any grammar compression with n variables can be transformed into such

---

[1] In practical sense, $\lg^* N$ is a constant for sufficiently large N.

a restricted CFG with at most $2n$ variables. Moreover, this restriction is useful for practical applications to compression algorithms, e.g., LZ78 [16], REPAIR [8] and LCA [10], and indexes e.g. SLP [2] and ESP [9].

The derivation tree of G is represented by a rooted ordered binary tree such that internal nodes are labeled by variables in V and the *yields*, i.e., the sequence of labels of leaves is equal to S. In this tree, any internal node $Z \in V$ has a left child labeled X and a right child labeled Y, corresponding to the $Z \to XY \in P$.

If a CFG is obtained from any other CFG by a permutation $\pi : \Sigma \cup V \to \Sigma \cup V$, they are identical to each other because the string derived from one is transformed to that from the other by the renaming. For example, $P = \{Z \to XY, Y \to ab, X \to aa\}$ and $P' = \{X \to YZ, Z \to ab, Y \to aa\}$ are identical each other. Thus, we assume the following canonical form of grammar compression.

Definition 1 ([6]). An SLP is a grammar compression over $\Sigma \cup V$ whose production rules are formed by either $X_i \to a$ or $X_k \to X_i X_j$, where $a \in \Sigma$ and $1 \le i, j < k \le |V|$.

2.2 ESP-index

Edit-Sensitive Parsing: We review the method for constructing a derivation tree by Cormode and Muthukrishnan [3], called ESP (Edit-Sensitive Parsing), satisfying the condition: When ESP trees T_S and T_P for strings S and P are given, P occurs in S iff there is a sequence of subtrees in T_P of length $O(\lg^* |P| \lg |P|)$ that are embedded adjacently into T_S. Using this characteristics, we can develop a self-index on grammar compression.

The basic idea is to (i) start from a string $S \in \Sigma^*$, (ii) replace as many as possible of the same digrams in common substrings by the same variables, and (iii) iterate this process until $|S| = 1$.

In each iteration, S is divided into the maximal non-overlapping substrings such that $S = S_1 S_2 \cdots S_\ell$ and each S_i belongs to one of three types: (1) a repetition of a symbol; (2) a substring not including a type1 substring and of length at least $\lg^* |S|$; (3) a substring being neither type1 nor type2 substrings. Substrings of S_i is parsed by $A \to XY$ (2-tree) or $A \to XYZ$ (2-2-tree), where $A \to XYZ$ is further transformed to $A \to XB$ and $B \to YZ$ to obtain a binary tree.

S_i is parsed according to its type. In case S_i is a type1 or type3 substring, it is parsed by the typical left aligned parsing where 2-trees are built from left to right in S_i and a 2-2-tree is built for the last three symbols if $|S_i|$ is odd, as follows:

- If $|S_i|$ is even, ESP builds $A \to S_i[2j-1, 2j]$, $j = 1, ..., |S_i|/2$,
- Otherwise, it builds $A \to S_i[2j-1, 2j]$ for $j = 1, ..., (\lfloor |S_i|/2 \rfloor - 1)$, and builds $A \to B S_i[2j+1]$ and $B \to S_i[2j-1, 2j]$ for $j = \lfloor |S_i|/2 \rfloor$.

In case S_i is a type2 substring, S_i is partitioned into several substrings such that $S_i = s_1 s_2 ... s_\ell$ $(2 \le |s_j| \le 3)$ using *alphabet reduction* [3], which is detailed below. ESP builds $A \to s_j$ if $|s_j| = 2$, or builds $A \to s_j[2,3]$, $B \to s_j[1]A$ otherwise for $j = 1, ..., \ell$. After transforming S_i to S_i', the concatenated string S_i' $(i = 1, ..., \ell)$ is parsed at the next level.

Alphabet Reduction: Given a type2 substring S, consider $S[i]$ and $S[i-1]$ as the binary integers. Let p be the position of the least significant bit in which $S[i]$ differs from $S[i-1]$, and let $bit(p, S[i]) \in \{0, 1\}$ be the value of $S[i]$ at the p-th position, where p starts at 0. Then, $L[i] = 2p + bit(p, S[i])$ is defined for any $i \geq 2$. Since S is type2, so is the resulted string $L = L[2]L[3] \ldots L[|S|]$. We note that if the number of different symbols in S is n which is denoted by $[S] = n$, clearly $[L] \leq 2 \lg n$. Setting $S := L$, the next label string L is iteratively computed until $[L] \leq \lg^* |S|$, where $\lg^* n = \min\{i | \lg^{(i)} n \geq 1\}$. We can consider $\lg^* n$ as constant in practical sense, since $\lg^* n \leq 5$ for $n \leq 2^{65536}$. At the final L^*, $S[i]$ of the original S is called *landmark* if $L^*[i] > \max\{L^*[i-1], L^*[i+1]\}$. After deciding all landmarks, if $S[i]$ is a landmark, it is parsed by a 2-tree or 2-2-tree.

T_P is divided into a sequence of maximal *adjacent* subtrees rooted by nodes $v_1, \ldots, v_k$ such that $yield(v_1 \cdots v_k) = P$, where $yield(v)$ denotes the string represented by the leaves of v and $yield(v_1 \cdots v_k)$ is analogous. If z is the lowest common ancestor of v_1 and v_k, denoted by $z = lca(v_1, v_k)$, the sequence $(v_1, \ldots, v_k)$ is said to be *embedded into* z, denoted by $(v_1, \ldots, v_k) \prec z$. When $yield(v_1 \cdots v_k) = P$, z is called an *occurrence node* of P.

Definition 2 ([9]). An evidence of P is defined as a string $Q \in (\Sigma \cup V)^*$ of length k satisfying the following condition: There is an occurrence node z of P iff there is a sequence $(v_1, \ldots, v_k)$ such that $(v_1, \ldots, v_k) \prec z$, $yield(v_1 \cdots v_k) = P$, and $L(v_1 \cdots v_k) = Q$, where $L(v)$ is the variable of node v and $L(v_1 \cdots v_k)$ is its concatenation.

This is well defined because a trivial Q with $Q = P$ always exists. An evidence Q transforms the problem of finding an occurrence of P into that of embedding a shorter string Q into T_S.

Evidence Extraction: The evidence Q of P is iteratively computed from the parsing of P as follows. Let $P = \alpha\beta$ for a maximal prefix α belonging to type1, 2 or 3. For i-th iteration of ESP, α and β of P are transformed into α' and β', respectively. In case α is not type2, define $Q_i = \alpha$ and update $P := \beta'$. In this case, Q_i is an evidence of α and β' is an evidence of β. In case α is type2, define $Q_i = \alpha[1, j]$ with $j = \min\{p \mid p \geq \lg^* |S|, P[p] \text{ is landmark}\}$ and update $P := x\beta'$ where x is the suffix of α' deriving only $\alpha[j+1, |\alpha|]$. In this case, Q_i is an evidence of $\alpha[1, j]$ and $x\beta'$ is an evidence of $\alpha[j+1, |\alpha|]\beta$. Repeating this process until $|P| = 1$, we obtain the evidence of P as the concatenation of all Q_i.

Counting, Locating, and Extracting: A node z in T_S is an occurrence node of P iff $\exists (v_1, \ldots, v_k) \prec z$ and $L(v_1 \cdots v_k) = Q$. It is sufficient to adjacently embed all subtrees of $v_1, \ldots, v_k$ into T_S. We recall the fact that the subtree of v_1 is left adjacent to that of v_2 iff v_2 is a leftmost descendant of $right\_child(lra(v_1))$ where $lra(v)$ denotes the *lowest right ancestor of* v, i.e., the lowest ancestor of v such that the path from v to it contains at least one left edge. Because $z = lra(v_1)$ is unique and the height of T_S is $O(\lg |S|)$, we can check whether $(v_1, v_2) \prec z$

in $O(\lg |S|)$ time. Moreover, $(v_1, v_2, v_3) \prec z'$ iff $(z, v_3) \prec z'$ (possibly $z = z'$). Therefore, when $|Q_i| = 1$ for each i, we can execute the embedding of whole Q in $t = O(\lg |P| \lg |S| \lg^* |S|)$ time. For general case of $Q_i \in q_i^+$, the same time complexity t is obtained.

Theorem 1 ([9,15]). *Let $|S| = N$, $|P| = m$, and $|G| = n$. Counting time of ESP-index is $O((m + occ \lg m \lg N) \lg \lg n \lg^* N)$ with $2n + n \lg n + o(n \lg n)$ bits of space. With auxiliary $n \lg N + o(n)$ bits of space, ESP-index supports locating in the same time complexity and also supports extracting in $O((m + \lg N) \lg \lg n)$ time, where occ is the frequency of the largest embedded subtree for P.*

3   Algorithm

We propose the algorithm for the compressed Hamming distance problem: given a grammar compressed string S, a pattern P, and $d \geq$, enumerating all similar substrings $P' = S[i, j]$ such that $d(P', P) \leq d$. The proposed algorithm is an improvement of the baseline method (Fig. 1).

3.1   Baseline Algorithm

The string S is preprocessed beforehand and represented by the ESP-index. Without loss of generality, we can assume $d < m$. For a pattern P and its evidence Q, if $P = S[i, j]$, we can embed Q into T_S. We express this embedding by $Q \prec T_S[i, j]$.

Algorithm 1. The baseline algorithm with ESP tree T_S for S, pattern P, and $d < m = |P|$.

Decompose P into $d + 1$ blocks $P_0, \ldots, P_d$ with $|P_k| = m/d$.
for $k := 0, \ldots, d$ **do**
Extract the evidence Q_k of P_k.
if $Q_k \prec T_S[i, j]$ and Q_ℓ $(\ell < k)$ $\not\prec T_S[i - \frac{m}{d}(k - \ell), j - \frac{m}{d}(k - \ell)]$ **then**
$(X, Y) \leftarrow (P_0 \cdots P_{k-1}, P_{k+1} \cdots P_d)$, $(S_X, S_Y) \leftarrow (S[i - |X|, i - 1], S[j + 1, j + |Y|])$.
if $d(X, S_X) + d(Y, S_Y) \leq d$ **then** return the matched position $i - |X|$ in S.

Theorem 2. *The time of the Algorithm 1 is $O(occ_d \lg \lg n \lg N (d \lg \frac{m}{d} \lg^* N + m))$ where occ_d is the frequency of the largest embedded subtree for a P_k $(0 \leq k \leq d + 1)$.*

Proof. Since $d < m = |P|$, if $d(S[i, j], P) \leq d$, then there is at least one P_k such that P_k exactly matches with the kth block in $S[i, j]$. By Theorem 1, after $t_1 = O(\frac{m}{d} \lg \lg n \lg^* N)$ preprocessing time for P_k, occurrences of P_k in S can be found in $t_2 = O(occ_d \lg \frac{m}{d} \lg N \lg \lg n \lg^* N)$ search time. For each candidate of $S[i, j] = P$ with $d(S[i, j], P) \leq d$, the time to check whether $P_0 \cdots P_{k-1}$ and $P_{k+1} \cdots P_d$ can be embedded together is $t_3 = O(m \cdot occ_d \lg N \lg \lg n)$. Since, the number of blocks is $d + 1$, the total time of preprocessing and searching for P is $O(d(t_1 + t_2) + t_3) = O(occ_d \lg \lg n \lg N (d \lg \frac{m}{d} \lg^* N + m))$. $\qquad \square$

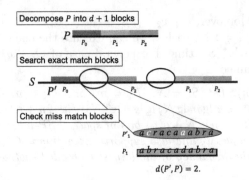

Fig. 1. An example for 3 blocks of $P = P_0P_1P_2$.

3.2 The Improvement

We show the computation of $d(X, S_X) + d(Y, S_Y) \leq d$ in Algorithm 1 can be reduced to the embedding of evidences of X as follows.

Lemma 1. *Given T_P and T_S and a substring P' of S with $m = |P| = |P'|$. The time for checking $d(P, P') \leq d$ is $O(d \lg \frac{m}{d} \lg \lg n \lg N \lg^* N)$.*

Proof. We assume $P' = e_1 x_1 e_2 x_2 \cdots e_d x_d e_{d+1}$ for the mismatch symbol x_i. Since each e_i appears in S in this order, there is an evidence of length $O(\lg \frac{m}{d} \lg^* N)$. Thus, the time of checking $d(P, P') \leq d$ is $t_4 = O(d \lg \frac{m}{d} \lg \lg n \lg N \lg^* N)$.   □

Theorem 3. *The time complexity of Algorithm 1 is $O(\lg \lg n \lg^* N(m + d \, occ_d \lg \frac{m}{d} \lg N))$.*

Proof. Lemma 1 shows the time for checking $d(X, S_X) \leq d$ for a single block $X = P_\ell$ ($0 \leq \ell \leq k - 1$) in Algorithm 1. Thus, taking account into the time for the preprocessing of the input pattern P and the frequency occ_d, the time bound is $O(d(t_1 + t_2) + occ_d \, t_4) = O(\lg \lg n \lg^* N(m + d \, occ_d \lg \frac{m}{d} \lg N))$.   □

4   Experiments

We implement the proposed algorithms: one is the baseline algorithm and the other is the improved algorithm presented in the previous section. We also compared proposed algorithms to FM-index (https://code.google.com/p/fmindex-lus-plus/) and LZ-index. We prepare LZ-index (http://pizzachili.dcc.uchile.cl/indexes/LZ-index/LZ-index-1.tar.gz) which is applied pigeonhole principle. Locating time of LZ-index is not obtained because LZ-index cannot return exact location.

We evaluate their performance on the following environment: one core of an eight-core Intel Xeon CPU E7-8837 (2.67GHz) machine with 1024 GB memory. The dataset of DNA sequence (E-Coli, 107MB) is obtained from the text collection: Pizza&Chili (http://pizzachili.dcc.uchile.cl/). The pattern length is set from 100 to 1000, and for each length, 1000 substrings are randomly extracted from the sequence as the test patterns.

Table 1 indecates index size and construction time. Memory consumption depends on index size. We show memory consumption of these algorithms (Fig. 2). Memory consumption of baseline altorithm is the same as improved algorithm. These indicate that our algorithm can operate on small memory. We show the number of extracted substrings for Hamming distance $2 \leq d \leq 4$ (Fig. 3). Next, we measured the locating time of all occurrences of patterns within Hamming distance $d \leq 4$ for a given pattern. The results are all average time for 1000 trials. Besides, we also show the computation time of checking mismatch blocks in comparison to the two algorithms. The locating time is not much faster than the previous algorithm (Figs. 4, 6 and 8). However, the time of checking mismatch blocks is up to 2.3 times faster than the naive algorithm (Figs. 5, 7 and 9). Especially, our algorithm is efficient for long patterns and the small Hamming distance because our algorithm can reduce searching nodes for such that settings.

Table 1. About Index

Algorithm	Index size[MB]	Construction time[sec]
ESP-index	27.1	16.9
FM-index	210.5	33.5
LZ-index	119.9	26.7

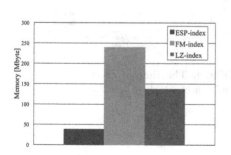

Fig. 2. Memory consumption

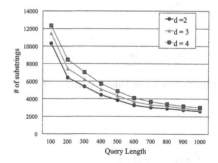

Fig. 3. The number of substrings

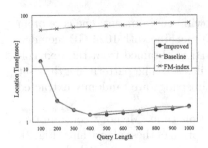

Fig. 4. Locating time when $d = 2$.

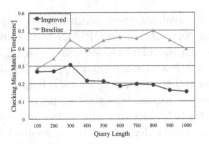

Fig. 5. The time of checking mismatch blocks when $d = 2$.

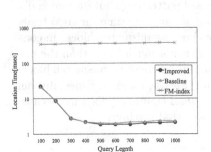

Fig. 6. Locating time when $d = 3$.

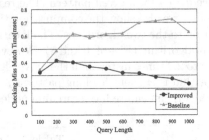

Fig. 7. The time of checking mismatch blocks when $d = 3$.

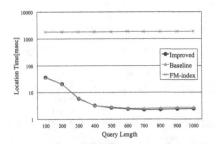

Fig. 8. Locating time when $d = 4$.

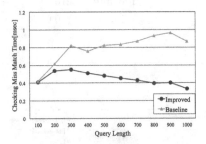

Fig. 9. The time of checking mismatch blocks when $d = 4$.

5   Conclusion

We proposed a framework of Hamming distance computation in grammar compressed strings. Our algorithm can locate a pattern faster than a naive extracting algorithm. By the experiments, it is expected that pattern matching for a long pattern with small Hamming distance is much faster. However, total locating time is not significantly faster than the naive algorithm because the locating time is mainly dominated by locating blocks not by embedding of trees. As an important future works, we would improve the bottleneck developing another data structure. On the other hand, computing harder edit distance problem on grammar compression is an interesting challenge.

References

1. Charikar, M., Lehman, E., Liu, D., Panigrahy, R., Prabhakaran, M., Sahai, A., Shelat, A.: The smallest grammar problem. IEEE Trans. Inf. Theory **51**(7), 2554–2576 (2005)
2. Claude, F., Navarro, G.: Self-indexed grammar-based compression. Fundam. Inf. **111**(3), 313–337 (2011)
3. Cormode, G., Muthukrishnan, S.: The string edit distance matching problem with moves. ACM Trans. Algor. 3(1): Article 2 (2007). doi:10.1145/1186810.1186812
4. Gagie, T., Gawrychowski, P., Kärkkäinen, J., Nekrich, Y., Puglisi, S.J.: A faster grammar-based self-index. In: Dediu, A.-H., Martín-Vide, C. (eds.) LATA 2012. LNCS, vol. 7183, pp. 240–251. Springer, Heidelberg (2012)
5. Jeż, A.: Approximation of grammar-based compression via recompression. In: Fischer, J., Sanders, P. (eds.) CPM 2013. LNCS, vol. 7922, pp. 165–176. Springer, Heidelberg (2013)
6. Karpinski, M., Rytter, W., Shinohara, A.: An efficient pattern-matching algorithm for strings with short descriptions. Nordic J. Comput. **4**(2), 172–186 (1997)
7. Kreft, S., Navarro, G.: On compressing and indexing repetitive sequences. Theor. Comput. Sci. **483**, 115–133 (2013)
8. Larsson, N.J., Moffat, A.: Offline dictionary-based compression. Proc. IEEE **88**(11), 1722–1732 (2000)
9. Maruyama, S., Nakahara, M., Kishiue, N., Sakamoto, H.: ESP-Index: a compressed index based on edit-sensitive parsing. J. Discrete Algorithms **18**, 100–112 (2013)
10. Maruyama, S., Sakamoto, H., Takeda, M.: An online algorithm for lightweight grammar-based compression. Algorithms **5**(2), 213–235 (2012)
11. Maruyama, S., Tabei, Y., Sakamoto, H., Sadakane, K.: Fully-online grammar compression. In: Kurland, O., Lewenstein, M., Porat, E. (eds.) SPIRE 2013. LNCS, vol. 8214, pp. 218–229. Springer, Heidelberg (2013)
12. Navarro, G.: Implementing the LZ-index: theory versus practice. ACM J. Exp. Algorithmics **13**, 25–65 (2008)
13. Rytter, W.: Application of Lempel-Ziv factorization to the approximation of grammar-based compression. Theor. Comput. Sci. **302**(1–3), 211–222 (2003)
14. Sakamoto, H.: A fully linear-time approximation algorithm for grammar-based compression. J. Discrete Algorithms **3**(2–4), 416–430 (2005)
15. Takabatake, Y., Tabei, Y., Sakamoto, H.: Improved ESP-index: a practical self-index for highly repetitive texts. In: Gudmundsson, J., Katajainen, J. (eds.) SEA 2014. LNCS, vol. 8504, pp. 338–350. Springer, Heidelberg (2014)
16. Ziv, J., Lempel, A.: Compression of individual sequences via variable-rate coding. IEEE Trans. Inf. Theory **24**(5), 530–536 (1978)

Detecting Anomalous Subgraphs on Attributed Graphs via Parametric Flow

Mahito Sugiyama[1,2]([✉]) and Keisuke Otaki[3]

[1] ISIR, Osaka University, 8-1, Mihogaoka, Ibaraki-shi, Osaka 567-0047, Japan
mahito@ar.sanken.osaka-u.ac.jp
[2] JST, PRESTO, Chiyoda-ku, Japan
[3] Graduate School of Informatics, Kyoto University, Yoshida-Honmachi, Sakyo-ku, Kyoto 606-8501, Japan

Abstract. Detecting anomalies from structured *graph* data is becoming a critical task for many applications such as an analysis of disease infection in communities. To date, however, there exists no efficient method that works on massive *attributed graphs* with millions of vertices for detecting anomalous subgraphs with an abnormal distribution of vertex attributes. Here we report that this task is efficiently solved using the recent graph cut-based formulation. In particular, the full hierarchy of anomalous subgraphs can be simultaneously obtained via the *parametric flow algorithm*, which allows us to introduce the size constraint on anomalous subgraphs. We thoroughly examine the method using various sizes of synthetic and real-world datasets and show that our method is more than five orders of magnitude faster than the state-of-the-art method and is more effective in detection of anomalous subgraphs.

1 Introduction

Anomaly detection is one of crucial tasks in data mining as anomalous objects (*outliers*) causes serious problems across applications [1]. Despite recent development of anomaly detection methods on multivariate datasets [5,6,25,27] that efficiently find sparsely populated objects, anomaly detection on structured data, in particular on *graphs*, is still developing. The main difficulty is accounting for *inter-dependencies* between objects to find anomalous regions in which objects are connected to each other, which makes the task of anomaly detection on graphs largely different from that on multi-dimensional feature vectors.

Rapid technological advances produce massive amount of *attributed graphs*, where each vertex is associated with a label/attribute, and an anomalous subgraph corresponds to a densely connected region in which a distribution of attributes is significantly different from the rest of the region. Moreover, in many cases, such vertex attribute directly shows whether or not the vertex is anomalous, that is, they can be used as partially supervised information of anomalous subgraphs. Thus, in this situation, our task is to *detect potentially (or hidden) anomalous regions* on the given graph structure using the attribute information (see Fig. 1). This task therefore corresponds to *transductive learning* [8] in the

© Springer-Verlag Berlin Heidelberg 2015
T. Murata et al. (Eds.): JSAI-isAI 2014 Workshops, LNAI 9067, pp. 340–355, 2015.
DOI: 10.1007/978-3-662-48119-6_26

field of machine learning, where we aim at predicting labels of unlabeled data in a given dataset. For example in an analysis of disease infection on a social network, people are annotated as whether or not they are already infected and, to understand the cause of disease infection and to find potentially infected people, the goal is to detect an anomalous local community with the high rate of infected people.

To date, several methods including the current state-of-the-art gAnomaly [18] have been proposed that try to solve the above task of detecting anomalous subgraphs on attributed graphs. However, the following two problems remain unsolved: (1) *scalability*; massive graphs with millions of vertices cannot be treated in a reasonable time; (2) *cardinality constraint*; the size of anomalous subgraphs cannot be specified by the user, which is an important requirement in real-world applications. In gAnomaly, we have to rerun it many times by changing its parameter in small steps until reaching at a subgraph with the desirable size.

Our goal in this paper is to overcome the above two issues. The key technique of our proposal is to use the recently proposed graph cut-based formulation [4], where the method, called SConES, has been proposed and used for feature selection on networks (weighted graphs). SConES uses the fact that the optimal features correspond to the minimum cut on a given graph, and hence it can be solved by a *maximum flow algorithm* in an efficient and exact manner. Although the original SConES cannot directly handle the size constraint on the resulting subgraph, we solve the problem by applying the *parametric flow algorithm* proposed by Gallo et al. [11], which gives the entire *regularization path* along with changes in a regularization parameter. Since the size of subgraphs depends on the regularization parameter, we can pick up the best solution that fulfills the size constraint from the set of possible solutions obtained by the parametric flow algorithm.

This paper is organized as follows: Sect. 2 describes our method; we first formulate our problem in Sect. 2.1 and introduce the cardinality constraint in Sect. 2.2, followed by achieving ranking and visualization of anomalous subgraphs in Sect. 2.3. Related work is discussed in Sect. 3, and our proposal is evaluated by experiments in Sect. 4. We conclude the paper with summarizing our contribution in Sect. 5.

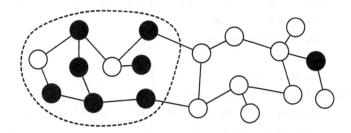

Fig. 1. Our problem setting. Open and filled circles denote normal and anomalous vertices, respectively, given by attributes. Our goal is to find an anomalous subgraph, denoted by a dotted line, where two normal vertices in the subgraph are potential anomalies.

2   Anomalous Subgraph Detection

Given an weighted graph $G = (V, E)$, where V is a set of vertices and E is a set of edges, and a weight $w(e)$ is assigned to each edge $e \in E$. We consider the situation in which the *degree of anomalousness* of each vertex is given through an *attribute function* A from V, where its range, denoted by range(A), can be either binary (range$(A) = \{0, 1\}$) or real-valued (range$(A) = \mathbb{R}$). In the binary case, a vertex v is *anomalous* if $A(v) = 1$ and is normal if $A(v) = 0$, while v is more and more anomalous if $A(v)$ gets a larger and larger value in the real-valued case. In the following, we treat the graph $G = (V, E)$ and an attribute function A as a triplet $G = (V, E, A)$ and call G an *attributed graph*. In this setting, the function A can be viewed as partial information of anomalies, and our objective is to recover potentially anomalous regions from together with A and the given graph structure G. Notations used in the paper is summarized in Table 1.

One of the most direct mathematical formulation of this problem is as follows: Find a subset $S \subset V$ which maximizes the sum of values $\sum_{v \in S} A(v)$ under two constraints that the vertices in S of G are *connected to each other* by edges and the cardinality $|S| = k$, which is specified by the user. Unfortunately, this problem is infeasible to solve in practice because the maximum-weight connected

Table 1. Notation

$G = (V, E, A)$	Attributed graph
V	Set of vertices of G
E	Set of edges of G
A	Attribute function from V to $\{0, 1\}$ or $\mathbb{R}$
S	Subset of V
$G[S]$	Subgraph of G induced by S
v, u	Vertex; $v, u \in V$
e	Edge; $e \in E$
$w(e)$	Weight of edge e
$C(S)$	Cut set of S, i.e., $C(S) = \{\{v, u\} \in E \mid v \in V \setminus S, u \in S\}$
λ	Parameter for connectivity
η	Parameter for sparsity
n	Number of vertices
m	Number of edges
$G' = (V', E')$	$s\text{-}t$ graph constructed from G
k	Size constraint (upper bound of the number of vertices of subgraph)
γ	Parameter of parametric network
$S_1, S_2, \ldots, S_l$	Optimal solutions obtained by the parametric flow algorithm
q	Scoring function from V to $[0, 1]$
$i(v)$	Natural number such that $v \notin S_{i(v)-1}$ and $v \in S_{i(v)}$ for all $v \in V$

graph (MCG) problem:

$$\max_{S \subset V} \sum_{v \in S} A(v) \text{ such that } G[S] \text{ is connected and } |S| = k$$

is known to be strongly NP-complete [17] ($G[S]$ denotes the subgraph of G induced by $S \subset V$). Thus, instead of tackling this hopeless problem, we focus on *local connectivity* rather than conducting an exhaustive search over all connected subgraphs. Our formulation, which is introduced in the next subsection, allows the user to pick up more than one subgraph at the same time.

2.1 Formulation

To achieve our objective and find anomalous subgraphs, here we define the following problem based on the SConES formulation [4]: Given an attributed graph $G = (V, E, A)$, the *anomalous subgraph finding problem* is to find the optimal subgraph $G[S]$ induced by a subset $S \subset V$, which is the solution of

$$\max_{S \subset V} \sum_{v \in S} A(v) - \lambda \sum_{e \in C(S)} w(e) - \eta |S|, \tag{1}$$

where $w(e)$ is the weight of the edge e,

$$C(S) = \{\{v, u\} \in E \mid v \in V \setminus S, u \in S\}$$

is the set of edges that have one of two endpoints in S (i.e., the *cut set* of S), and λ and η are two real-valued regularization parameters. The first term is for quantifying the anomalousness of a subgraph $G[S]$, which coincides with the cardinality of the set $\{v \in S \mid A(S) = 1\}$ in the binary case. The second and third terms are penalties, where the second is to enforce the *connectivity* of S as it penalizes selecting a vertex without selecting all of its neighbors, and the third is to enforce the *sparsity* of the subgraph. Note that SConES has been originally proposed for supervised feature selection on graphs, and we transfer it into the problem of anomalous subgraph detection, where anomalous subgraphs correspond to selected features.

The notable advantage of this formulation is that this is exactly and efficiently solved by the *maximum flow algorithm*. For a graph $G = (V, E, A)$, we construct an s-t graph $G' = (V', E')$ as follows: $V' = V \cup \{s, t\}$ by adding a source node s and a sink node t, $E' = E \cup \{\{u, v\} \mid u \in \{s, t\}, v \in V\}$, and the capacity $c : E' \to \mathbb{R}$ is given as

$$c(\{u, v\}) = \begin{cases} A(v) - \eta & \text{if } u = s \text{ and } v \in V, \\ \eta - A(v) & \text{if } u = t \text{ and } v \in V, \\ \lambda w(\{v, u\}) & \text{if } u, v \in V. \end{cases}$$

An example of transformation into an s-t graph is shown in Fig. 2. For mathematical convenience, a capacity of an edge can be negative in the above definition.

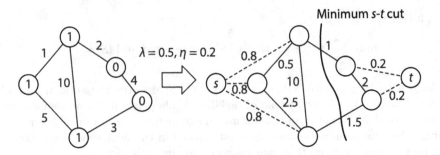

Fig. 2. Example of a graph (left) and its corresponding s-t graph (right) for the maximum flow problem. Numbers in circles denote attribute values $A(v)$ and those on edges denote weights (left) and capacities (right). In this example, $\lambda = 0.5$ and $\eta = 0.2$.

Such edges with negative capacities are ignored in the maximum flow algorithm. We have the following powerful property for this s-t graph.

Theorem 1 ([4]). *Given an attributed graph G. Let $(S \cup \{s\}, V \setminus S \cup \{t\})$ be the minimum s-t cut of the s-t graph G'. Then the set S coincides with the solution of Problem (1) on G.*

Since the s-t minimum cut problem is solved as a *maximum flow problem*, thanks to the famous max-flow min-cut theorem [23, Chap. 6.1], the optimal subgraph $G[S]$ is exactly obtained by simply applying a maximum flow algorithm to the transformed s-t graph G', whose time complexity is $O(nm \log(n^2/m))$ [13], where $n = |V'|$ and $m = |E'|$.

2.2 Parametric Flow on Anomalous Subgraphs

In Formulation (1), which is fundamentally the same as SConES, it is not intuitive how to set two parameters, in particular η that controls the size of S. However in anomalous subgraph detection, it is desirable to allow the user to input the constraint on the size of subgraphs. Here we achieve this requirement by solving the following modified problem:

$$\max_{S \subset V,\ \eta \in \mathbb{R}} \sum_{v \in S} A(v) - \lambda \sum_{e \in C(S)} w(e) - \eta |S|, \tag{2}$$

$$\text{subject to: } |S| \leq k,$$

where k is a natural number specified by the user.

Interestingly, we can obtain all possible minimum cuts simultaneously along with changes of the parameter η without increasing the time complexity of the maximum flow algorithm $O(nm \log(n^2/m))$. This is achieved by applying the

---

Algorithm 1. paraAnomaly

Input: Attributed graph $G = (V, E, A)$, size constraint k, connectivity parameter λ
Output: Anomalous subgraph of G
 1: Construct the s-t graph G' from G
 2: Apply the parametric flow algorithm to G' and obtain the set of optimal solutions $S_1, S_2, \ldots, S_l$ along with changes of the sparsity parameter η
 3: Output $G[S_i]$, where $|S_i| \leq k$ and $|S_{i+1}| > k$

---

parametric flow algorithm presented by Gallo et al. [11][1] since the s-t graph G' always becomes a parametric network.

A *parametric network* is a specific type of networks equipped with a real-valued parameter γ satisfying the following three conditions:

1. The capacity $c(\{s, v\})$ is a non-decreasing function of γ for all $v \in V \setminus \{t\}$.
2. The capacity $c(\{t, v\})$ is a non-increasing function of γ for all $v \in V \setminus \{s\}$.
3. The capacity $c(\{u, v\})$ is constant for all $u, v \in V$ with $u \neq s$ and $v \neq t$.

From the definition of the s-t graph G', we can easily confirm the following fact by letting $\gamma = -\eta$.

Lemma 1. *The s-t graph G' is a parametric network with respect to* $(-\eta)$.

Notice that the weight $w(e)$ of edges $e \in E$ is used to construct the s-t graph G', while it is treated as a constant in the parametric network because it is independent from η.

For a parametric network, it is known that the maximum flow value takes a continuous piecewise linear function of $\gamma = -\eta$. Then there must be a finite number of breakpoints $\gamma_1 < \gamma_2 < \cdots < \gamma_{l-1}$, and for each interval $[\gamma_{i-1}, \gamma_i)$, the optimal solution S_i does not change for any $\gamma \in [\gamma_{i-1}, \gamma_i)$. Hence a finite sequence of optimal solutions (subsets of vertices) $S_1, S_2, \ldots, S_l$ is produced by the parametric flow algorithm, where $l - 1$ is the number of breakpoints uniquely determined from the property of a given graph.

Here an important property of the sequence of solutions is that they always have the *nesting property*:

$$\emptyset \subset S_1 \subset S_2 \subset \cdots \subset S_{l-1} \subset S_l \subset V$$

with increasing the corresponding parameter values $\gamma_1, \gamma_2, \ldots, \gamma_l$ (i.e., decreasing the parameter values $\eta_1, \eta_2, \ldots, \eta_l$ such that $\gamma_i = -\eta_i$). The optimal solution of Problem (2) is therefore computed by simply choosing S_i such that $|S_i| \leq k$ and $|S_{i+1}| > k$. The entire process, which we call *paraAnomaly*, is summarized in Algorithm 1.

---

[1] This fact is pointed out in [26] but has not been used in any applications. A related result is theoretically analyzed in [16].

2.3   Parametric Flow to Rank

The proposed method paraAnomaly can go one step further: It achieves not only binary discrimination of anomalous subgraphs from the entire graph but also *ranking of anomalous subgraphs*, which is often desirable in anomaly detection. This is directly achieved from the hierarchical structure of the optimal solutions $S_1 \subset S_2 \subset \cdots \subset S_l$, that is, a smaller subgraph with a larger regularization parameter η is more anomalous than a larger subgraph.

Moreover, this ranking can be visualized by designing a scoring function for vertices by focusing on the difference between consecutive subgraphs. Let us denote by $i(v)$ a natural number such that $v \notin S_{i(v)-1}$ and $v \in S_{i(v)}$ for any vertices $v \in V$. Define a scoring function $q : V \rightarrow [0,1]$ as

$$q(v) := \frac{l - i(v) + 1}{l},$$

where the numerator $l - i(v) + 1$ is the number of solutions containing v and the denominator l is the normalizer so that the resulting value $q(v) \in [0,1]$. Then vertices in a highly anomalous subgraph receives a higher score than those in a low anomalous subgraph. This visualization reveals the hierarchical structure of anomalous regions. We will show an example of visualization of a real-world dataset in Sect. 4.

3   Related Work

Anomaly detection on graphs is roughly divided into two settings: on plain (unlabeled) graphs and attributed (labeled) graphs. In this section we briefly discuss related work about anomaly detection on graphs, mainly focusing on anomaly detection on attributed graphs, and point out the difference between our method and the existing methods. A comprehensive survey is given by [3].

On plain graphs, the objective is to detect regions that have rare structural patterns. Various approaches have been proposed [2,7,14,15,19], for example, Akoglu et al. [2] introduced the concept of an *egonet*, which is a subgraph with its neighbors, and measured the abnormality of vertices by checking whether their egonets obey some power-low extracted from real-world graph data.

Followed by studies on plain graphs, anomaly detection on attributed graphs have been also heavily studied. Noble and Cook [22] were the first to investigate anomaly detection on attributed graphs, where the Minimum Description Length (MDL) principle was used to define abnormal substructures through measuring the compression quality of frequent subgraphs. Eberle and Holder [10] tried to define the degree of anomalousness based on the structure of subgraphs with their attributes. Gao et al. [12] introduced *community outliers*, which significantly deviate from the rest of the local community members, and proposed an algorithm CODA to find them, but the algorithm strongly depends on the initialization step and the convergence is not guaranteed [3]. A node outlier ranking technique GOutRank was proposed by Müller et al. [20], although it does not aim at finding densely connected subgraphs. The concept of *focused* clustering and outlier detection was introduced by Perozzi et al. [24], where only clusters and outliers focused by the user through their exemplars are detected.

Despite the detailed studies of anomaly detection on attributed graphs, none of the above methods aggressively treat attributes as supervision of anomalousness. This means that their setting is basically *unsupervised*, and hence attributes are not directly associated with the degree of anomalousness. In contrast, recently, Li et al. [18] have considered the transductive setting and tried to recover anomalous subgraphs from partially labeled vertices by estimating probability distributions of anomalous attributes by the EM algorithm. This is the problem setting that we are considering in this paper, and their method, called gAnomaly, is compared to our proposed method paraAnomaly in the next section as it is the current state-of-the-art.

Clustering techniques on attributed graphs, which do not focus on detecting anomalies, can be used for anomaly detection since the task of anomaly detection can be achieved by dividing the whole vertices into two clusters of normal and abnormal vertices. A representative method GBAGC [29] is also compared to our method in our experiments.

4 Experiments

In this section, we examine our method paraAnomaly on synthetic and real-world graph datasets. First we describe our experimental setting, followed by discussing the results.

4.1 Experimental Methods

Environment. We used Mac OS X version 10.9.4 with a 3.5 GHz Intel Core i7 CPU and 32 GB of memory. Our method paraAnomaly is implemented in R, version 3.1.1, which calls the parametric flow algorithm[2] written in C++ and compiled by gcc version 4.9.0.

Comparison Partners. Our main comparison partner is the state-of-the-art method gAnomaly [18]. We re-implemented gAnomaly in R since the official code is not available. Note that the most expensive optimization part of gAnomaly is done by an R function `optim`, in which the core part is implemented in C. Thus comparison of running time between paraAnomaly and gAnomaly is fair. The function $R_N^{(2)}$ (see [18, Eq. (3.7)] for its definition) was used as a network regularizer because the author claims that it is more robust to the parameter setting than the other regularize $R_N^{(1)}$.

---

[2] Source code is available at http://research.microsoft.com/en-us/downloads/d3adb 5f7-49ea-4170-abde-ea0206b25de2/. Since the code can handle only integers for parameters, we first transform every parameter to an integer by multiplying some constant value.

In addition, a clustering method GBAGC is also included as a comparison partner because it is used as the solo comparison partner of gAnomaly in [18]. The official implementation[3] was used.

Datasets. We generated various sizes of attributed graph datasets in the following manner: First, we generated graphs according to the Watts-Strogatz network model [28] using the R `igraph` package. Second, we took the largest dense subgraph using a method proposed by Clauset et al. [9] and assumed that this community is an anomalous subgraph, that is, a vertex is labeled as 1 if it is in the subgraph and 0 otherwise. Although our method can handle real-valued attributes, we systematically examine only the binary case as gAnomaly cannot treat real-valued attributes.

We used three real-world datasets: CORA[4], DBLP, and Amazon. DBLP and Amazon were obtained from SNAP[5]. In CORA, we used the largest cluster "Neural Networks" as an anomalous subgraph, which is the same protocol as in [18]. In DBLP and Amazon, we chose the largest community given by [30] and assigned it as anomalous. Moreover, we used a small subset of DBLP (denoted as DBLP(s)) by taking the four largest communities. Statistics of datasets are summarized in Table 2.

Evaluation. To investigate the performance of detection methods, in each synthetic and real-world graph dataset, we randomly chose 20 % of vertices from the anomalous subgraph and assigned the label 0 to them. Hence the task is to recover those hidden anomalous vertices from the rest of 80 % anomalous vertices. Precision and recall were computed, and the F-measure was also computed from them to summarize the performance. In addition, we used the gain of the modularity [21] to evaluate the goodness of division of resulting subgraphs without label information. We report the mean and the standard deviation of these values in 20 repeats in every case.

4.2   Results and Discussion

Efficiency. First we examine the efficiency using synthetic datasets. Figure 3 shows results of running time of each method with respect to the number of vertices. The number of edges are fixed as twice the number of vertices. We could not finish GBAGC when the number of vertices is larger than 10^5 as it run out of memory. This means that GBAGC cannot treat large graphs, although such graphs are now emerging and needed to be analyzed.

We can clearly see that paraAnomaly is much faster than gAnomaly, and it is the only method that can be applied to large graphs with more than 10^5 vertices in a reasonable time. The running time scales sub-quadratically with the number

---

[3] http://www.cais.ntu.edu.sg/~chi/software.html.

[4] http://www.cs.umd.edu/~sen/lbc-proj/LBC.html.

[5] http://snap.stanford.edu/index.html.

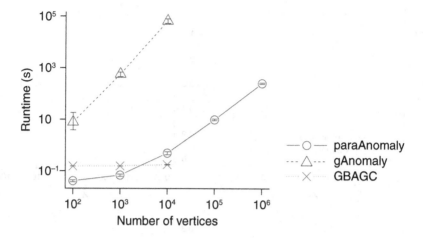

Fig. 3. Running time (seconds). GBAGC run out of memory and broke down when the number of vertices is larger than 10^5.

Table 2. Statistics of real-world datasets. In the table, "Ratio" denotes the ratio of the number of anomalous vertices in each graph

Data	$|V|$	$|E|$	Average degree	Ratio
CORA	2708	5429	4.010	0.302
DBLP(s)	3194	8714	5.456	0.202
Amazon	334863	925872	5.530	0.160
DBLP	317080	1049866	6.622	0.024

of vertices in paraAnomaly, while quadratically in gAnomaly. In real datasets in Table 3, our method is also the fastest and more than five orders of magnitude faster than gAnomaly. We can therefore say that *paraAnomaly is the first method that can handle massive graphs with millions of vertices in detecting anomalous subgraphs.*

Sensitivity. Next we analyze the sensitivity of our method with respect to changes in the parameter λ. Since gAnomaly also has a regularization parameter λ, we also analyze the sensitivity of gAnomaly and compare it to our method. We used synthetic graphs of 1000 vertices and 2000 edges. We also applied GBAGC to synthetic data, but it could not find any anomalous subgraph in any setting, i.e., all vertices are always in the same cluster. The reason might be that their framework is too sensitive to labels of neighbors and cannot handle our setting.

Results are plotted in Fig. 4. These plots show that our method is robust to changes in the parameter λ if it is smaller than 1, and is not stable if it gets larger than 1. This is why the penalty with respect to the connectivity is too

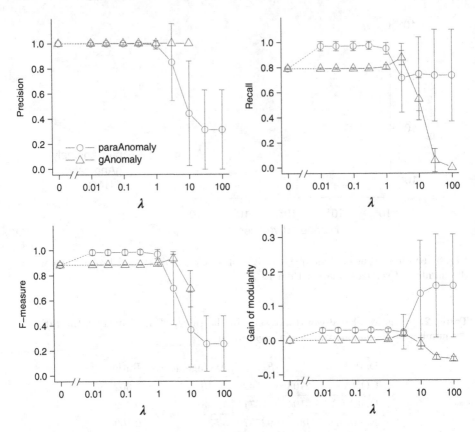

Fig. 4. Performance with respect to changes in regularization parameter λ.

strong, resulting in choosing too small anomalous subgraphs if λ is large. Thus both precision and recall become low values while the modularity increases. If λ is sufficiently small (around 0.1), the performance of paraAnomaly in terms of both the F-measure and the modularity is always better than gAnomaly. In addition, if $\lambda = 0$, precision stays high while recall gets lower again. The reason is that, in such a case, there is no regularization and the method just picks up all given 80 % of vertices and does not pick up any hidden anomalous vertices. To summarize, these results indicate that we do not need to be carefully tune the parameter λ and just set to a small value. In the following, we always set $\lambda = 0.01$ in paraAnomaly.

In gAnomaly, if the parameter λ is small (from 0.01 to 1), it cannot regularize the detection, that is, it just chooses vertices labeled as anomalous like the case of $\lambda = 0$ in paraAnomaly. But once λ gets larger than 1, regularization effect becomes suddenly too strong and the performance gets worse in terms of both the F-measure and the modularity. When λ is larger than 30, it did not find any anomalous subgraph. This result indicates that in practice it is not easy to find a good setting of λ. In the following, we always set $\lambda = 1$ in gAnomaly.

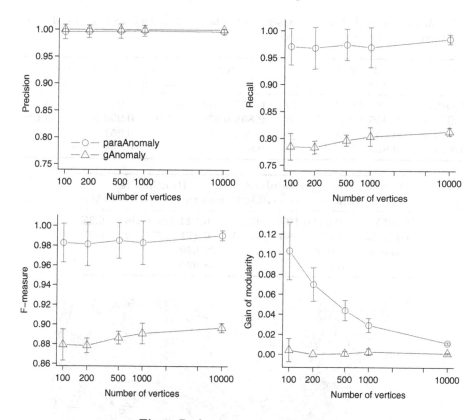

Fig. 5. Performance on synthetic data

Effectiveness. We investigate the performance on various sizes of synthetic graphs and real-world graphs. Results on synthetic data are shown in Fig. 5 and those on real data are in Table 3.

In synthetic data, we can see that our method paraAnomaly is always superior to gAnomaly in terms of both the F-measure and the modularity on every data size. In real data, paraAnomaly shows the best scores on all datasets in the F-measure, while gAnomaly is the best in the modularity. However, note that gAnomaly is not scalable and takes more than five orders of magnitude slower than paraAnomaly, thereby we could not finish gAnomaly on Amazon and DBLP. The clustering method GBAGC shows the worst score on every dataset. From those results, we can again confirm that our paraAnomaly is the only method that can efficiently and effectively find anomalous subgraphs from large scale graph data.

Visualization. Finally, we demonstrate visualization on the CORA dataset. The original anomalous vertices, corresponds to the cluster "Neural Networks", are shown in Fig. 6 and the resulting visualization by paraAnomaly is shown in

Table 3. Results on real-world datasets. In the table, "paraAno" and "gAno" denote paraAnomaly and gAnomaly, respectively.

Data	Precision			Recall			F-measure		
	paraAno	gAno	GBAGC	paraAno	gAno	GBAGC	paraAno	gAno	GBAGC
CORA	0.969	**0.977**	0.20	**0.867**	0.822	0.078	**0.915**	0.892	0.112
DBLP(s)	**0.955**	0.918	0.16	**0.858**	0.670	0.108	**0.904**	0.775	0.129
Amazon	**0.951**	—	—	**0.951**	—	—	**0.951**	—	—
DBLP	**0.868**	—	—	**0.828**	—	—	**0.848**	—	—

Data	Gain of modularity			Runtime (s)		
	paraAno	gAno	GBAGC	paraAno	gAno	GBAGC
CORA	0.062	**0.107**	−0.272	**0.124**	32861.436	0.358
DBLP(s)	0.059	**0.085**	−0.031	**0.171**	39450.032	0.279
Amazon	**0.078**	—	—	**26.649**	—	—
DBLP	**0.011**	—	—	**48.626**	—	—

Fig. 6. CORA dataset. Anomalous vertices are colored by red (Color figure online).

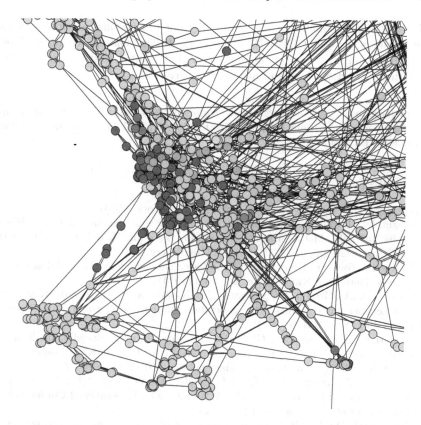

Fig. 7. CORA dataset. Vertices are colored according to the score q obtained by paraAnomaly.

Fig. 7. Here we can confirm that vertices that are close to (i.e., densely connected to) anomalous vertices are claimed to be anomalous according to their anomalous scores. Thus, by our method paraAnomaly, one can visualize interesting anomalous communities from the given attributed graphs, which are simultaneously ranked according to their degree of anomalousness.

5 Conclusion

In this paper we have presented a scalable method paraAnomaly, which detects anomalous subgraphs from attributed graphs. This method is based on the SConES formulation [4], thereby it is exactly and efficiently solved by the maximum flow algorithms through a minimum cut reformulation. Moreover, using the parametric flow algorithm [11], we have achieved to introduce the cardinality constraint, that is, the user can specify the desirable number of vertices. Experiments have shown that our method is much faster than the state-of-the-art method gAnomaly and is more effective on synthetic and real graph datasets.

Currently, paraAnomaly can handle only one-dimensional attributes, while some methods including GBAGC can use *multi-dimensional* attributes. Thus extending our formulation to multi-dimensional attributes, that is, how to design the attribute function A, is an interesting future work.

Acknowledgment. The authors thank Yoshinobu Kawahara for insightful discussions. This work was partially supported by JSPS KAKENHI 26880013 and Grand-in-Aid for JSPS Fellows 26-4555.

References

1. Aggarwal, C.C.: Outlier Analysis. Springer, New York (2013)
2. Akoglu, L., McGlohon, M., Faloutsos, C.: oddball: spotting anomalies in weighted graphs. In: Zaki, M.J., Yu, J.X., Ravindran, B., Pudi, V. (eds.) PAKDD 2010. LNCS, vol. 6119, pp. 410–421. Springer, Heidelberg (2010)
3. Akoglu, L., Tong, H., Koutra, D.: Graph based anomaly detection and description: a survey. Data Min. Knowl. Disc. **29**, 1–63 (2014)
4. Azencott, C.A., Grimm, D., Sugiyama, M., Kawahara, Y., Borgwardt, K.M.: Efficient network-guided multi-locus association mapping with graph cuts. Bioinformatics **29**(13), i171–i179 (2013)
5. Bhaduri, K., Matthews, B.L., Giannella, C.R.: Algorithms for speeding up distance-based outlier detection. In: Proceedings of the 17th ACM SIGKDD Conference on Knowledge Discovery and Data Mining, pp. 859–867 (2011)
6. Breunig, M.M., Kriegel, H.P., Ng, R.T., Sander, J.: LOF: identifying density-based local outliers. In: Proceedings of the ACM SIGMOD International Conference on Management of Data, pp. 93–104 (2000)
7. Chakrabarti, D.: AutoPart: parameter-free graph partitioning and outlier detection. In: Boulicaut, J.-F., Esposito, F., Giannotti, F., Pedreschi, D. (eds.) PKDD 2004. LNCS (LNAI), vol. 3202, pp. 112–124. Springer, Heidelberg (2004)
8. Chapelle, O., Schölkopf, B., Zien, A.: A discussion of semi-supervised learning and transduction. In: Chapelle, O., Schölkopf, B., Zien, A. (eds.) Semi-Supervised Learning, Chap. 25, pp. 473–478. MIT Press, Cambridge (2006)
9. Clauset, A., Newman, M.E.J., Moore, C.: Finding community structure in very large networks. Phys. Rev. E **70**(6), 066111 (2004)
10. Eberle, W., Holder, L.: Discovering structural anomalies in graph-based data. In: IEEE International Conference on Data Mining (ICDM) Workshop, pp. 393–398 (2007)
11. Gallo, G., Grigoriadis, M.D., Tarjan, R.E.: A fast parametric maximum flow algorithm and applications. SIAM J. Comput. **18**(1), 30–55 (1989)
12. Gao, J., Liang, F., Fan, W., Wang, C., Sun, Y., Han, J.: On community outliers and their efficient detection in information networks. In: Proceedings of the 16th ACM SIGKDD International Conference on Knowledge Discovery and Data Mining, pp. 813–822 (2010)
13. Goldberg, A.V., Tarjan, R.E.: A new approach to the maximum-flow problem. J. ACM **35**(4), 921–940 (1988)
14. Henderson, K., Eliassi-Rad, T., Faloutsos, C., Akoglu, L., Li, L., Maruhashi, K., Prakash, B.A., Tong, H.: Metric forensics: a multi-level approach for mining volatile graphs. In: Proceedings of the 16th ACM SIGKDD International Conference on Knowledge Discovery and Data Mining, pp. 163–172 (2010)

15. Henderson, K., Gallagher, B., Li, L., Akoglu, L., Eliassi-Rad, T., Tong, H., Faloutsos, C.: It's who you know: graph mining using recursive structural features. In: Proceedings of the 17th ACM SIGKDD International Conference on Knowledge Discovery and Data Mining, pp. 663–671 (2011)
16. Kawahara, Y., Nagano, K.: Structured convex optimization under submodular constraints. In: Proceedings of Uncertainty in Artificial Intelligence (UAI), pp. 459–468 (2013)
17. Lee, H.F., Dooly, D.R.: Algorithms for the constrained maximum-weight connected graph problem. Naval Res. Logistics **43**(7), 985–1008 (1996)
18. Li, N., Sun, H., Chipman, K., George, J., Yan, X.: A probabilistic approach to uncovering attributed graph anomalies. In: Proceedings of SIAM International Conference on Data Mining (SDM), pp. 82–90 (2014)
19. Lin, C.Y., Tong, H.: Non-negative residual matrix factorization with application to graph anomaly detection. In: Proceedings of SIAM International Conference on Data Mining (SDM), pp. 143–153 (2011)
20. Müller, E., Sanchez, P.I., Mülle, Y., Böhm, K.: Ranking outlier nodes in subspaces of attributed graphs. In: ICDE Workshop, pp. 216–222 (2013)
21. Newman, M.E.J., Girvan, M.: Finding and evaluating community structure in networks. Phys. Rev. E **69**(2), 026113 (2004)
22. Noble, C.C., Cook, D.J.: Graph-based anomaly detection. In: Proceedings of the 9th ACM SIGKDD International Conference on Knowledge Discovery and Data Mining, pp. 631–636 (2003)
23. Papadimitriou, C.H., Steiglitz, K.: Combinatorial Optimization: Algorithms and Complexity. Dover, New York (1998)
24. Perozzi, B., Akoglu, L. Sánchez, P.I., Müller, E.: Focused clustering and outlier detection in large attributed graphs. In: Proceedings of the 20th ACM SIGKDD International Conference on Knowledge Discovery and Data Mining (2014)
25. Pham, N., Pagh, R.: A near-linear time approximation algorithm for angle-based outlier detection in high-dimensional data. In: Proceedings of the 18th ACM SIGKDD Conference on Knowledge Discovery and Data Mining, pp. 877–885 (2012)
26. Sugiyama, M., Azencott, C.A., Grimm, D., Kawahara, Y., Borgwardt, K.M.: Multi-task feature selection on multiple networks via maximum flows. In: Proceedings of SIAM International Conference on Data Mining (SDM), pp. 199–207 (2014)
27. Sugiyama, M., Borgwardt, K.M.: Rapid distance-based outlier detection via sampling. In: Advances in Neural Information Processing Systems, pp. 467–475 (2013)
28. Watts, D.J., Strogatz, S.H.: Collective dynamics of 'small-world' networks. Nature **393**(6684), 440–442 (1998)
29. Xu, Z., Ke, Y., Wang, Y., Cheng, H., Cheng, J.: GBAGC: a general Bayesian framework for attributed graph clustering. ACM Trans. Knowl. Disc. Data **9**(1), 1–43 (2014)
30. Yang, J., Leskovec, J.: Defining and evaluating network communities based on ground-truth. In: Proceedings of the 2012 IEEE International Conference on Data Mining (ICDM), pp. 745–754 (2012)

Author Index

Abzianidze, Lasha   66

Baral, Chitta   259
Barker, Chris   184
Bekki, Daisuke   23, 83

Castroviejo, Elena   114
Chatzikyriakidis, Stergios   172

de Groote, Philippe   53

Gaur, Shruti   259
Gehrke, Berit   114
Goebel, Randy   244
Goto, Tetsuji   227

Hirata, Kouichi   296, 310, 317

Ishizaka, Yuma   296

Jin, Quming   310
Jirakunkanok, Pimolluck   274

Kashihara, Kazuaki   259
Kawahara, Yoshinobu   293
Kim, Mi-Young   244
Kiselyov, Oleg   99
Kuboyama, Tetsuji   293, 310, 317

Liefke, Kristina   6
Luo, Zhaohui   172

Maeda, Koji   331
McCready, Elin   23

Mery, Bruno   144
Mineshima, Koji   3, 83
Moot, Richard   144
Mori, Yoshiki   160

Nakashima, Masaya   310
Nakayama, Yasuo   37

Okano, Shinya   160
Otaki, Keisuke   340

Retoré, Christian   144

Sakamoto, Hiroshi   293, 331
Sano, Katsuhiko   274
Shinohara, Takeshi   310
Sugiyama, Mahito   340

Tabei, Yasuo   331
Takabatake, Yoshimasa   331
Tanaka, Ribeka   83
Tancredi, Christopher   200
Tojo, Satoshi   225, 227, 274

Vo, Nguyen H.   259

Winter, Yoad   53

Xie, Zhiguo   130
Xu, Ying   244

Yoshino, Takuya   296, 317

nited States
aylor Publisher Services